Bimal Krishna Banik and Sangeeta Bajpai (Eds.)
Tellurium Chemistry

Also of interest

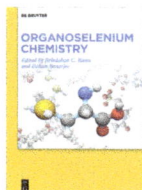

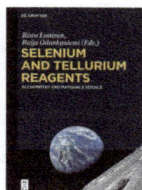

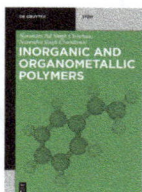

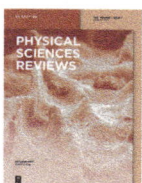

Tellurium Chemistry

Edited by
Bimal Krishna Banik
and Sangeeta Bajpai

DE GRUYTER

Editors
Bimal Krishna Banik, Ph. D.
Prince Mohammad Bin Fahd University
College of Sciences and Human Studies
Deanship of Research Development
1664 P. O. Box
Al Khobar 31952
Saudi-Arabia
bbanik@pmu.edu.sa

Dr. Sangeeta Bajpai
Department of Chemistry
Amity School of Applied Sciences
Amity University, Lucknow, 226028
Uttar Pradesh
India
sbajpai1@amity.edu

ISBN 978-3-11-073930-5
e-ISBN (PDF) 978-3-11-073584-0
e-ISBN (EPUB) 978-3-11-073587-1

Library of Congress Control Number: 2022940106

Bibliographic information published by the Deutsche Nationalbibliothek
The Deutsche Nationalbibliothek lists this publication in the Deutsche Nationalbibliografie;
detailed bibliographic data are available on the internet at http://dnb.dnb.de.

© 2022 Walter de Gruyter GmbH, Berlin/Boston
Cover image: Gettyimages/sanches812
Typesetting: TNQ Technologies Pvt. Ltd.
Printing and binding: CPI books GmbH, Leck

www.degruyter.com

Preface

The unique properties of 'Tellurium' (Te) have aggrandized this silvery white metalloid, from a rare to well-known species. Apart from its less abundance around 1pbb in the Earth's crust, the increasing research interest in the field of tellurium compounds has been documented. Since its discovery in 1782, research findings have unfolded the chemistry of tellurium compounds. The increase in the number of inorganic tellurium compounds and organic/organometallic tellurium compounds by 40% to 70%, envisages their importance. The emerging significance of tellurium compounds is evident from their diverse and potential applications in various fields of chemistry. This is due to its unique property of three-centered bonding, hyper valence and secondary bonding interactions. Basic knowledge of chemistry is very essential for the novel applications of organotellurium compounds in diverse fields.

The main purpose of this book is to provide an authentic and comprehensive account of current materialistic applications of organotellurium compounds. This purpose is fulfilled by assembling a knowledgeable team of contributing authors having considerable interest and skill in this field. The combined efforts of the authors and their expertise in the relevant field have explored and produced almost complete coverage of unique properties and modem applications of organotellurium compounds. All twelve chapters of this book are designed in a way to impart synthetic and practical knowledge of tellurium and organotellurium compounds.

Sahoo *et al.* have focused on various synthetic and applications of organotellurium compounds in chapter 1. Kamboj has explored the chemistry of tellurium containing macrocycles and its applications in chapter 2. The semiconductor properties and applications of organotellurium compounds in carbohydrate synthesis, cyclization reactions, solar cells and chemical sensors have been discussed and elaborated by Das and Banik *et al.* in chapters 3, 4, 5, 9 and 11.

Chapter-6 contributed by Ashraf has thrown light on the toxic toxic behaviour of tellurium which is the utmost important while working with organotellurium compounds.

Chapter-7 authored by Pandey *et al.* have covered the detailed study of tellurium existence and its effects on various environmental sections.

Banerjee *et al.* have explored the latest developments in the synthesis of bioactive organotellurium scaffolds in chapter-8.

A detailed study of tellurium and its novel low-dimensional derivatives offering intriguing nonlinear optical responses, making them promising candidates for design of various photonic devices, has been identified by Rose *et al.* in chapter-10.

The reactivity of organotellurium compounds as catalysts, reagents and sensors through functional group activation has been investigated by Ray *et al.* in chapter-12.

https://doi.org/10.1515/9783110735840-201

The editors of this book express their sincerest gratitude to all the authors for contributing excellent chapters on tellurium. The work of Ms. Stella Muller and Ms. Christene Smith in realization of this endeavour has been remarkable. It is impossible to publish this book without the timely support from the authors, Ms. Muller, Ms. Smith and the production team. We expect that this book on tellurium will be used extensively by the scientific community. Thank you ALL.

Contents

Bubun Banerjee, Aditi Sharma, Gurpreet Kaur, Anu Priya, Manmeet Kaur and Arvind Singh

Anjaly Das, Aparna Das and Bimal Krishna Banik

List of contributing authors

Sara Ali A Aldawood
Department of Mathematics and
Natural Sciences
College of Sciences and Human Studies
Prince Mohammad Bin Fahd University
Al Khobar 31952
Kingdom of Saudi Arabia

Muhammad Waqar Ashraf
Mathematics & Natural Sciences
Prince Mohammad Bin Fahd University
Azizeyah
Al-Khobar, 31952
Saudi Arabia
E-mail: mashraf@pmu.edu.sa

Sangeeta Bajpai
Department of Applied Chemistry,
Amity School of Applied Sciences
Amity University, Uttar Pradesh
Malhour
Lucknow 226028
India
E-mail: sbajpai1@amity.edu

Bubun Banerjee
Department of Chemistry
Akal University
TalwandiSabo
Bathinda
Punjab 151302
India
E-mail: banerjeebubun@gmail.com

Bimal Krishna Banik
Department of Mathematics and
Natural Sciences
College of Sciences and Human Studies
Prince Mohammad Bin Fahd University
Al Khobar 31952
Kingdom of Saudi Arabia
E-mail: bimalbanik10@gmail.com

Preetismita Borah
CSIR-Central Scientific Instruments Organization
Chandigarh
India

Anjaly Das
National Institute of Electronics & Information
Technology
Calicut 673601
Kerala
India

Aparna Das
Department of Mathematics and Natural Sciences
College of Sciences and Human Studies
Prince Mohammad Bin Fahd University
Al Khobar, 31952
Kingdom of Saudi Arabia
E-mail: aparnadasam@gmail.com

Syed Iqleem Haider
Chemistry
University College Hyderabad
Hyderabad
Pakistan

Adya Jain
Department of Chemistry
MRK Educational Institutions
IGU
Rewari
Haryana
India

Monika Kamboj
Department of Applied Chemistry
Amity School of Applied Sciences
Amity University Uttar Pradesh
Lucknow Campus
Lucknow
226028, UP
India
E-mail: mkamboj@lko.amity.edu

Gurpreet Kaur
Department of Chemistry
Akal University
TalwandiSabo
Bathinda
Punjab 151302
India

https://doi.org/10.1515/9783110735840-202

Manmeet Kaur
Department of Chemistry
Akal University
TalwandiSabo
Bathinda
Punjab 151302
India

Suman Mazumdar
Department of Scientific and Industrial Research
Ministry of Science & Technology
Government of India
Technology Bhawan
New Mehrauli Road
110016 New Delhi
Delhi
India

Almas Fatima Memon
Chemistry
Government College University
Hyderabad
Pakistan

Garima Pandey
Department of Chemistry
SRM Institute of Science and Technology
Delhi NCR Campus
Modinagar 201204
Ghaziabad
Uttar Pradesh
India
E-mail: garimapandey.pandey8@gmail.com

Anu Priya
Department of Chemistry
Akal University
TalwandiSabo
Bathinda
Punjab 151302
India

Devalina Ray
Amity Institute of Biotechnology
Amity University
Noida 201313
UP
India
E-mail: dray@amity.edu

Biswa Mohan Sahoo
Roland Institute of Pharmaceutical Sciences
Berhampur-760010
Odisha
India
E-mail: drbiswamohansahoo@gmail.com

Aditi Sharma
Department of Chemistry
Akal University
TalwandiSabo
Bathinda
Punjab 151302
India

Arvind Singh
Department of Chemistry
Akal University
TalwandiSabo
Bathinda
Punjab 151302
India

Amber Rehana Solangi
Center of Excellence in Analytical Chemistry
University of Sindh
Jamshoro
Pakistan

Priya Rose Thankamani
International School of Photonics
Cochin University of Science and Technology
Cochin 682022
Kerala
India
and
Inter University Center for Nanomaterials and Devices (IUCND)
Cochin University of Science and Technology
Cochin 682022
Kerala
India
E-mail: priyarose@cusat.ac.in

Sheenu Thomas
International School of Photonics
Cochin University of Science and Technology
Cochin 682022
Kerala
India

Abhishek Tiwari
Faculty of Pharmacy
IFTM University
Moradabad
Uttar Pradesh, 244102
India

Varsha Tiwari
Faculty of Pharmacy
IFTM University
Moradabad
Uttar Pradesh, 244102
India

Biswa Mohan Sahoo*, Bimal Krishna Banik*, Abhishek Tiwari, Varsha Tiwari, Adya Jain and Preetismita Borah

1 Synthesis and application of organotellurium compounds

Abstract: Organotellurium compounds define the compounds containing carbon (organic group) and tellurium bond (C–Te). The first organic compound containing tellurium was prepared by Wohler in 1840 after the discovery of the metal by the Austrian chemist F. J. Muller von Reichenstein in the year 1782. The term tellurium was derived from Latin tellus. Tellurium was observed first time in ores mined in the gold districts of Transylvania. Naturally occurring tellurium compounds are present in various forms based on their oxidation states such as TeO_2 (+4) and TeO_3 (+6). These oxidation states of tellurium compounds are more stable as compared to the other oxidation states. Tellurium is a rare element and is considered a non-essential, toxic element. Tellurium possesses only one crystalline form which consists of a network of spiral chains similar to that of hexagonal selenium. Tellurium is used for the treatment and prevention of microbial infections prior to the development of antibiotics. Hence, the utilization of organotellurium compounds plays a significant role as reagents and intermediates in various organic syntheses.

Keywords: application; compounds; organic group; synthesis; tellurium.

1.1 Introduction

Tellurium is an element represented with the symbol Te. It has an atomic number of 52 with an atomic mass of 127.60 g·mol^{-1}. It is considered under the chalcogen (group 16) family of elements in the periodic table. Chalcogen refers to the oxygen-family elements of p-block [1]. Te exhibits two allotropes as crystalline and amorphous. In the case of crystalline form, Te is generally silvery-white with a metallic luster. While the amorphous type of tellurium is black-brown powder. The crystalline form of tellurium possesses parallel helical chains of Te atoms with three atoms per turn. It has the properties of both metals and non-metals. The tellurium from natural sources contains

*Corresponding authors: **Biswa Mohan Sahoo**, Roland Institute of Pharmaceutical Sciences, Berhampur-760010, Odisha, India; and **Bimal Krishna Banik**, Department of Mathematics and Natural Sciences, College of Sciences and Human Studies, Prince Mohammad Bin Fahd University, Al Khobar 31952, Kingdom of Saudi Arabia, E-mail: drbiswamohansahoo@gmail.com (B.M. Sahoo)
Abhishek Tiwari and Varsha Tiwari, Faculty of Pharmacy, IFTM University, Moradabad, Uttar Pradesh-244102, India
Adya Jain, Department of Chemistry, MRK Educational Institutions, IGU, Rewari, Haryana, India
Preetismita Borah, CSIR-Central Scientific Instruments Organization, Chandigarh, India

As per De Gruyter's policy this article has previously been published in the journal Physical Sciences Reviews. Please cite as: B. M. Sahoo, B. K. Banik, A. Tiwari, V. Tiwari, A. Jain and P. Borah "Synthesis and application of organotellurium compounds" *Physical Sciences Reviews* [Online] 2022. DOI: 10.1515/psr-2021-0105 | https://doi.org/10.1515/9783110735840-001

Table 1.1: List of natural isotopes.

Name of isotopes	Relative abundance (%)
^{120}Te	0.09
^{122}Te	2.55
^{123}Te	0.89
^{124}Te	4.74
^{125}Te	7.07
^{126}Te	18.84
^{128}Te	31.74
^{130}Te	34.08

eight isotopes [2]. Out of which, six isotopes are stable such as ^{120}Te, ^{122}Te, ^{123}Te, ^{124}Te, ^{125}Te, and ^{126}Te. Whereas the remaining two isotopes (^{128}Te and ^{130}Te) are slightly radioactive (Table 1.1) [3].

Tellurium exits in different oxidation states such as −2, +2, +4, and +6. The +2 oxidation state is displayed by the dihalides including $TeCl_2$, $TeBr_2$, and TeI_2 (Figure 1.1, Table 1.2). Te is considered as one of the least available elements in the earth's lithosphere [4]. It was discovered by Austrian chemist F. J. Muller von Reichenstein in 1782. It was reported that the first organotellurium compound was identified more than 150 years ago with the synthesis of diethyl tellurides by the scientist Wohler in 1840. Taniyama and co-workers reported the strong antibacterial activity of diaryltellurium-dihalides and the inhibitory activity of cyclic-tellurium substances against the growth of bacteria in 1922 [5].

Haiduc and Edelmann reported the chemistry of organotellurium compounds in 1999. X-ray diffraction study is performed to confirm the geometry of these compounds. When Te is present in the ground state, it exists two unpaired electrons and hence the oxidation state (II) is well known in such compounds one 's' and three 'p' orbital hybridized to give four sp^3 hybrid orbitals (Table 1.3). Out of which, two are occupied by lone pair and the remaining two have bonding electron pairs. So, the geometry is

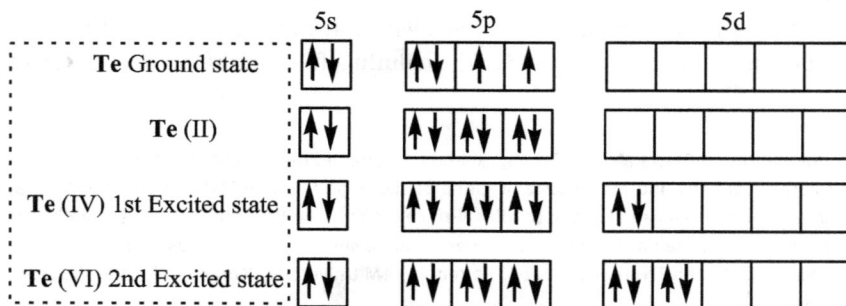

Figure 1.1: Oxidation states of tellurium.

Table 1.2: List of organotellurium compounds.

Type	Examples
Organotellurium(II) compounds	
R_2Te and $RR'Te$	Diphenyltellurides
	Dialkyltellurides
	Diaryltellurides(p-MeOC$_6$H$_4$TePh)
	Methyl vinyl telluride
	Telluracyclopentane
	Tellurium azamacrocycles
R2Te$_2$ and $RR'Te_2$	Te$_2$(CF$_3$)$_2$, Te$_2$(C$_6$F$_5$)$_2$
	R_2Te_2 (R = Me, Et, Me$_2$CH)
RTeX	RTeX (X = Cl, Br, I, CN)
Organotellurium(IV) compounds	
RTeX$_3$	2-Biphenyl tellurium tri-bromides
	Phenyl tellurium tri-iodide
	Phenyl tellurium tri-chloride
	Phenyltellurium tri-bromide
R_2TeX_2	Biphenylene tellurium(IV) dichloride
	Diphenyl tellurium(IV) monothiocarbamates
	Diaryl tellurium dihalides *bis*(Ferrocenylcarboxylato)telluranes
R_3TeX	Triphenyltelluronium thiocyanate (C$_6$F$_5$)$_3$TeCl
	(CF$_3$C$_6$F$_4$)$_3$TeCl
	(CF$_3$C$_6$F$_4$)$_3$TeBr
R_4Te	Tetraphenyl tellurium
	Tetramethylchalcogens
Organotellurium(VI) compounds	
RTeX$_5$	(C$_2$F$_5$)TeF$_4$Cl
	CH$_3$TeF$_5$
	(CH$_3$O)TeF$_4$Cl
R_2TeX_4	(C$_2$F$_5$)$_2$TeF$_4$
R_6Te	Ar$_6$Te (Ar = 4CF$_3$C$_6$H$_4$, C$_6$H$_5$)

V-shaped as presented in Figure 1.2 [6]. In the first excited state, the coordination number greater than four is attained by accepting electrons with suitable donors in the empty d-orbital. The coordination bonds are generally studied by X-ray diffraction study and NMR spectroscopy.

1.2 Synthesis of organotellurium compounds

There are different synthetic routes for producing organotellurium compounds that involve the use of elemental tellurium (Te), tellurium tetrachloride (TeCl$_4$), and (iii)

Table 1.3: List of tellurium compounds with coordination number and geometry.

Valency	Coordination number	No. of bonds	No. of lone pairs	Geometry	Hybridization	Examples
II	2	2	2	Angular	sp^3	R_2Te
						RTeX
	3	3	2	Pyramidal	sp^3d	$RTeX_2$
	4	4	2	Square planar	sp^3d^2	Te $[SC(NH_2)_2]_2Cl_2$
IV	4	4	1	Distorted trigonal bipyramidal	sp^3d	R_2TeX_2 $RTeX_3$ R_4Te
	5	5	1	Square pyramidal	sp^3d^2	$RTeX_4$ $RTeX_3L$ (L = Monodentate)
	6	6	1	Distorted octahedral	sp^3d^3	$[RTeX_4]^{2-}$ $RTeX_3L$ (L = Bidentate)
	7	7	1	Distorted pentagonal bipyramidal	sp^3d^4	$RTe(Et_2NCS_2)_3$
	8	8	1	Distorted dodecahedral	sp^3d^5	$Te(Et_2NCS_2)_4$
VI	6	6	0	Octahedral	sp^3d^2	$C_2F_5TeF_4Cl$ $(C_2F_5)_2TeF_4$

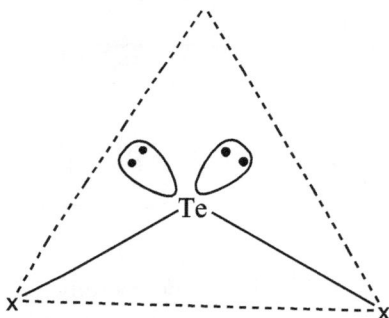

Figure 1.2: V-shaped structure of Te(II) compounds.

intermediates obtained by the above methods during the preparation of organo-tellurium compounds (Figures 1.3 and 1.4).

Junk et al. reported the synthesis of 2-acylamino and 2-arylamino-1,3-benzo-tellurazoles and also evaluate their properties (Figure 1.5). The reaction mechanism

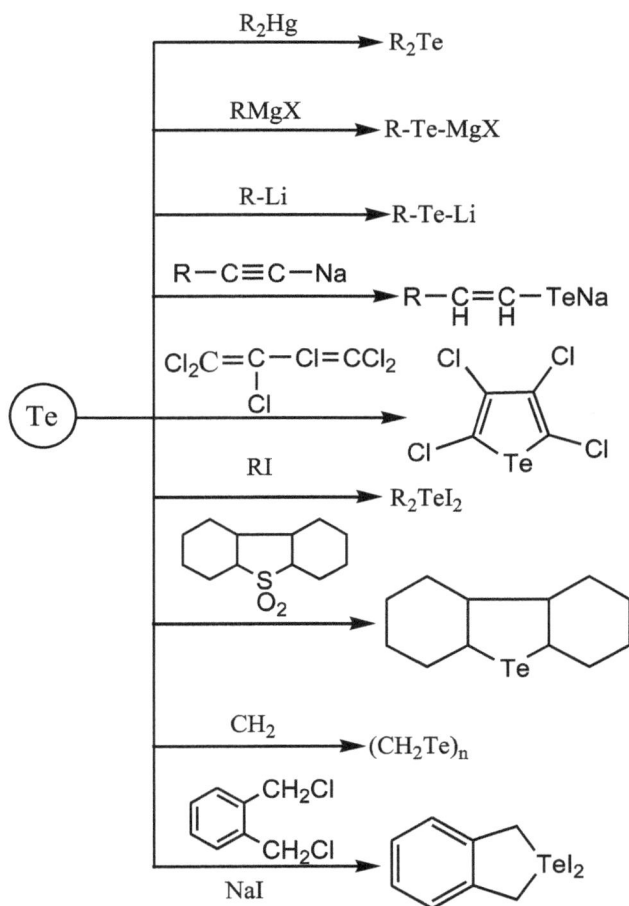

Figure 1.3: Synthetic routes for organotellurium compounds using elemental tellurium (Te).

involves the insertion of mercury into the Te-Te bond (**1**), followed by an intramolecular nucleophilic attack of the thiocarbonyl moiety by the resulting insertion product. The targeted compounds were produced with 44–67% yield by reacting bis(2-aminophenyl)ditelluride with acyl and aryl isothiocyanates respectively, and the subsequent reductive cyclization of the resulting thiourea analogs. The synthesized compound, 2-acylamino-1,3-benzotellurazoles (**3**) are found to be crystalline solids. These compounds are stable at ambient light, air, and moderate heat [7].

Al-Fregi et al. demonstrated several quinolone-based organotellurium compounds. (Figure 1.6). These compounds were obtained by the reaction of 8-(quinolyl) mercuric chloride (**5**) with tellurium tetrabromide (TeBr$_4$) in the presence of dry dioxane to produce 8-(quinolyl)tellurium tribromide(**6**) and *bis*[8-(quinolyl)]tellurium

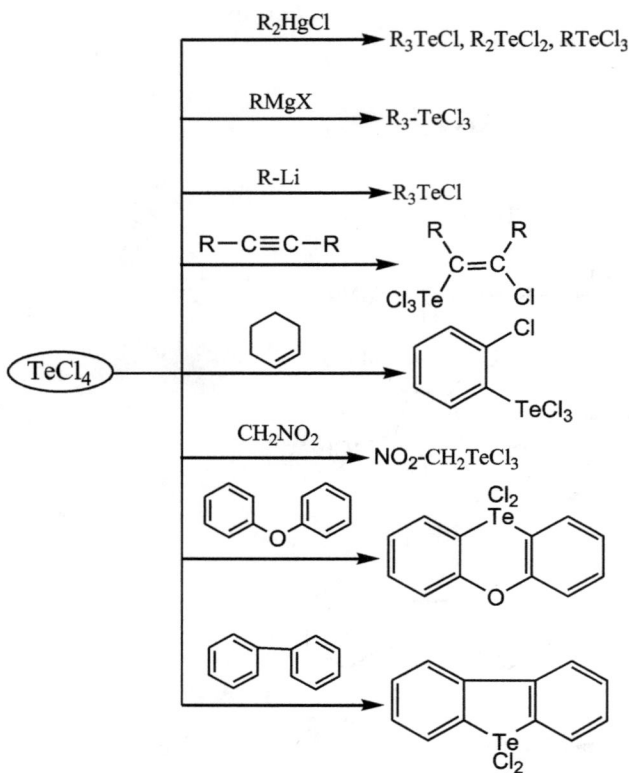

Figure 1.4: Synthetic routes for organotellurium compounds using tellurium tetrachloride (TeCl₄).

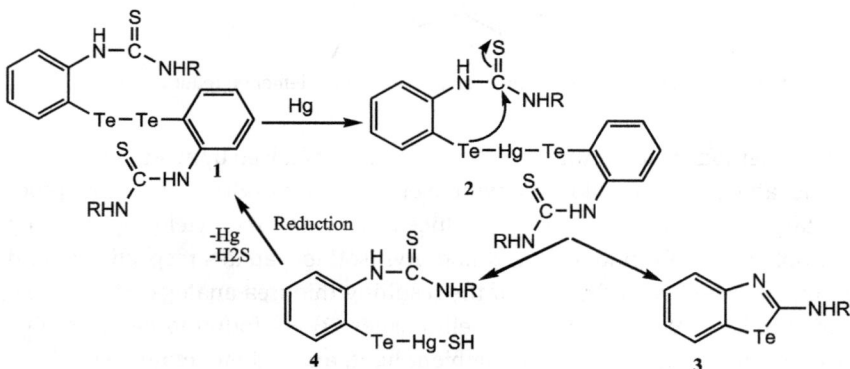

Figure 1.5: Synthetic route to benzotellurazoles.

dibromide(**7**), respectively. Further, the reaction of compounds **6** and **7** by ethanolic hydrazine hydrate yields *bis*[8-(quinolyl)]ditelluride (**8**) and *bis*[8-(quinolyl)]telluride (**9**)respectively [8].

Netellurolate is used as the tellurenating reagent to prepare optically active tellurium-containing binaphthyl (Figure 1.7). The chirality inducers like ditellurides are used in the asymmetric 1,4-addition reactions of α,β-unsaturated ketones. The optically active (*S,S*)-bis[1-(1′-naphthyl)-2-naphthyl]ditelluride (**10**) was produced with a high product yield (93%) by the sequential reaction of (*S*)-2,2′-dibromo-1,1′-binaphthyl with an equimolar quantity of *t*-butyllithium, elemental tellurium, another equimolar *t*-butyllithium and methanol followed by oxidation [9].

The monotellurated derivative is produced by the reaction between benzothiophene (**11**) and two equivalents of lithium diisopropylamide (LDA) followed by the

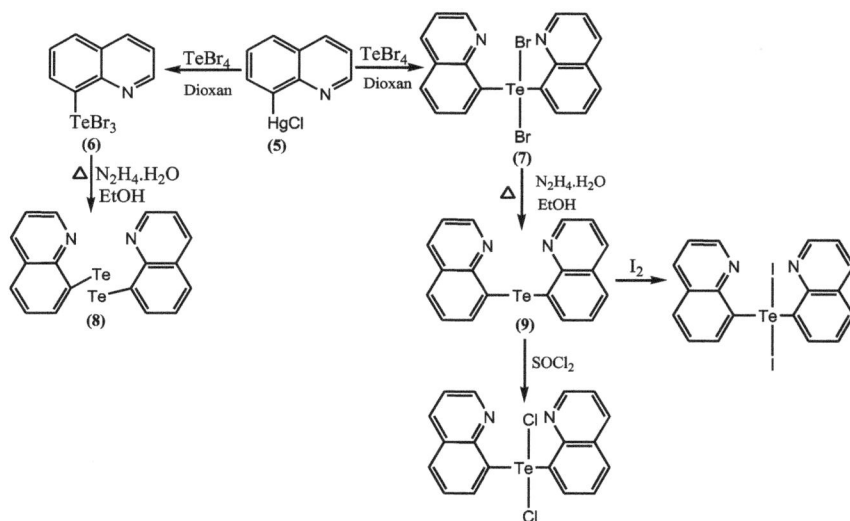

Figure 1.6: Synthesis of organotellurium compounds based on quinolone.

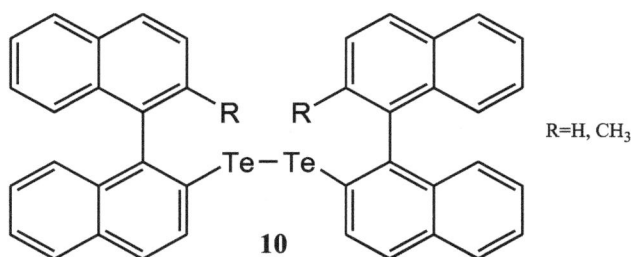

Figure 1.7: Synthesis of tellurium-containing binaphthyl.

addition of tellurium. The reaction of lithium benzothiophene-tellurolate (**12**) with 1,2-dibromoethane afford di-2-benzo[*b*]thienylditelluride (**13**) (Figure 1.8). The crystal structure of the compounds is determined by an X-ray diffraction study [10].

The synthesis of the ditellurides **14** and **15** is carried out by the *ortho*-lithiation scheme (Figure 1.9). The single-crystal X-ray study is performed to determine the presence of intramolecular interactions in case of ditelluride [11].

The treatment of compound **16** with *tert*-butyllithium (3.5 equivalent) at a temperature of −78 °C in diethyl ether, followed by the addition of tellurium powder and the oxidation of the resulting tellurolate, produces ditelluride **17** (Figure 1.10) [12].

Tellurium oxide (TeO2) reacts with non-conjugated dienes (**18–20**) in the presence of acetic acid and lithium halide to produce corresponding bis(2-acetoxyalkenyl) ditellurides (**21–23**) (Figure 1.11) [13].

An efficient synthetic scheme was developed to prepare tellurophene (**25**). The synthetic methodology involves the ring-closure addition-elimination reaction between 2,3-dimethoxy-1,3-butadiene (**24**) and Tellurium dichloride (TeCl$_2$) in the presence of NaOAc (Figure 1.12) [14].

Huang et al. reported the condensation reaction of telluronium salts **26** with aldehydes and dibutyl telluride **29**, bromide **30** with aldehyde **27**. This reaction proceeds efficiently in the presence of ionic solvent [bmim][BF4], to produce *(E)-α,β*-unsaturated compounds **28** with high purity, excellent yields, and high stereo-selectivity (Figure 1.13) [15].

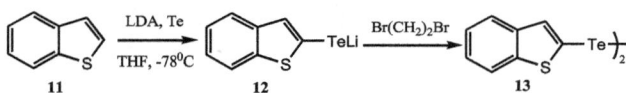

Figure 1.8: Synthesis of di-2-benzo[*b*]thienylditelluride.

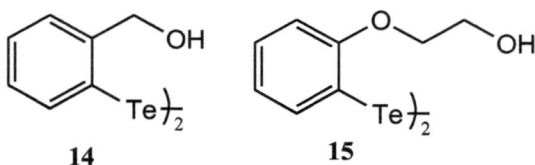

Figure 1.9: Synthesis of ditellurides.

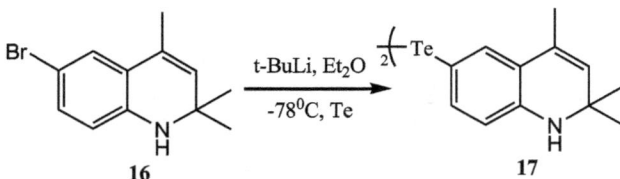

Figure 1.10: Synthesis of ditelluride (**17**).

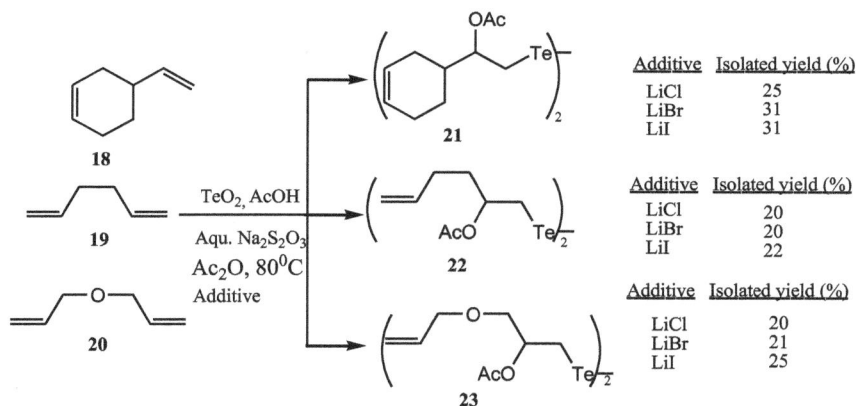

Figure 1.11: Synthesis of bis(2-acetoxyalkenyl)ditellurides (**21–23**).

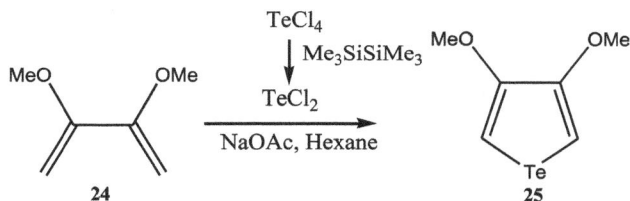

Figure 1.12: Synthesis of tellurophene.

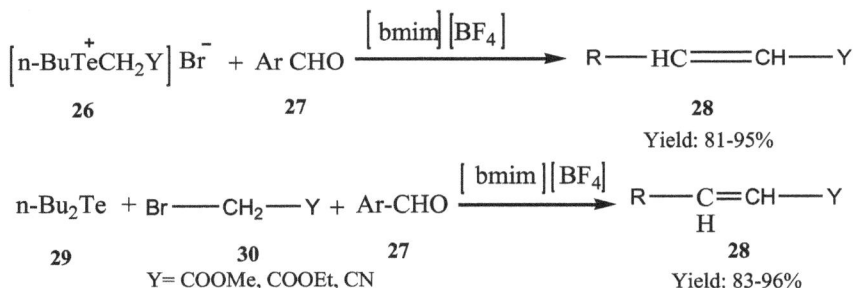

Figure 1.13: Synthesis of α,β-unsaturated compounds.

Hameed et al. reported the synthesis, characterization, and computational study of several organotellurium compounds linked with azomethine groups (Figure 1.14). Schiff bases undergo direct telluration in the presence of tellurium tetrahalides that result in the production of the ionic products which arise due to hydrolysis of the tellurium tetrahalides. The reaction of Ar_2TeBr_2 with hydrazine hydrate (N_2H_4) produces tellurides (Ar_2Te) with excellent yield [16].

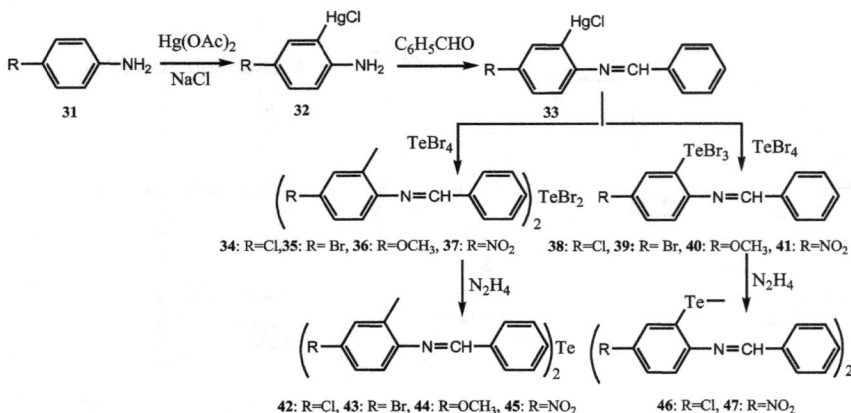

Figure 1.14: Synthesis of organotellurium compounds containing azomethine groups.

4-Bromo aniline (**48**) undergoes telluration in the presence of potassium tell-rocyanate (KTeCN) to produce 4-aminobenzeno-tellurocyanate (**49**) which is successively treated with sodium hydroxide (NaOH) and ammonium chloride (NH$_4$Cl) to afford bis(4-aminophenyl)ditelluride (**50**). The ditelluride is further treated with formalin (CH$_2$O) to obtain the target compound (**51**) with a 90% yield (Figure 1.15) [17].

The preparation of bis(2-halo-3-pyridyl)ditelluride (**56**) is carried out by the aerial oxidation of 2-halo-3-pyridyl telluride (**55**) with a 60% yield (Figure 1.16). It involves the reaction between one equivalent of n-BuLi and one equivalent of diisopropylamine (DIA) in tetrahydrofuran (THF) under a nitrogen environment at 20 °C. The solution was subjected to stirring at 0 °C for 1 h. A little amount of lithium-diisopropylamine (LDA) was added to 2-halopyridine (**52**) in tetrahydrofuran (THF) with continuous stirring at 40 °C and LDA induced at the C-3 position. After that, fine tellurium powder was added and stirred for 1 h for surface oxidation of the metal. The resulting product was hydrolyzed and the reaction was performed under aerial oxidation for 12 h [18].

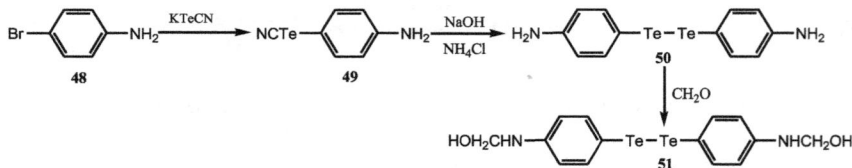

Figure 1.15: Synthesis of the title compound (**51**).

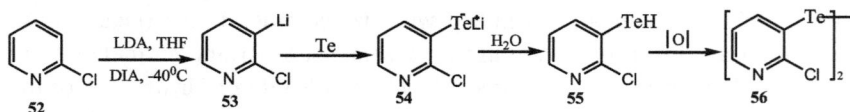

Figure 1.16: Synthesis of ditelluride via aerial oxidation.

1.3 Application of organotellurium compounds

Organotellurium compounds are applied as a key component in different fields that include medicine, antibiotics, cancer drug development, biomarker, X-ray, biochemical analysis, etc.

1.3.1 Applications of tellurium compounds in diagnosis and therapy

In 1926, it was reported that organ tellurium compounds are used for the treatment of leprosy and syphilis. In 1984, it was proposed that TeO_3^{2-} can be used as a potential antisickling agent of red blood cells (RBC) for the management and treatment of sickle cell anemia [19, 20]. Compounds like AS-101 inhibit the formation of IL-10, IFN-γ, IL-2R, and IL- 5. It was demonstrated that AS-101 also protects the bone marrow stem cells during chemotherapy. Tellurite (TeO_3^{2-}) and tellurate (TeO_4^{2-}) have been suggested for utilize in the selective medium for the evaluation of *Streptococci* of feces. Similarly, Cefixime-tellurite media has been used to isolate the organisms from minced beef and rectal swabs of cattle. It was studied that Tellurate is about 2- to 10-fold less toxic than tellurite in most organisms. Diphenylditelluride (PhTeTePh) is used extensively in toxicological studies. The toxicological profile of organotellurium compounds is determined by evaluating the relative toxicity in animals or by the inhibition of cellular growth [21–23].

1.3.2 Applications of tellurium compounds in biology

Chasteen et al. demonstrated the evaluation of the toxicity of a series of ditellurides (**57–61**, Figure 1.17) against HL-60 cells by assessing the induction of apoptosis using cell cytometry. The ditellurides exhibited a significant apoptosis induction with doses of 1 μM [24, 25].

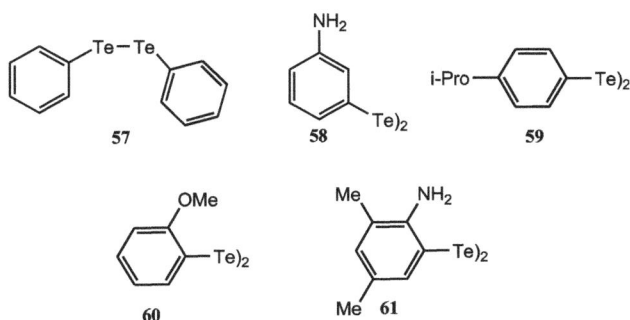

Figure 1.17: Structures of ditellurides with anticancer activity.

Figure 1.18: First organotellurium compounds with antioxidant property.

In 1994, Andersson et al. reported the *in vitro* antioxidant potential of the organotellurium compounds. It was reported that tellurides **62** and **63** are the first organotellurium compounds that exhibit antioxidant properties. They inhibit lipid peroxidation of cells caused by oxidative conditions (Figure 1.18) [26].

The utility of organotellurium compounds plays a significant role in Photodynamic therapy (PDT). PDT is designed as an alternative cancer remedy for the treatment of head, neck, lung, digestive tract, genitourinary tract, and pancreas carcinomas [27]. During the synthesis of organic compounds, the use of organotellurium compounds is mainly categorized into two types such as carbon–carbon bond (C–C) forming reactions and different types of functional group inter-conversions. In the case of carbon-carbon bond-forming reactions, tellurium is easily introduced into a variety of organic compounds. The aromatic compounds (**64**) undergo an electrophilic aromatic substitution reaction with tellurium tetrachloride (TeCl$_4$) to produce diaryl-tellurium dichlorides (**65**). These compounds can be converted into symmetrical biaryls (**66**) on treatment with degassed Raney nickel in high-boiling ether solvents (Figure 1.19) [28].

Similarly, the aryl-aryl coupling is achieved starting from an aryltelluriumtrichloride (**67**). Example: Naphthayl-tellurium trichloride undergoes a coupling reaction to produce a 2,2′-binaphthyl compound (**68**). Whereas phenoxatellurine (**69**) afford dibenzofuran (**70**) *via* intramolecular C–C bond formation (Figure 1.20) [29].

The synthesis of biaryl compounds is carried out by thermal decomposition of tetra-aryltellurium species. Tetraphenyl-tellurium (**71**) on heating in the presence of toluene at 140 °C afford biphenyl (**72**) and diphenyl telluride (**73**) (Figure 1.21) [30].

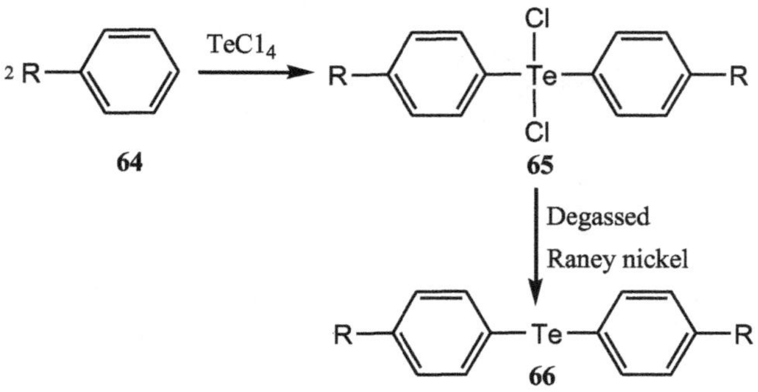

Figure 1.19: Synthesis of symmetrical biaryls.

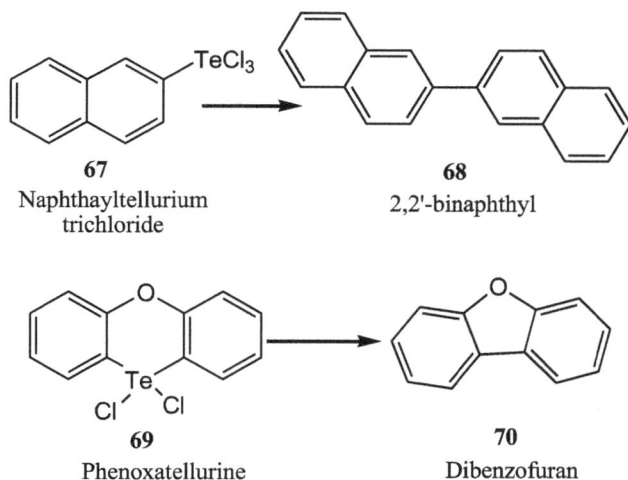

67
Naphthayltellurium
trichloride

68
2,2'-binaphthyl

69
Phenoxatellurine

70
Dibenzofuran

Figure 1.20: Synthesis of 2,2'-binaphthyl compound and dibenzofuran.

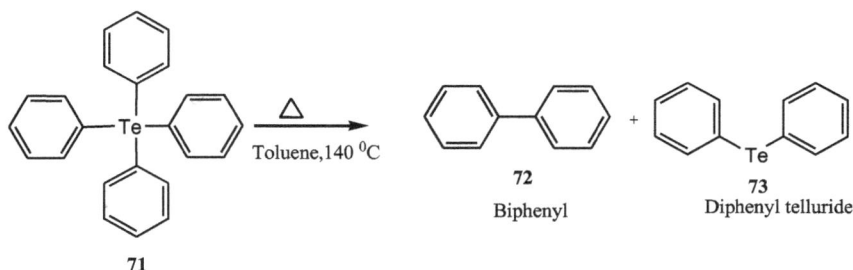

72
Biphenyl

73
Diphenyl telluride

71

Figure 1.21: Synthesis of biphenyl (**72**) and diphenyl telluride (**73**).

Tellurium tetrachloride (TeCl$_4$) is the reagent used to synthesize several organo-tellurium compounds. TeCl$_4$ is prepared by the reaction of elemental tellurium with a stream of chlorine at high temperature, followed by distillation of the TeCl$_4$ into glass ampoules. Anisole (**74**) reacts with tellurium tetrachloride under microwave irradiation at 100 W for 3 min to produce p-methoxy-phenyl-tellurium-trichloride (**75**) with a good yield (86%) (Figure 1.22). The synthesis is carried out in absence of organic solvents [31].

Engman et al. reported the scheme for the synthesis of a series of ditellurides and evaluated their glutathione peroxidase-like (GPx) activity (Figure 1.23). Both the ¹H-NMR and coupled reductase assay methods were used to determine the GPx activity of the synthesized compounds. Diarylditellurides **77–82** were obtained from aryl bromides (**76**) by sequential reaction including lithiation, tellurium insertion, and ferrocyanide oxidation [32].

$$SO_2Cl_2 + Te \xrightarrow[65^0C]{Microwave} TeCl_4$$

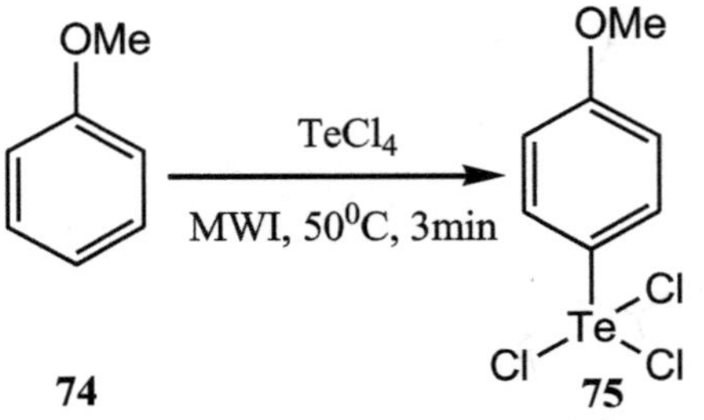

Figure 1.22: Synthesis of *p*-methoxy-phenyl-telluriumtrichloride.

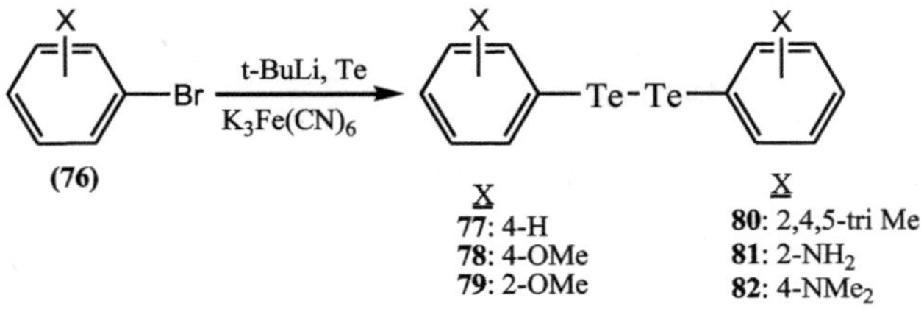

Figure 1.23: Synthesis of diarylditellurides.

Vazquez-Tato et al. reported the microwave-assisted synthesis of organotellurium compound ammonium trichloro(dioxoethylene-O,O′)tellurate (AS101). AS101 is found to be a potent immunomodulator with several therapeutic potentials including anti-tumor, antibacterial, antioxidant, anti-inflammatory and anti-apoptotic, etc. MAOS offers an efficient, clean, and faster process for the synthesis of AS101 under solvent-free conditions. The reaction is based on Albeck's synthesis that involves heating of TeCl$_4$ (**83**) and NH$_4$Cl (**84**) in ethylene glycol (**85**) to produce AS101 (**86**) (Figure 1.24) [33].

Savegnag et al. performed the synthesis, characterization, and antioxidant activity of chrysin-based organotellurium compounds (**89**) (Figure 1.25). Chrysin is chemically known as 5,7-dihydroxyflavone. It is a flavonoid and is present commonly in various plant extracts, honey, fruits, vegetables, etc. It is reported that chrysin exhibits biological activities such as antiviral, anticancer, antibacterial, anti-inflammatory, anti-allergic, anti-mutagenic, anti-anxiolytic and antioxidant, etc. The chemical modifications of the natural products play a vital role to improve their biological

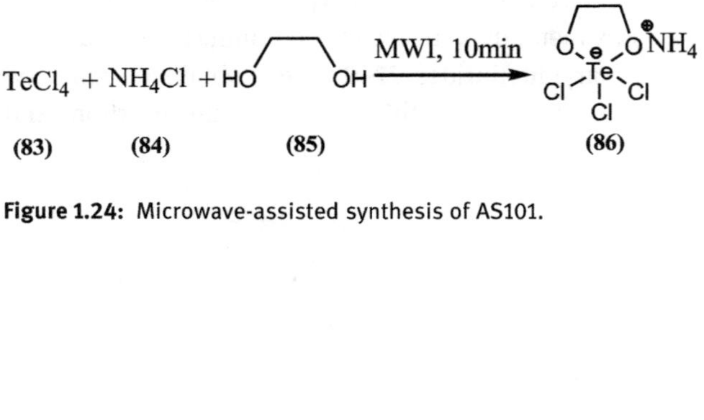

Figure 1.24: Microwave-assisted synthesis of AS101.

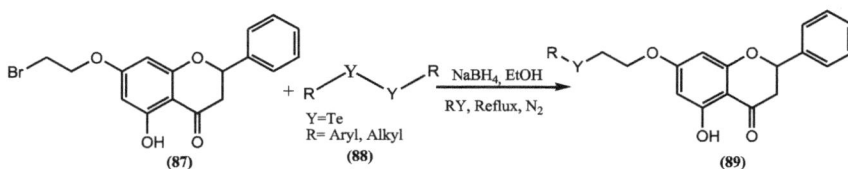

Figure 1.25: Synthesis of tellurium-containing chrysin derivatives.

activities. The different types of modifications include cyclization reaction, dehydration, reduction, oxidation reactions, etc. These reactions involve the insertion of organo-chalcogen moieties. The optimization of the pysico-chemical properties of organo-chalcogens makes an attractive synthetic target for selective chemical reactions [34].

Cadmium telluride (CdTe) nanoparticles are fluorescent. So it can be used as quantum dots in imaging and diagnosis. It was investigated that CdTe is applied as a biomarker in the diagnosis of tumors [35]. 2-amino-5-carboxyphenyl mercury chloride (**90**) reacts with tellurium tetrabromide (TeBr$_4$) in chloroform (CHCl$_3$) to produce 2-amino-5-carboxyphenyl tellurium tribromide (**91**). Compound **91** further reacts with 4-hydroxyphenyl mercury chloride (**92**) in the presence of argon environment to afford 4-hydroxyphenyl-2-amino-5-carboxyphenyl tellurium dibromide (**93**) followed by reduction with hydrate hydrazine (N$_2$H$_4$.H$_2$O) to obtain 4-hydroxyphenyl-2-amino-5-carboxyphenyl telluride (**94**) (Figure 1.26). Compound **94** exhibits antitumor and antioxidant activities [36].

Butterfield et al. performed the antioxidant activity of the organotellurium com-pound, 3-[4-(N,N-dimethylamino)benzenetellurenyl]propane sulfonic acid (NDBT, **95**) against oxidative stress in synaptosomal membrane systems and neuronal cultures (Figure 1.27) [37].

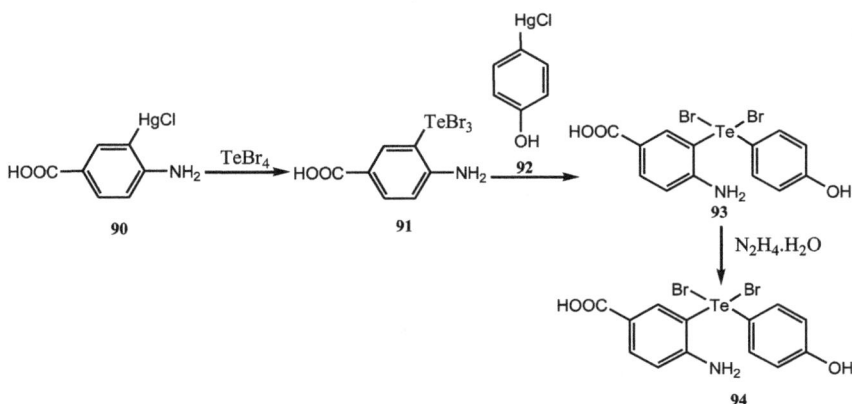

Figure 1.26: Synthesis of 4-hydroxyphenyl-2-amino-5-carboxyphenyl telluride.

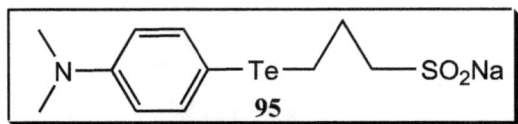

Figure 1.27: Structure of NDBT.

Kwon et al. reported benzo[b]tellurophenes as a potential histone H3 lysine 9 demethylase (KDM4) inhibitor (Figure 1.28). Among the carbamates, alcohol, and aromatic derivatives, tert-butylbenzo[b]tellurophen-2-ylmethylcarbamate (**96**) exhibited KDM4 specific inhibitory activity in cervical cancer HeLa cells (IC$_{50}$: 30.24 ± 4.60 µM) [38].

Giorgio et al. reported a novel organotellurium compound (RT-01) as a new anti-leishmanial agent (Figure 1.29). The empirical formula of RT-01 (**97**) is $C_{13}H_{22}N^+ \cdot C_3H_3Cl_4OTe^-$.The organotellurane (RT-01) is evaluated to determine the effects *in vitro* against *L. amazonens* is and *in vivo* in *L. amazonens* is infected mice. The screening results revealed that the intralesional administration (720 µg/kg/day) of RT-01 in mice displayed a significant delay in the development of cutaneous lesions and decreased the number of parasites obtained from the lesions [39].

Tellurium compounds are found to prevent and reverse type-1 diabetes in NOD Mice by modulating the $\alpha_4\beta_7$ integrin activity, IL-1β, and T Regulatory Cells Tellurium-based compounds such as AS101 and SAS considerably elevate the number of T regulatory cells in the pancreas and thereby potentially control the autoimmune system. Chemically, AS101 (**86**) is ammonium trichloro(dioxoethylene-o,o′)tellurate and SAS (**98**) is chemically known as octa-O-bis-(R, R)-tartarateditellurane (Figure 1.30) [40].

AS101 generally inactivates the cysteine proteases by interacting with the thiol group. It also inhibits the enzyme caspases and thereby down-regulating the caspase-1 inflammatory products such as IL-18 and IL-1β. The direct inhibition of anti-inflammatory cytokine IL-10 induces the up-regulation of glial cell line-derived

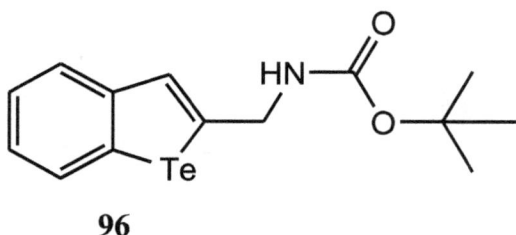

Figure 1.28: Structures of tert-butylbenzo[b]tellurophen-2-ylmethylcarbamate.

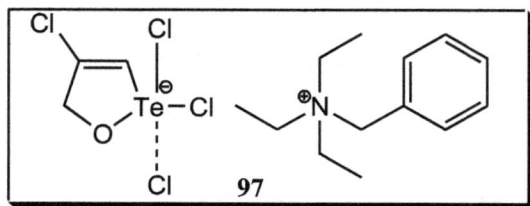

Figure 1.29: Structure of organotellurium compound (RT-01).

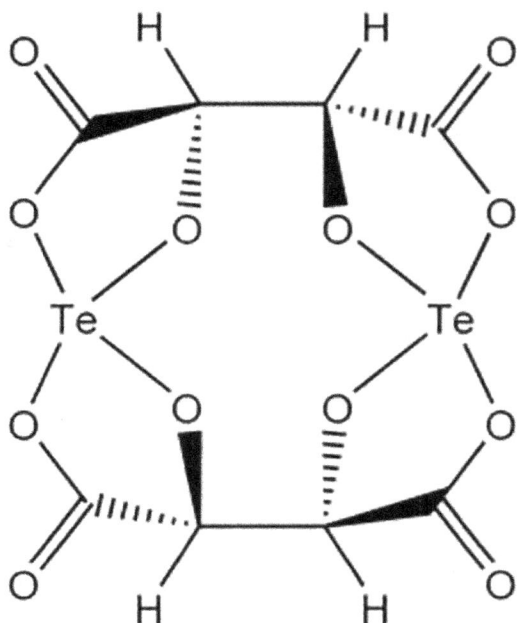

Figure 1.30: Structure of SAS (**98**).

neurotrophic factor (GDNF), which in turn induces the activation of Akt and the associated cell survival pathways (Figure 1.31).

Similarly, the compound RT-04 mainly inhibits the cathepsin B and thereby induces apoptosis in HL60 cells with no significant toxic effects observed in normal bone marrow cells. The possible biochemical mechanism of action includes the regulation of Bcl proteins in the cancer cells. Chemically, RT-04 is (3E)-4-chloro-3-[dichloro(4-methoxyphenyl)tellanyl]-2-methylbut-3-en-2-ol (Figure 1.32).

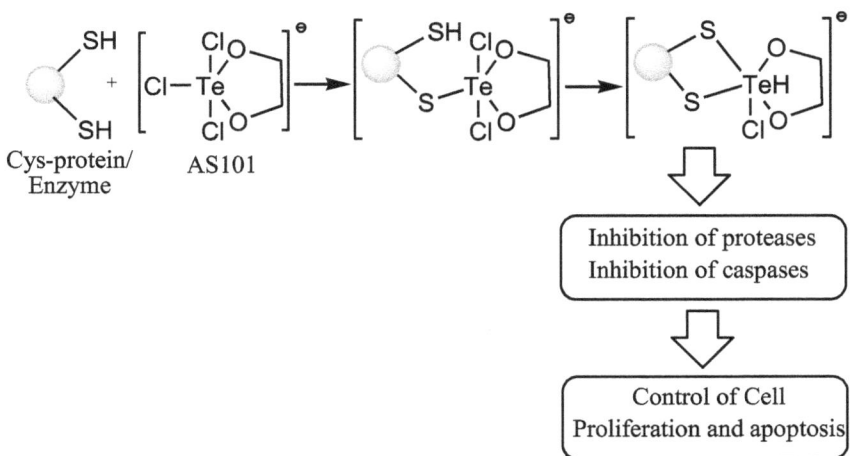

Figure 1.31: Mode of action of AS101.

Abbas et al. performed the preparation of tetrazole-based organotellurium compounds (Figure 1.33). The tetrazoles are the most popular five-membered heterocyclic compound that contains four nitrogen atoms and one carbon atom in the ring. The tetrazole moiety possesses a carbon-hydrogen bond with an acidic character (pKa = 23).

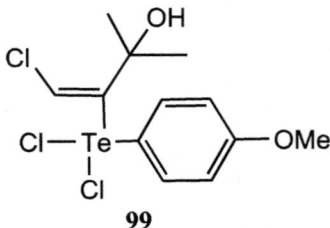

Figure 1.32: Structure of RT-04.

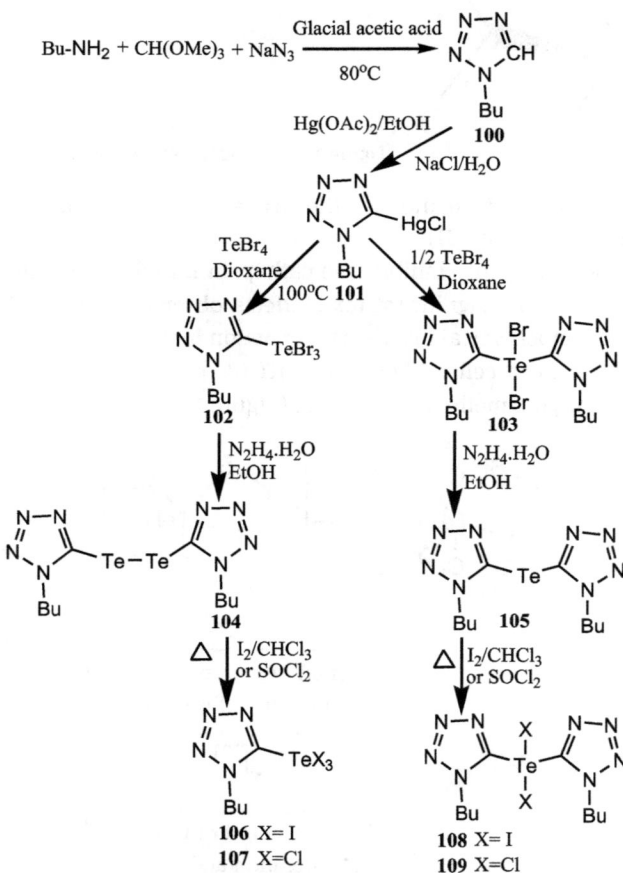

Figure 1.33: Synthesis of tetrazole-based organotellurium compounds.

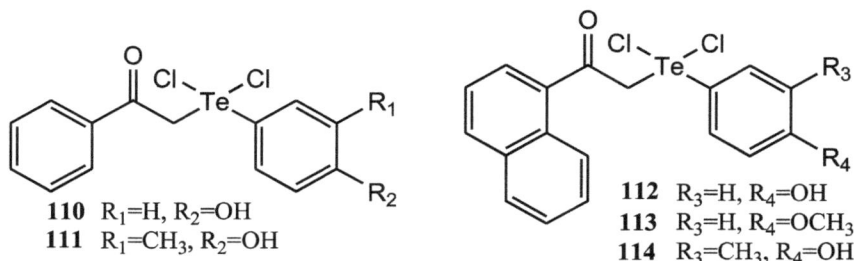

Figure 1.34: Organotelluranes with gram-negative antibacterial effect.

110 R_1=H, R_2=OH
111 R_1=CH$_3$, R_2=OH

112 R_3=H, R_4=OH
113 R_3=H, R_4=OCH$_3$
114 R_3=CH$_3$, R_4=OH

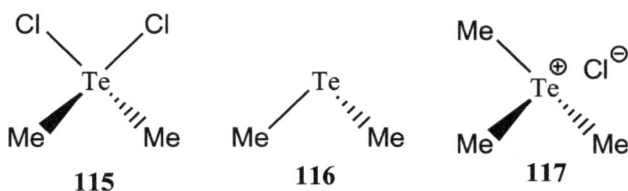

115   116   117

Figure 1.35: Tellurium-based inhibitors of cholesterol biosynthesis.

The synthetic route for generating organotellurium compounds involves the trans-metalation of organomercury compounds containing tetrazole scaffolds with tellurium tetrabromide. These are useful due to several synthetic routes and a wide range of applications [41].

Soni et al. reported the anti-bacterial potential of a series of unsymmetrical diorganyltellurium dichlorides (**110–114**) (Figure 1.34). It was observed that the target compounds exhibited anti-bacterial activity against both gram-positive (*Bacillus subtilis, Staphylococcus aureus*) and gram-negative bacteria (*Escherichia coli, Pseudomonas aeruginosa*, and *Salmonella sp.*). Among the series, naphthyl derivatives (**112–114**) were found most effective [42].

Laden et al. reported that the compounds **115, 116**, and **117** (Figure 1.35) exhibit potential inhibitory action on squalene monooxygenase (SM) that leads to the dramatic reduction in cholesterol biosynthesis. The inactivation of SM is due to the interaction of the telluranes and a pair of vicinal thiols from the catalytic cysteine in the enzyme system. Dimethyltellurium dichloride **115** is found to be the potential inhibitor of SM [43].

Mono-telluride has successfully been applied for the synthesis of chiral monotelluride **119** (Figure 1.36). This monotelluride is an optically active compound and its synthesis involves the reaction between one equivalent of finely ground powdered tellurium and two equivalents mixture of sodium tetrahydroborate, ethanol, and dimethylformamide. The reaction mixture is heated at 70 °C for 1 h till the formation of a colorless solution. Further, the resulting solution is cooled to 25 °C, and two equivalents of terphenyltosylate **118** are added in the presence of dimethylformamide [44].

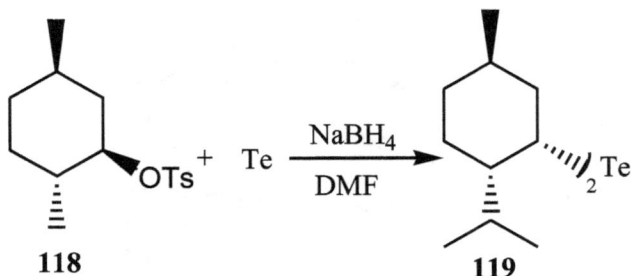

Figure 1.36: Synthesis of chiral monotelluride.

Tellurobistocopherols **121** (Figure 1.37) is prepared by the reaction between bromotocopherols **120**, t-butyl lithium, and tellurium tetrachloride followed by reduction. Tellurium is introduced into a carbon lithium bond (C–Li). The compound **120** is found to have antioxidant properties due to the presence of the tellurium atom [45].

The trichloro(4-methoxyphenyl)tellurium **123** reacts with *N*-methylbenzothiazole-2-(3*H*)-thione **124** to afford trichloro(4-methoxyphenyl)-tellurium-*N*-methylbenzothiazole-2-(3*H*)-thione complex **125** (Figure 1.38). This complex displays square pyramidal geometry due to the presence of lone pairs of electrons in the central atom of tellurium. First of all, anisole **122** reacts with tellurium tetrachloride refluxed in dry chloroform to yield trichloro(4-methoxyphenyl)tellurium **123**. Then five equivalents of compound **123** are treated with one equivalent of compound **124** in tetrahydrofuran. The resulting mixture is stirred at room temperature for 1 h to produce yellow precipitates of compound **125** with 87% yield [46].

Compound **128** exhibits antioxidant activity due to electron transfer by tellurium which is attained by the high reactivity of the tellurium molecules. For the synthesis of compound **128**, one equivalent of compound **126** in ethanol is added into two and a half equivalents of sodium borohydride (NaBH$_4$). Then the resulting solution is stirred at room temperature and refluxed in the presence of a nitrogen environment till the appearance of colorless solution. Further, two equivalents of compound **127** are added to the colorless solution. The resulting solution is further refluxed to obtain yellow-colored product **128** with 87% yield (Figure 1.39) [47].

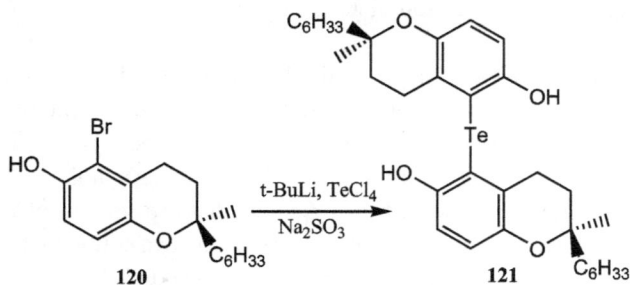

Figure 1.37: Synthesis of tellurobistocopherol.

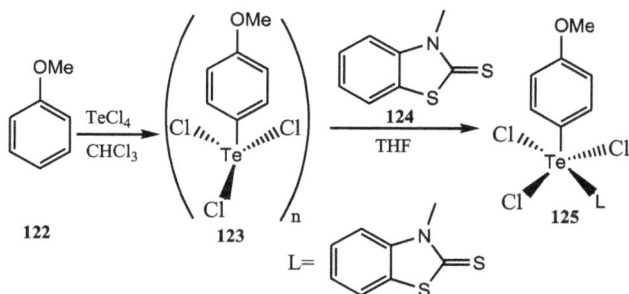

Figure 1.38: Synthesis of tellurium-N-methylbenzothiazole-2-(3H)-thione complex.

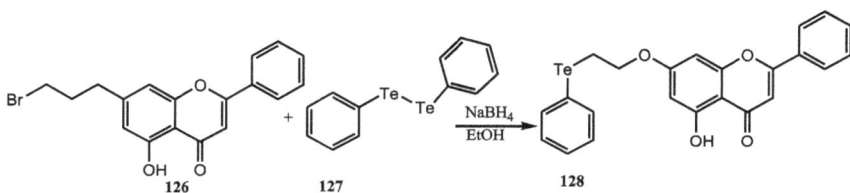

Figure 1.39: Synthesis of compound **128**.

4-amino-N-(5-methyl isoxazole-3-yl)benzenesulfonamide (sulphamethoxazole) **129** reacts with mercuric acetate in the presence of sodium chloride to produce 2-amino-(N-(5-methyl isoxazole-3-yl)sulfamoyl)mercuric chloride **130**. Compound **130** further reacts with tellurium tetrabromide to afford organotellurium compound **131** which on reduction in presence of hydrazine hydrate to produce diorganylditelluride **132** in a good yield (Figure 1.40). These compounds are found to have antibacterial activity [48].

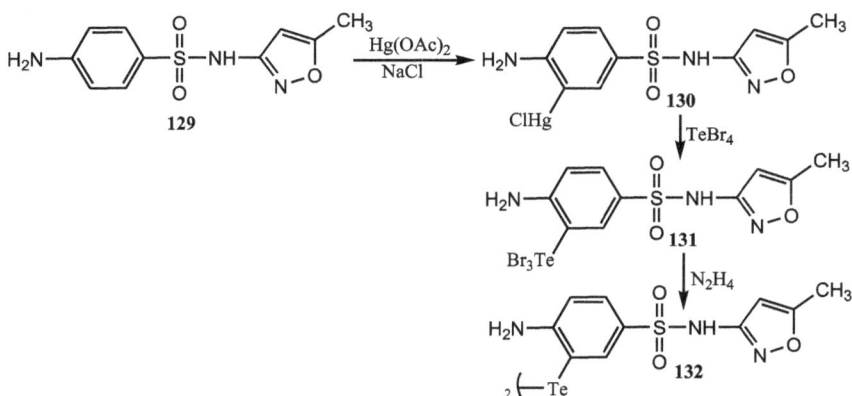

Figure 1.40: Synthesis of diorganylditelluride.

Al-Fregi et al. performed the synthesis and antimicrobial evaluation of organo-tellurium compounds based on pyrazole scaffold with a general molecular formula of $ArTeBr_3$ and Ar_2TeBr_2 [Ar = 2-(3-(4-substituted phenyl)-5-(2-chlorophenyl)-1*H*-pyrazol-1-yl)-3,5-dinitrophenyl] (Figure 1.41). First of all, organomercuric chloride-containing pyrazole moiety **135** is attained by reaction between substituted chalcones **133** and 2-hydrazinyl-3,5-dinitrophenylmercury chloride **134**. Then, the corresponding aryl mercuric chlorides react **135** with $TeBr_4$ in two different mole ratios of 1:1 and 2:1 to produce $ArTeBr_3$ **136** and Ar_2TeBr_2 **137** respectively. The synthesized compounds are evaluated for their antimicrobial activity against both Gram-negative and Gram-positive bacteria based on the agar diffusion method. It is observed that the presence of substituents like bromo, methoxy, and methyl on aryl rings potentiate antimicrobial activity [49].

Andersson et al. reported that the diaryl tellurides **138** are found to exhibit inhibitory action efficiently on the peroxidation in hepatocytes and liver microsomes of rats (Figure 1.42). The mechanism action of the thiol-peroxidase activity of organo-tellurides was further demonstrated by Engman (Figure 1.43). It was observed that the organotelluroxide in the hydrate form reacts with thiols to produce disulfides and regenerate the initial diorganotelluride [50].

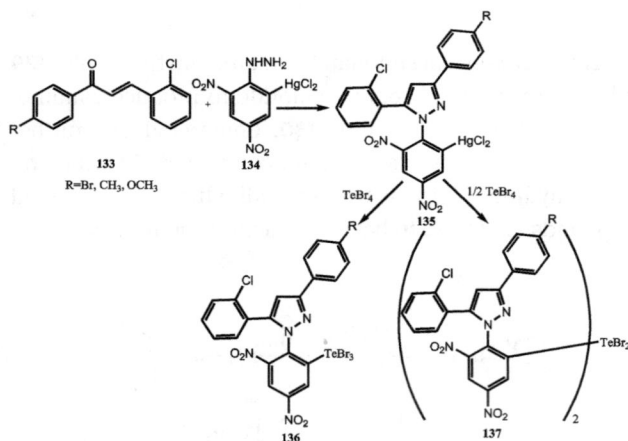

Figure 1.41: Synthesis of organotellurium compounds based on pyrazole derivatives.

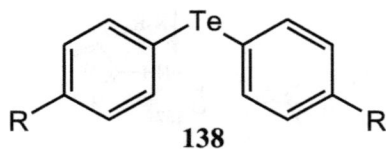

138

R= NH_2, OH, OMe, F, Me, CF_3, Cl, Br, NO_2

Figure 1.42: Synthesis of diaryltellurides.

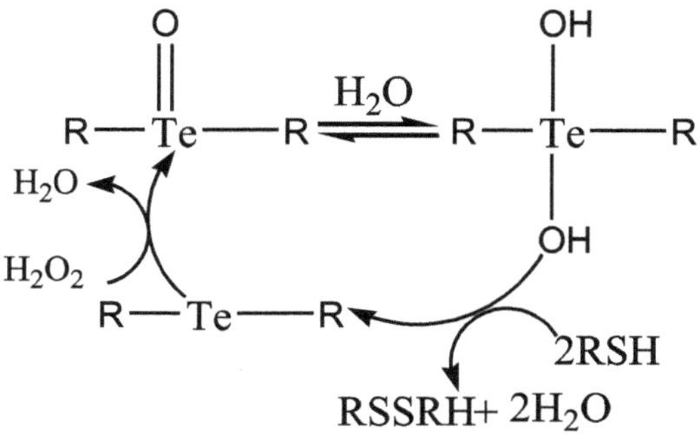

O
‖
R—Te—R $\rightleftharpoons$ R—Te—R
H_2O

OH
|
R—Te—R

OH

H_2O
H_2O_2

R—Te—R

2RSH

RSSRH+ $2H_2O$

Figure 1.43: Mechanism action of diorganotellurides towards the GPx-like activity.

1.4 Conclusions

The synthesis and applications of organotellurium compounds provide wide opportunities for the research and development of medicinally active agents. Organotellurium compounds are handled easily and purified because these compounds are crystalline. These compounds display their applications in the different chemical reactions including oxidation, reduction, functionalization, addition, cyclization, elimination, rearrangement reactions, polymerization reactions, etc. The utility of organotellurium compounds is increased in industrial applications, synthetic transformations, and the production of pharmaceutical products. Organochalcogen compounds also act as a potential component of various biologically active compounds. So, the knowledge of the toxicity profile, physicochemical properties, mode of action, and synthetic methodologies of the organotellurium compounds is essential for the development of potential medicinally active agents.

References

1. Aicha Ba L, Doring M, Jamier V, Jacob C. Tellurium: an element with great biological potency and potential. Org Biomol Chem 2010;8:4203–16.
2. Lars E. Synthetic applications of organotellurium chemistry. Acc Chem Res 1985;18:274–9.
3. Irfan M, Rehman R, Razali MR, Rehman S, Rehman A, Iqbal MA. Organotellurium compounds: an overview of synthetic methodologies. Rev Inorg Chem 2020;40:193–232.
4. Larner AJ. Biological effects of tellurium: a review. Trace Elem Electrolytes 1995;12:26–31.
5. Dittmer DC. Tellurium. Chem Eng News 2003;81:128.
6. Bajpai S, Shamsi M. Organotellurium compounds: from molecular to supramolecular chemistry. Asian J Chem 2018;30:1183–9.
7. Smith WE, Franklin DV, Goutierrez KL, Fronczek FR, Mautner FA, Junk T. Organotellurium chemistry: synthesis and properties of 2-acylamino- and 2-arylamino-1,3-benzotellurazoles. Am J Het Chem 2019;5:49–54.
8. Al-Fregi AA, Abdul-Sattar J, Abdulsahib HT. Synthesis and characterization of some new organotellurium compounds based on quinolone. Eur J Chem 2017;8:218–23.

9. Irie M, Doi Y, Ohsuka M, Aso Y, Otsubo T, Ogura F. Synthesis of Optically Active Tellurium-Containing Binaphthyls and Their Use in the Asymmetric 1,4-Addition Reaction of α,β- Unsaturated Ketones. Tetrahedron Asymmetry 1993;4:2127.
10. Kumar SK, Singh HB, Das K, Sinha UC. Synthesis and structure of di-2-benzo[b]thienyl ditelluride. J Organomet Chem 1990;397:161.
11. Tripathi SK, Patel D, Roy D, Sunoj RB, Singh HB, Wolmershauser G, et al. o-Hydroxylmethylphenylchalcogens: Synthesis, Intramolecular Nonbonded Chalcogen,,,OH Interactions, and Glutathione Peroxidase-like Activity. J Org Chem 2005;70:9237.
12. Kumar S, Engman L, Valgimigli L, Amorati R, Fumo MG, Pedulli GF. Antioxidant Profile of Ethoxyquin and Some of Its S, Se, and Te Analogues. J Org Chem 2007;72:6046.
13. Yoshimori Y, Cho CS, Uemura S. Diacetoxylation of nonconjugated dienes with TeO2 and the isolation of intermediate organotellurium compounds. J Organom Chem 1995;487:55.
14. Patra A, Wijsboom YH, Leitus G, Bendikov M. Synthesis, Structure, and Electropolymerization of 3,4-Dimethoxytellurophene: Comparison with Selenium Analogue. Org Lett 2009;11:1487.
15. Wanga L, Huang Z. First application of ionic liquid to reactions involving organotellurium compounds as intermediates. J Chem Res 2005;7:446–8.
16. Al-Rubaie A, Al-Masoudi W, Al-Jadaan SAN, Jalbout AF, Hameed AJ. Synthesis, characterization, and computational study of some new organotellurium compounds containing azomethine groups. Heteroatom Chem 2008;19:307–15.
17. Bhasin KK, Gupta V, Sharma RP. Synthesis of alkali metal tellurides and ditellurides in THF and their relative reactivities towards alkali bromides: a convenient synthesis of dialkyl tellurides and dialkyl ditellurides. Ind J of Chem Section A: Inorganic, Physical, Theoretical and Analytical 1991; 30:632–4.
18. Bhasin K, Singh N, Doomra S, Arora E, Ram G, Singh S, et al. Regioselective synthesis of bis(2-halo-3-pyridyl)dichalcogenides (E= S, Se, and Te): directed ortholithiation of 2-halopyridines. Bioinorg Chem App 2007;3:1–9.
19. De Meio RH, Henriques FC. Tellurium IV, excretion and distribution in tissues studied with a radioactive isotope. J Biol Chem 1947;169:609–23.
20. Asakura T, Shibutani Y, Reilly MP. Antisickling effect of tellurite: a potent membrane acting agent *in vitro*. Blood 1984;64:305–7.
21. Shohat M, Mimouni D, Ben-Amitai D, Sredni B, Sredni D, Shohat B, et al. *In vitro* cytokine profile in childhood alopecia areata and the immunomodulatory effects of AS-101. Clin Exp Dermatol 2005; 30:432–4.
22. Guest I, Uetrecht J. Bone marrow stem cell protection from chemotherapy by low molecular weight compounds. Exp Hematol 2001;29:123–37.
23. Dogan HB, Kuleasan H, Cakir I, Halkman AK. Evaluation of increased incubation temperature and cefixime-tellurite treatment for the isolation of *Escherichia coli* O157:H7 from minced beef. Int J Food Microbiol 2003;87:29–34.
24. Sailer BL, Liles N, Dickerson S, Chasteen TG. Cytometric determination of novel organotellurium compound toxicity in a promyelocytic (HL-60) cell line. Arch Toxicol 2003;77:30–6.
25. Chasteen TG, Bentley R. Biomethylation of selenium and tellurium: microorganisms and plants. Chem Rev 2003;103:1–26.
26. Andersson CM, Brattsand R, Hallberg A, Engman L, Persson J, Moldéus P, et al. Diaryltellurides as inhibitors of lipid peroxidation in biological and chemical systems. Free Radic Res 1994;20: 401–10.
27. Leonard KA, Nelen MI, Simard TP, Davies SR, Gollnick SO, Oseroff AR, et al. Synthesis and evaluation of chalcogenopyrylium dyes as potential sensitizers for the photodynamic therapy of cancer. J Med Chem 1999;42:3953–64.
28. Engman L. Synthetic applications of organotellurium chemistry. Acc Chem Res 1985;18:274–9.

29. Bergman J. In: Berry FJ, Mc Whinnie WR, editors. Proceedings of the Fourth International Conference on the Organic Chemistry of Selenium and Tellurium. The University of Aston in Birmingham; 1983. p. 215.
30. Cuthbertson E, MacNicol DD. Tellurium extrusion: synthesis of benzocyclobutene and naphto[b] cyclobutene. Tetrahedron Lett 1975;16:1893–4.
31. Princival C, Alcindo A, Santos D, Joao V. Comasse to solventless and mild procedure to prepare organotellurium (IV) compounds under microwave irradiation. J Braz Chem Soc 2015;26:832–6.
32. Engman L, Stern D, Cotgreave IA, Andersson CM. Thiol peroxidase activity of diarylditellurides as determined by a proton NMR method. J Am Chem Soc 1992;114:9737.
33. Vazquez-Tato MP, Mena-Menéndez A, Feás X, Seijas JA. Novel microwave-assisted synthesis of the immunomodulator organotellurium compound ammonium trichloro(dioxoethylene-O,O′)tellurate (AS101). Int J Mol Sci 2014;15:3287–98.
34. Fonseca SF, Lima DB, Alves D, Jacob RG, Perin G, Eder JL, et al. Synthesis, characterization and antioxidant activity of organoselenium and organotellurium compounds derivatives of chrysin. New Journal of Chem 2015;39:3043–50.
35. Chen HY, Wang YQ, Xu J, Ji JZ, Zhang J, Hu YZ, et al. Quantum dots for biological imaging. J Fluoresc 2008;18:801–11.
36. Al-Asadi RH, Al-Masoudi WA, Abdual-RassolSynthesis KS. Biological activity and computational study of some new unsymmetrical organotellurium compounds derived from 2-amino-5-carboxyphenyl mercury (II) chloride. Asian J Chem 2016;28:1171–6.
37. Kanski J, Drake J, Aksenova M, Engman L, Butterfield DA. Antioxidant activity of the organotellurium compound 3-[4-(N,N-dimethylamino)benzenetellurenyl]propane sulfonic acid against oxidative stress in synaptosomal membrane systems and neuronal cultures. Brain Res 2001;911:12–21. PMID: 11489439.
38. Yoon-Jung K, Dong HL, Yong-Sung C, Jin-Hyun J, Kwon SH. Benzo[b]tellurophenes as a potential histone H3Lysine 9 Demethylase (KDM4) Inhibitor. Int J Mol Sci 2019;20:5908.
39. Lima CBC, Arrais-Silva WW, Cunha RLOR, Giorgio S. A novel organotellurium compound (RT-01) as a new antileishmanial agent. Korean J Parasitol 2009;47:213–8.
40. Yossipof TE, Bazak ZR, Kenigsbuch-Sredni D, Caspi RR, Kalechman Y, Sredni B. Tellurium compounds prevent and reverse type-1 diabetes in NOD mice by modulating a4β7 integrin activity, IL-1β, and T regulatory cells. Front Immunol 2019;10:1–14.
41. Abbas SH, Al-Fregi AA, Al-Yaseen AA. Synthesis of some new organotellurium compounds based on 1-substituted tetrazole. J Phys Conf Ser 2021;3:1–27.
42. Soni D, Gupta PK, Kumar Y, Chandrashekhar TG. Antibacterial activity of some unsymmetrical diorganyltellurium(IV) dichlorides. Indian J Biochem Biophys 2005;42:398–40.
43. Laden BP, Porter TD. Inhibition of human squalene monooxygenase by tellurium compounds: evidence of interaction with vicinal sulfhydryls. J Lipid Res 2001;42:235–40.
44. Ścianowski J, Pacuła AJ, Wojtczak A. New and efficient methodology for the synthesis of chiral mono-and ditellurides. Tetrahedron Asymmetry 2015;26:400–3.
45. Poon J-F, Yan J, Singh VP, Gates PJ, Engman L. Regenerable radical-trapping tellurobistocopherol antioxidants. J Org Chem 2016;81:12540–4.
46. O'Quinn GK, Rudd MD, Kautz JA. Synthesis and characterization of complexes of Te (IV) with sulfur and selenium containing ligands: crystal and molecular structure of trichloro(4-methoxyphenyl) tellurium (IV)·N-methylbenzothiazole-2-(3H)-thione. Phosphorus Sulfur Silicon Relat Elem 2002; 177:853–62.
47. Fonseca SF, Lima DB, Alves D, Jacob RG, Perin G, Lenardao EJ, et al. Synthesis, characterization and antioxidant activity of organoselenium and organotellurium compound derivatives of chrysin. New J Chem 2015;39:3043–50.

48. AL-Jadaan S. Synthesis and characterization of some new organotellurium compounds derived from sulphamethoxazole. Res J Pharm Biolog Chem Sci 2014;5:594–8.
49. Sabti AB, Al-Fregi AA, Yousif MY. Synthesis and antimicrobial evaluation of some new organic tellurium compounds based on pyrazole derivatives. Molecules 2020;25:3439.
50. Andersson CM R, Hallberg A, Engman L, Persson J, Moldeus P, Cotgreave I. Diaryltellurides as inhibitors of lipid peroxidation in biological and chemical systems. Free Radic Res 1994;20: 401–10.

Monika Kamboj*

2 Chemistry of tellurium containing macrocycles

Abstract: The chemistry of Tellurium containing macrocycles has received great attraction and developed rapidly. Recently inorganic chemists are fascinated by ligands containing macrocycles having tellurium as soft donor and N and O as hard donor atoms. The tellurium atom is more electropositive than carbon due to its large size that resulted in polarisation of Te–C bond. So, tellurium containing macrocycles are explored due to their high reactivity and toxicity. Well-designed macrocycles containing different metals is an interesting field of chemistry as macrocycle with mixed donor atoms can bind two different metal atoms with different nature within the same cavity and thereby ion selectivity increases. Chemistry of macrocycles with tellurium as soft donor atoms also gives rise to very interesting coordination behaviour as addition of Tellurium in macrocycle adds an additional probe (^{125}Te NMR help to monitor their structures in solutions). The chemistry of hard and soft donors in macrocyclic framework makes interesting coordination chemistry and need to be explore. The discussion includes different types of tellurium macrocycles and their chemistry.

Keywords: coordination behaviour; hard donor; macrocycles; soft donor; tellurium.

2.1 Introduction

Tellurium is a semi -metallic, silver-white element, shiny, crystalline and hard. Tellurium forms many compounds like its group members (sulphur and selenium). Recently, chemists have shown their curiosity in Tellurium chemistry otherwise it was unnoticed for several years. Chemists concentrated their research on tellurite, tellurate and organic tellurides. After 1970, the ligand chemistry of tellurium was explored. During the last decade, several findings have fuelled a new interest in this element. After 1970, the ligand chemistry of tellurium was explored. Before 1970 there was delusion that they are toxic in nature, foul smelling, air-sensitive and non-accessibility of a various organotellurium ligands commercially. Now the new applications of tellurium compounds in material science have given impetus in this field [1]. The chemistry of Tellurium containing macrocycles has received great attraction and developed rapidly. Recently inorganic chemists are fascinated by tellurium ligands containing macrocycles having tellurium as soft donor and N and O

*Corresponding author: Monika Kamboj, Department of Applied Chemistry, Amity School of Applied Sciences, Amity University Uttar Pradesh, Lucknow Campus, Lucknow, 226028, UP, India, E-mail: mkamboj@lko.amity.edu. https://orcid.org/0000-0002-3367-3995

As per De Gruyter's policy this article has previously been published in the journal Physical Sciences Reviews. Please cite as: M. Kamboj "Chemistry of tellurium containing macrocycles" *Physical Sciences Reviews* [Online] 2022. DOI: 10.1515/psr-2021-0106 | https://doi.org/10.1515/9783110735840-002

as hard donor atoms. The chemistry of hard and soft donors with a metal centre [2] in framework make an interesting coordination chemistry and need to be explore [2, 3]. Such assemblies play significant role in transition metal catalysed asymmetric synthesis [4, 5] MOCVD processes [6–8] and in mimicking as models for proteins and enzymes [9–13]. The tellurium atom is more electropositive than carbon due to its large size that resulted in polarisation of Te–C bond. So, tellurium containing macrocycles are explored due to their high reactivity and toxicity. Well-designed macrocycles containing different metals is an interesting field of chemistry. The chemistry of macrocycles with tellurium as soft donor atoms stimulate interest in coordination chemistry. Amalgamation of large Tellurium atom in macrocycle alters the size of the cage cavity. Due to high σ-donating ability of tellurium, complexation with a variety of metal ions can occur easily [14]. This amalgamation of Te in macrocycle with donor atom N, S makes some interesting coordination behaviour. Macrocycles are useful in Supramolecular chemistry, which provide recognition sites in their cavity to bind the guest and modified their properties, is an emerging branch of chemistry with enormous application.

In recent years, synthesis of macrocyclic Schiff bases with phenol, pyridine, pyrrole, furan and thiophene have been studied along with their ligating chemistry [15–20]. Synthesis of polyazamacrocyclic complexes with hard and soft metal ions and numerous metallocene groups had been reported by Beer et al. [21]. The adverse non-bonding electron pair interaction between nitrogen atoms in the ring are reduced due to intramolecular Te -N interaction that gives impetus for the formation of the macrocyclic ring. The first example of macrobicyclic ligand with tellurium/selenium (Te_3N_8 or Se_3N_8) have been reported by A. Panda et al. [14].

2.2 Advantage of adding tellurium in macrocycle

- Addition of Tellurium in macrocycle adds an additional probe. [125]Te-NMR help to monitor their structures in solutions
- It has better sigma donating properties as a ligand which makes an important and rich coordination chemistry
- Introduction of Tellurium in macrocycles help to obtain macrocycle with mixed donor atoms. So, they can bind two different metal atoms with different nature within the same cavity and ion selectivity increases.
- Such complexes can be used for changing the oxidation-reduction abilities of the transition metal cation
- Promising tool for an allosteric effects and bimetallic catalysis [22]

2.3 Types of tellurium macrocycles

2.3.1 Telluraporphyrinoid

Porphyrinoid are core modified pyrrole containing macrocycles (**1**). They are also known as heteroporphyrinoids where one or more nitrogen atom(s) of the pyrrole ring in macrocycles are substituted by chalgogen atoms such as Oxygen, Sulphur, Selenium, and Tellurium through core modification method [23].

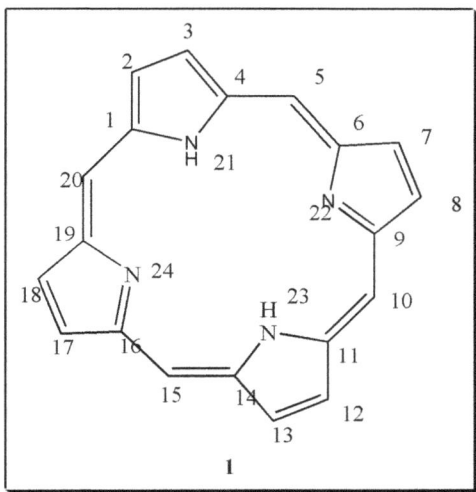

1

Telluraporphyrinoid are the core modified pyrrole containing macrocycles in which Te is the hetero atom at position 21 (**2**). They are Planar and aromatic in nature. Telluraporphyrinoids are unique in its physico-chemical properties owing to large size of tellurium atom. Telluraporphyrinoids are the interesting class of macrocycles as compared to normal as well as other chalcogen containing porphyrinoid macrocycles as the presence of tellurium atom alters the electronic properties of porphyrinoids [24]. The Telluraporphyrinoid chemistry is not investigated in depth and the progress in this field is at slow pace when compared to the porphyrinoids with oxygen and sulphur atoms. The chemistry of Telluraporphyrinoids is relatively less explored as compared to the other group-16 chalcogen due to the:
- Higher reactivity of the Te–C bond as compared to S–C and Se–C bonds,
- Toxic effect of organotellurium compounds [25].

1978 was the year of inception of Telluraporphyrinoids when porphyrin containing tellurium was published by A. Ulman [26]. Not much work was done in this area for almost two decades. The porphyrin macrocycles whose core has been modified by replacing NH at 21 positions by Te are known as Telluraporphyrins (**2**).

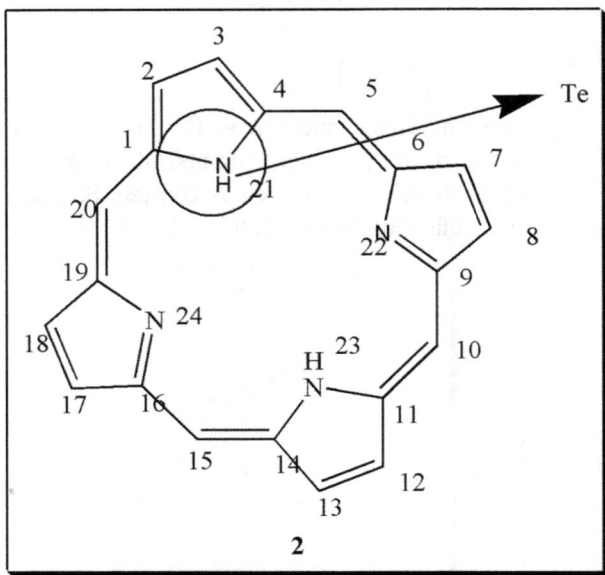

2

So there is a contact between Te and the 23- *trans* heteroatom (N or S) (**3a**, **3b**) due to tellurium's large van der Waal's radii [27].

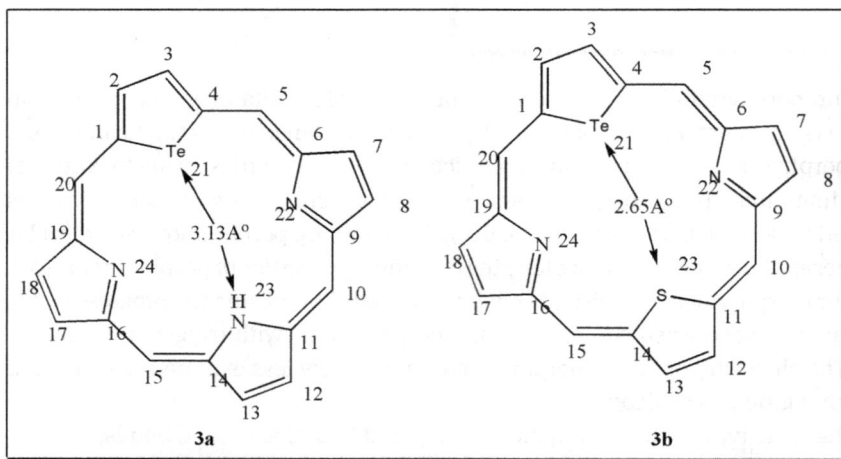

3a   **3b**

Due to large size of tellurium atom, two Te atoms cannot be accommodated in the macrocyclic core cavity [23]. If simultaneously two tellurium atoms are placed inside the macrocyclic cavity, then one of the tellurophene units is over turn. (**4**). Tellurium's large size also inhibit the metal ions in the cavity of macrocyclic core for coordination.

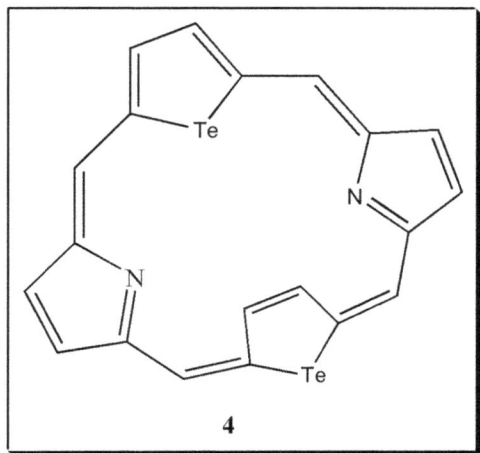

4

Due to this trans interaction, lone pair of heteroatoms bring about the oxidation of tellurium and can be utilised for the catalysis of H_2O_2 [28]. The first core modified telluroporphyrin with meso-substituted tellurium, **(5)** (tetraphenyl-21-tellura-23-thiaporphyrin) (Scheme 2.1) was reported by Ullman et al. in 1978 [26].

As shown in Scheme 2.1, Trace amount of structure A was obtained by refluxing 2,5-bis (α-phenylpyrrylmethylene) tellurophene with 2,5-bis(phenylhydroxymethyl) thiophene in dioxane

Tetraphenyl-21-tellura-23-thiaporphyrin **(5)** is first core modified telluroporphyrin. X-ray structure confirm the short distance of 2.65 A° between Te and S. The first core modified telluraporphyrin was reported by Ullman, with abnormal short Te …. S bond (at position 21 and 23) [26] and chemical bonding interaction between Te and S. The Porphyrin's inside and outside aromatic path varies due to this interaction, that changes shielding and deshielding zone of H atom at the periphery [29]. The downfield shift of the peripheral proton is not influence by electronegativity of heteroatom but due to Te …. S interaction, electron density is reduced for inner aromatic pathway. It has been observed that 21- Telluraporphyrins is distorted but planar, due to large size of Tellurium that resulted in longer interaction between Nitrogen atoms [30]. Since the macrocycle is planar and involve the participation of tellurophene ring in aromatic delocalisation. These 21-telluroporphyrins are easily oxidised as the large size of tellurium makes it distorted planar. They are used as catalysts as the Te is easily oxidised. The trans-atom at 23 positions donates the lone pair of electrons that bring the oxidation of Te.

Meso-substituted 21-telluraporphyrins **(6)** synthesis was reported by Latos-Grażyński et al. [30]. 2,5-bis(phenylhydroxymethyl)tellurophene, aromatic aldehyde and pyrrole in dichloromethane in (1:2:3 ratio) was condensed in presence of acid catalyst, followed by oxidation with *p*-chloranil as shown in (Scheme 2.2). 21-telluroporphyrins, **(6)** (Scheme 2.2) shows longer distance between N(1) and N(3) in

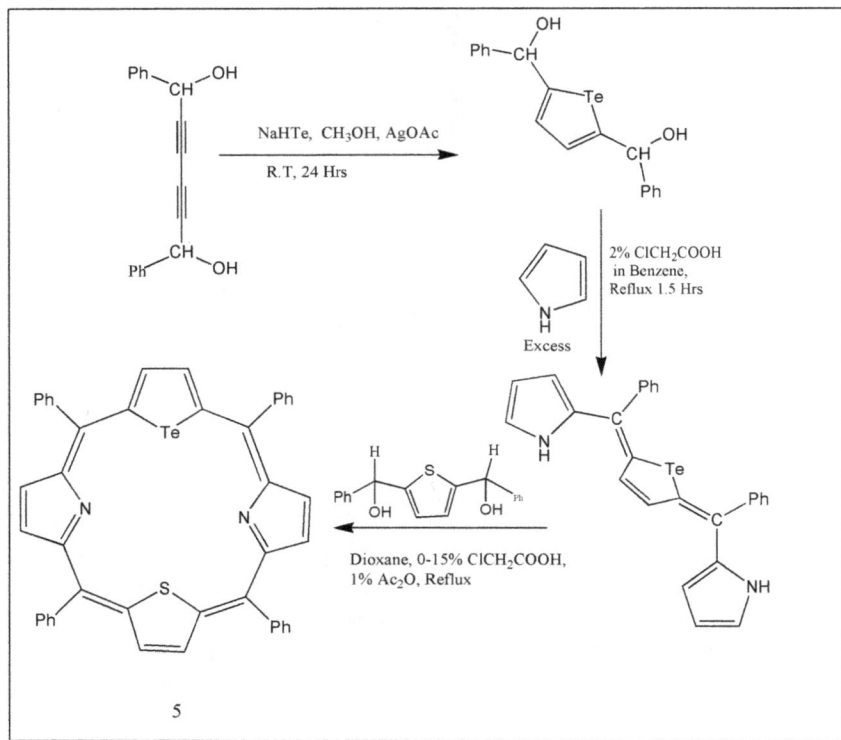

Scheme 2.1: Synthesis of tetraphenyl-21-tellura-23-thiaporphyrin (5) by Ulman et al.

comparison to normal porphyrin due to presence of large Tellurium atom in macrocycle that make it distorted. This has been corroborated by X-ray single crystal structure of 21-telluraporphyrin (Figure 2.1). This macrocyclic porphyrin is planar although the distance between Te and N(2) is short. The planar nature of this molecule is due to the participation of tellurophene ring in aromatic delocalisation unlike its selena counterpart.

Also, the 21-telluroporphyrins (6) is air sensitive and on reaction with *m*-chloroperoxybenzoic acid is rapidly oxidised to reddish-brown 21-oxaporphyrin (7). This renovation occurs through the isolation of intermediate (8) (tellurium hydroxyl compound) (Scheme 2.2). This novel tellurium compound, (8) has hydroxyl group attached to tellurium atom. Such types of conversion in tellurophene chemistry are rare. X-ray crystal structure of tellurium hydroxyl compound, (8) shows that NH of pyrrole trans to tellurophene ring protonate the oxygen of telluroxide and thereby produces a zwitterion in which hydroxyl group is attached to tellurium atom. A deprotonated nitrogen of trans pyrrole ring is weakly hydrogen bonded to this hydroxyl group. Ewa Pacholska *et* al [31] synthesised planar, novel macrocycle vacataporphyrin (aza deficient

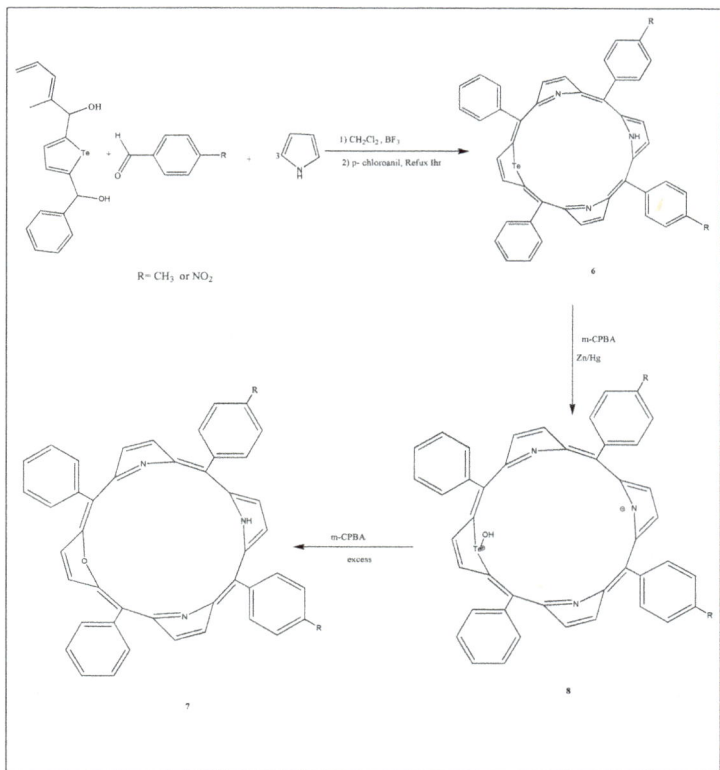

Scheme 2.2: (a) Synthesis of 21-telluraporphyrins(meso-substituted monotelluraporphyrins) (b) Oxidation of 21-telluraporphyrins.

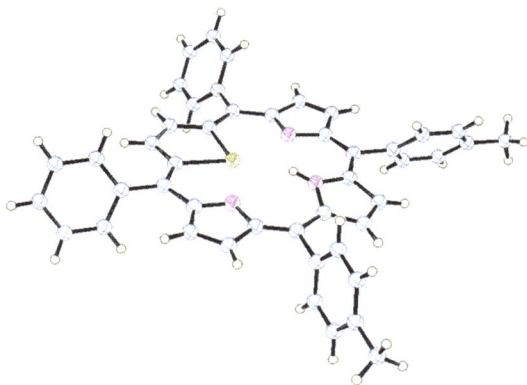

Figure 2.1: Crystal structure of 21-telluraporphyrin (6). Taken from ref [30] with permission from John Wiley & Sons, Inc.

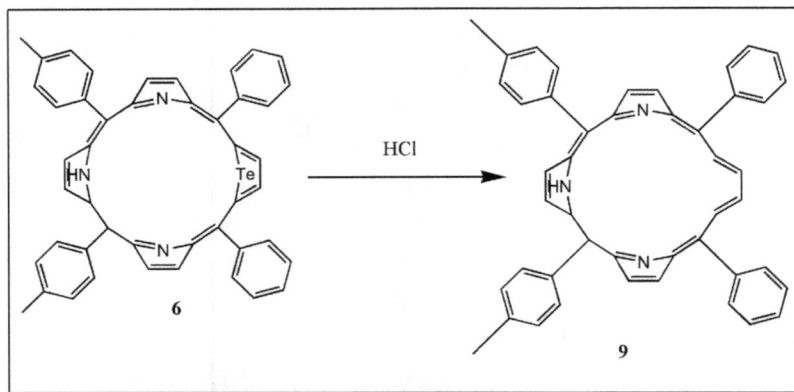

Scheme 2.3: Conversion of 21-telluraporphyrin (6) into 21-vacataporphyrin (9).

porphyrin) **(9)** from compound (5,20-diphenyl-10,15-di(p-tolyl)-21-telluraporphyrin) by refluxing it in 20% HCl and o-dichlorobenzene. This step expels the tellurium and enlarges the coordination core from 16 to 17 atoms (Scheme 2.3) and expand the macrocycle. Extrusion of Tellurium is possible as C-Te-C bond is fragile and electrophile H$^+$ attack carbon of tellurophene unit so this compound (5,20-diphenyl-10,15-di(p-tolyl)-21-telluraporphyrin) is considered as good substrate to synthesise new compounds.

Latos-Grażyński and his research team [32] had reported the extrusion of tellurium from ditelluraporphyrin. 5,10,15,20-tetraphenyl-21,23-ditelluraporphyrin was transformed into 21-tellura-23-vacataporphyrin **(10)** (by refluxing with HCl/Toluene) and 21,23-divacataporphyrins **(11)** (refluxing with HCl/o-dichlorobenzene) respectively, based on the reaction's conditions used (Scheme 2.4). Ewa Pacholska-Dudziak et al. [33] reported Pt(II) and Pt(IV) complexes, synthesised from 21,23- ditelluraporphyrin.

21,21-dichloro-21-telluraporphyrin **(12)** was synthesised by Detty and co-workers [34] from 21-telluraporphyrin.

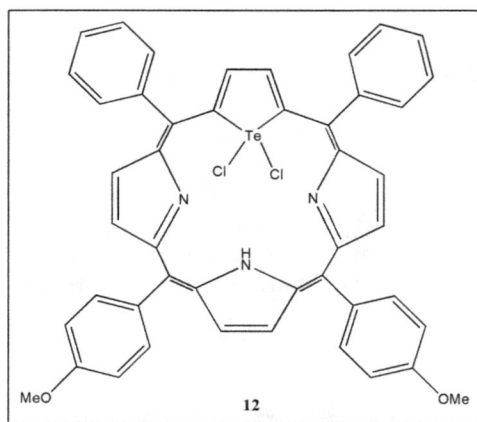

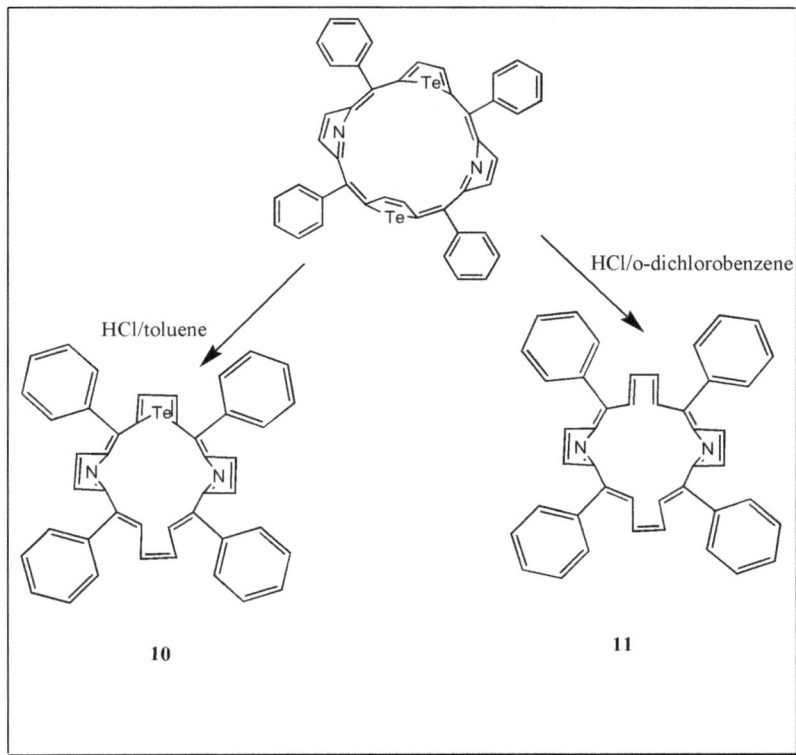

Scheme 2.4: Tellurium extrusion reactions of ditelluraporphyrin.

21-telluraporphyrin was first oxidised to telluroxide, followed by exchange of ligand with dichloromethane/HCl to give black, parallelepiped crystals of 21,21-dichloro-21-telluraporphyrin. 21-telluraporphyrin yielded monohalogenated derivatives on direct oxidation with halogen [28]. Formation of 21,21-dihalo-21-telluraporphyrins was unsuccessful by this method. They are outside the coordination core and so the distance between Te and trans N is longer. The Te atom with IV oxidation state, is in the centre of the distorted trigonal bipyramidal geometry with axial position occupied by two Cl atoms. Te-Cl bonds are of unequal length (one is 2.58 Å and other is 2.49 Å) and the Cl-Te-Cl bond angle was found to be 168.65°. 21,23-ditelluraporphyrin and 21-tellura-23-vacataporphyrin show ability to bind with palladium (II). 21,23-ditelluraporphyrins react with palladium (II) salts to form the products that strongly depend on the conditions of reaction and the source of palladium. On reaction with $Pd(OCOCH_3)_2$ in $CH_2Cl_2/(C_2H_5)_3N$, one Te atom in inverted tellurophene ring of 21,23-ditelluraporphyrin is substituted by a Pd atom. So, a regular palladacyclopenatadiene ring takes the position of an inverted tellurophene ring in 21,23-ditelluraporphyrin. Whereas on refluxing with $PdCl_2$ in $CH_2Cl_2/(C_2H_5)_3N$, Te insertion occurs with inverted tellurophene ring is preserved. In this $PdCl_2$ macrocyclic complex, Pd form the coordinate bond with the Te of normal tellurophene ring. 21-tellura-23-vacataporphyrin on reaction with $Pd(PhCN)_2Cl_2$ in dichloromethane formed the brown red coloured Pd(II) complex [29]. This Pd complex has square planar geometry and its X-ray structure showed that Pd is coordinated to tellurium and nitrogen (neighbouring porphyrin heteroatoms) and to two chlorides in a cis arrangement.

Scheme 2.5 depicts the unique features of Telluraporphyrins and its transformation journey. The salient features of this scheme [35] is:

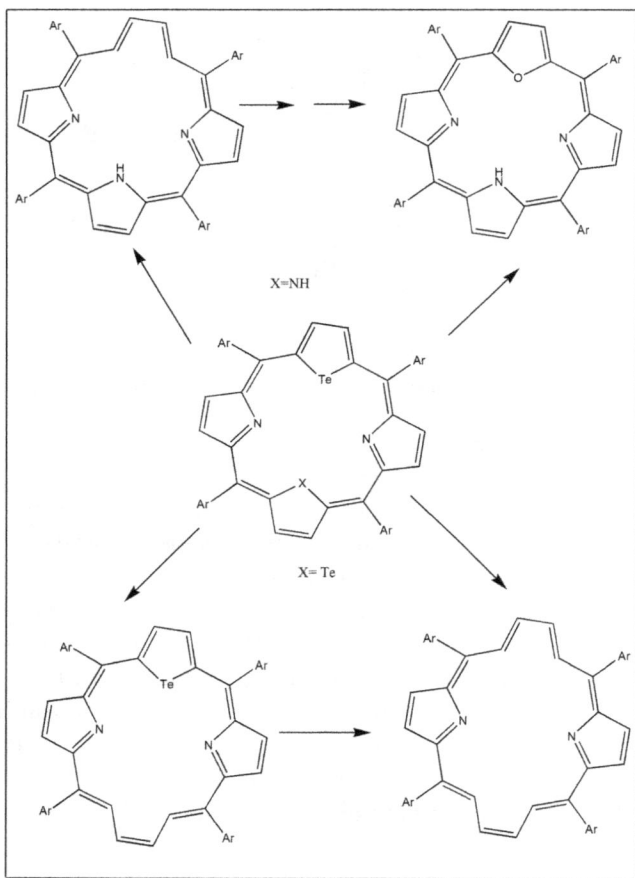

Scheme 2.5: Transformation of Telluraporphyrins. Reprinted with permission from ref. [35]. Copyright (2004) American Chemical society.

- The integrity of the porphyrin skeleton in 21-telluraporphyrin was maintained even after the removal of Te atom from it
- Expansion of macrocycle ring by removal of a tellurium atom from 1 by oxidation or by under acidic conditions to form 21-oxaporphyrin three and vacataporphyrin four respectively.
- 21,23-ditelluraporphyrin two can be transformed to 21-Tellura-23-vacataporphyrin 5, and 21,23-divacataporphyrin six by refluxing with acid
- Also 21-telluraporphyrin can be transformed to 21,21-dichloro-21-telluraporphyrin

2.3.2 Cryptand

Cryptands are a member of synthetic bicyclic and polycyclic multidentate ligands that binds a variety of cations in a cryst in three dimensions [36].

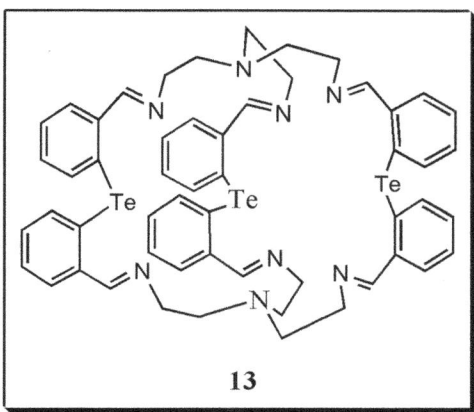

13

(**13**) is the only known tellurium-containing cryptand reported by A Panda et al. [37] It was synthesised by template condensation of 2 mol of tris(2-aminoethyl) amine (tren) and 3 mol of bis (*o*-formylphenyl)telluride in presence of templating cation, cesium ion (Scheme 2.6).

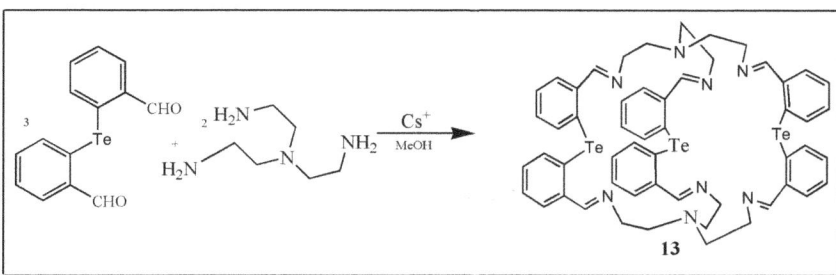

Scheme 2.6: Synthesis of Tellurium containing cryptand.

2.3.3 Macrocyclic Telluroether

These macrocyclic ligands are homoleptic compound. They are also known as Telluroether crown and are designated as (n- Te –m). Where n is the ring size, m is the number of Tellurium atoms in the ring. 1,5-ditelluracyclooctane, (8Te2), novel eight membered macrocyclic compound with two tellurium atoms (14), have reported by Furukawa and coworkers (Scheme 2.7) [38]. 8Te2 behaves as oxidising agent as on two-electron oxidation produces the novel ditelluride dication (15)

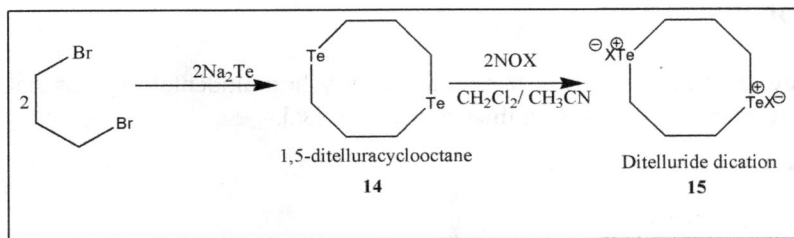

Scheme 2.7: Synthesis of macrocyclic Telluroether (8Te2) and ditelluride dication.

In 1996, the same research team reported the noval chlorine adduct of telluro-macrocycle. This 12-membered macrocyclic ring consist of three hypervalent tellurium (IV) moiety (17), resulted from 8-membered ring compound (16) by pyrolysis, through a ring expansion reaction (Scheme 2.8) [39]. In solid state, this 12-membered macrocycle (17) through intermolecular chlorine bridges have polymeric networks arrangement and is converted to 12Te3 (18).

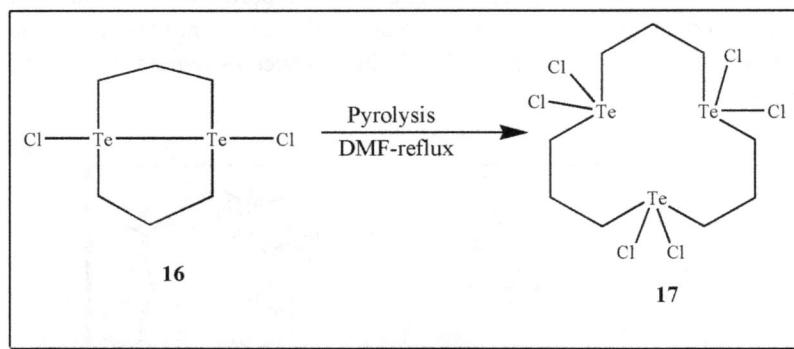

Scheme 2.8: Synthesis of chlorine adduct of telluro-macrocycle.

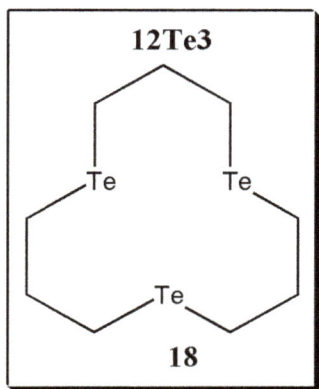

A distorted trigonal bipyramid geometry of Te(IV) atom (Figure 2.2) has been proposed in which chlorine atoms takes axial positions and equatorial position at corners of trigonal are occupied by the two alkyl C-atoms and a lone pair. Because of reactive nature of Te-C bonds, it is hard to prepare complexes of telluracrown ethers with metal.

2.3.4 Metallomacrocycles

McWhinnie & coworkers [40] reported the first 13 membered macrocyclic chelate involving tellurium (**19**). In the complex, two atoms of Cl and two Te atoms are bonded to tetrahedral Mercury (II). Tellurium (II) behave as Lewis's acid due to interaction

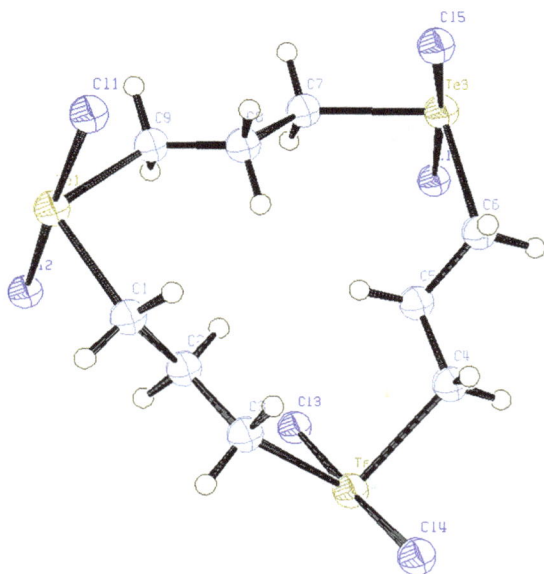

Figure 2.2: ORTEP plot for the structure of **17**. Reprinted with permission from [39]. Copyright (1996) American Chemical Society.

between Hg and Te and at the same time due to interaction of Te and lone pair of nitrogen, it acts as Lewis's base. The tellurium ligands have strong affinity towards 'soft' acids. The nitrogen's lone pair are involved in intramolecular Te·· ·N bond that result in decease in denticity of ligand.

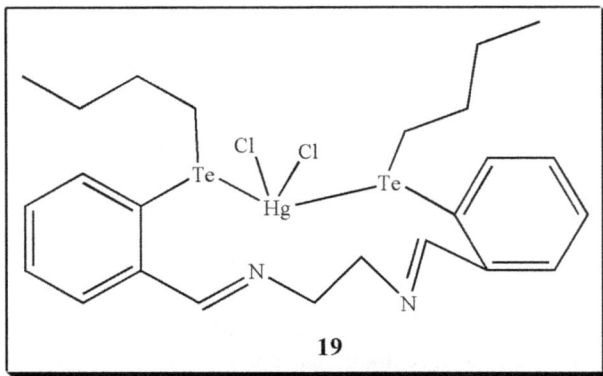

19

The 24-membered metallomacrocyclic Ag(I) complex (**20**) (Figure 2.3) has four Ag(I) atom and each one is forming coordinate bond with four tellurium atoms of four bridging $CH_3.Te(CH_2)_3Te-CH_3$ tetrahedrally [41].

The synthesis of 20 membered metallomacrocyclic ring (21,22) [42, 43] are reported by Singh and coworkers. This is the first example of organotellurium ligand forming 20 membered metallomacrocycle. The geometry of Pd and Pt are square planar. The two Chlorine atoms have trans position (Figures 2.4 and 2.5).

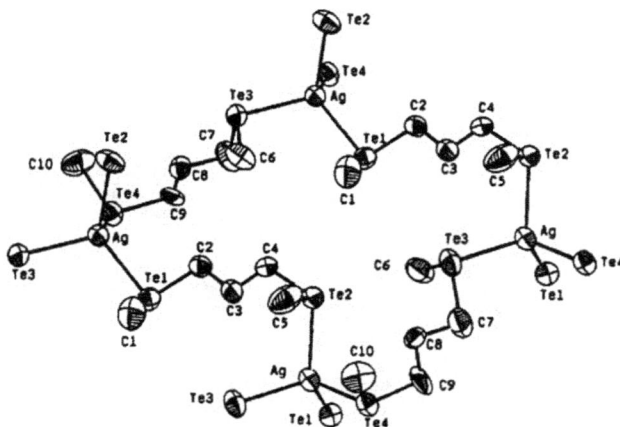

Figure 2.3: ORTEP drawing of polymeric [Ag-$CH_3Te(CH_2)_3TeCH_3)_2$]n cation (**20**). Reproduced by permission from ref. [41]. Copyright (1995) American Chemical Society.

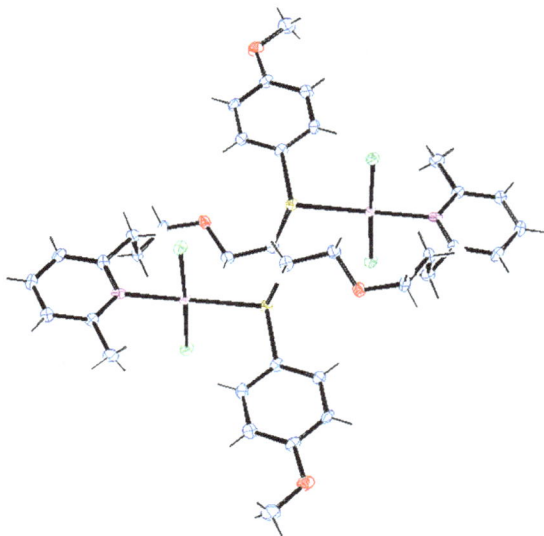

Figure 2.4: ORTEP diagram of (**21**). Taken from ref. [42]. Reproduced by permission from Elsevier.

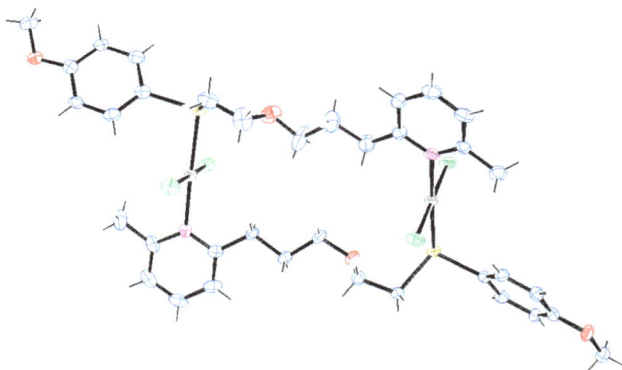

Figure 2.5: ORTEP diagram of (**22**). Taken from ref. [43]. Reproduced by permission from Elsevier.

2.3.5 Mixed Donor (O/N/S/Te) Macrocycles

Macrocycles with different cavity size and mixed donor atoms shows amazing coordination chemistry. The synthesis and coordination strength of macrocycles with heterocyclic units such as pyrrole, thiophene, phenol, pyridine and furan are studied in depth till date [15–17, 44–46]. The design and complexation studies of schiff base macrocycles with Te (**23**) was reported by S.C. Menon et al. [47]

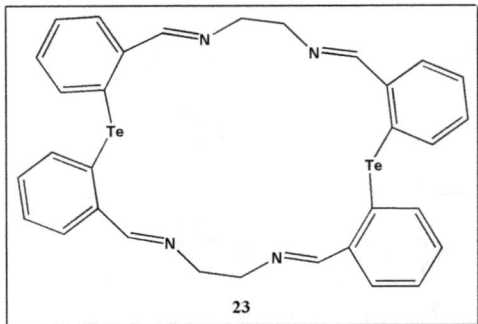

23

Intramolecular interaction between Te .... N (hypervalent bond) facilitates the formation of the macrocycle by template free method. This intramolecular interaction reduces the interaction that exist between lone pair on nitrogen atoms in the ring. The purity of the solvent is the sole factor on which the yield of the reaction depends. The solubility of ligand is found in chloroform & dichloromethane. In solvents like methanol and DMSO, it was found insoluble.

Since then, a template free synthesis of many tellurium azamacrocyclic ring has been isolated in high yield by condensing two moles of bis (2-formylphenyl) telluride with two moles of diamines in one step (Scheme 2.9) [47–51]. Single crystal X-ray structure (**24**) of the Pd (II) complex of a first telluraaza macrocyclic Schiff base cation had been reported by S.C. Menon et al. [49].

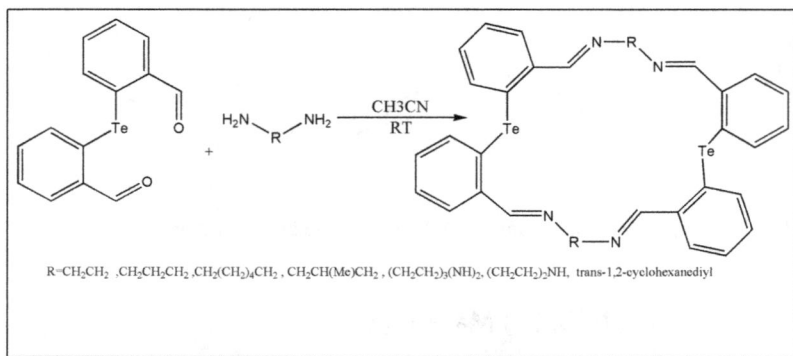

R=CH₂CH₂ ,CH₂CH₂CH₂ ,CH₂(CH₂)₄CH₂ , CH₂CH(Me)CH₂ , (CH₂CH₂)₃(NH)₂, (CH₂CH₂)₂NH, trans-1,2-cyclohexanediyl

Scheme 2.9: Synthesis of tellurium azamacrocycles.

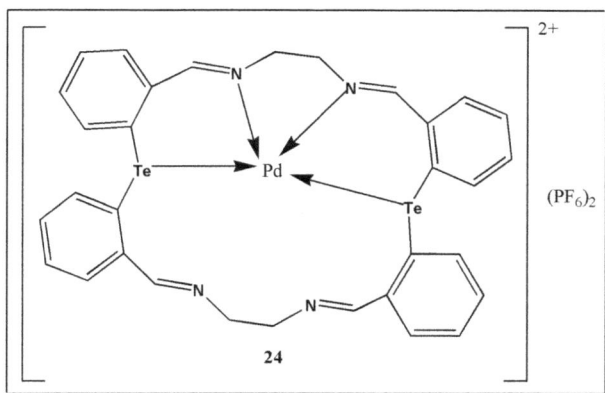

24

The metal complexes of this tellurium azamacrocycles with $HgCl_2$ (**25**), $PdCl_2$ (**26**) and $NiCl_2$ (**27**) have also been reported.

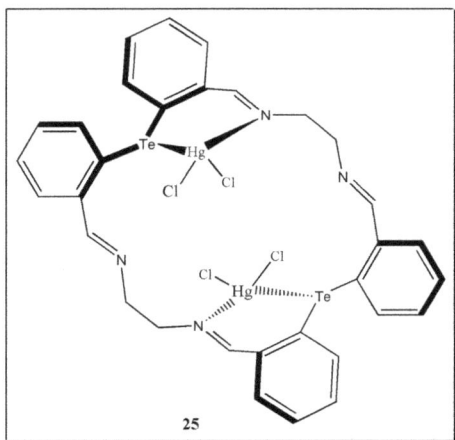

25

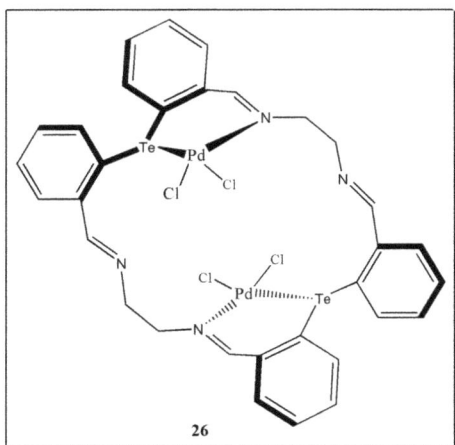

26

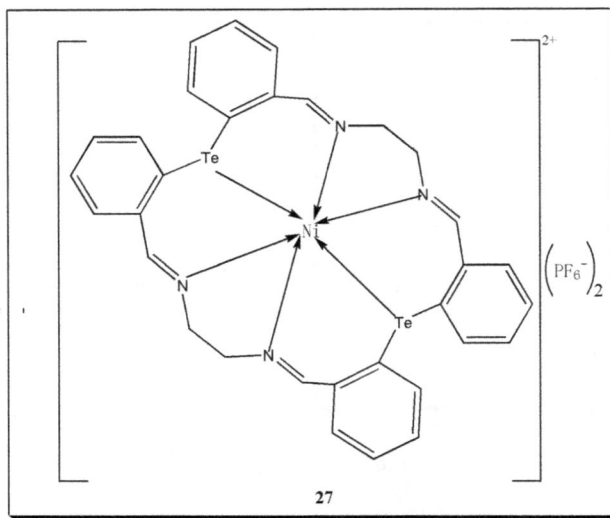

27

The tellurium azamacrocycles (23), react with

(a) PdCl$_2$ to form stable complex (26) (1:2 ratio) [48].

(b) HgCl$_2$ to form unstable complex (1:2 ratio) so undergo dismutation reaction to gives a mixture of (28) and (29) (Scheme 2.10) [48].

(c) NiCl$_2$·6H$_2$O with excess of PF$_6$–to form stable complex which is stable to oxidation and is paramagnetic in nature and reddish brown (27) [50]. Nickel ion is coordinated to six (two Te and four N) donor atoms in a distorted octahedral geometry.

2 moles of bis(aminoalkyl)tellurides and 2 moles of 2,6-diacetyl-4- methylphenol undergoes condensation via non template method to form macrocyclic ring with cavity size of 24 & 28 having two tellurium, four nitrogen and two oxygen as donor atoms. (30) [51]

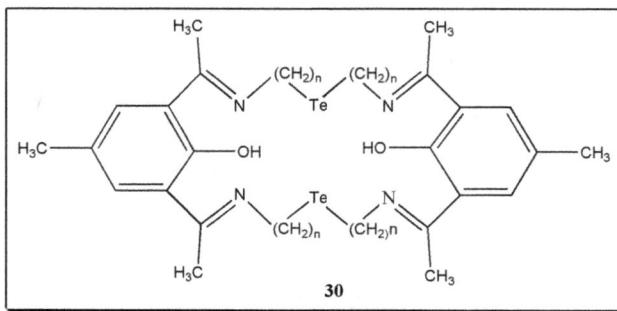

30

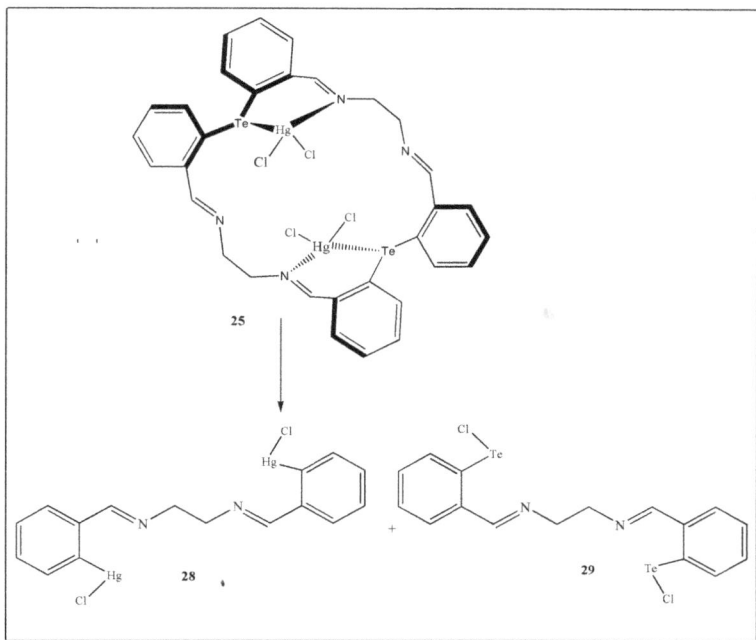

Scheme 2.10: Dismutation reaction of Mercury complex of tellurium azamacrocycles.

Light yellow coloured 32-Membered tellurium-containing Schiff base mercuraazamacrocycles (**31**) are isolated by reacting 2 moles each of bis(6-formyl-{2,3,4-tri-methoxymethylphenyl})mercury(II) and 3,3'-telluorobis-1-propanamine. Secondary intramolecular Hg···N interaction also reduces the repulsion between lone pairs on nitrogen, thereby play important role to architecture macrocycles [52].

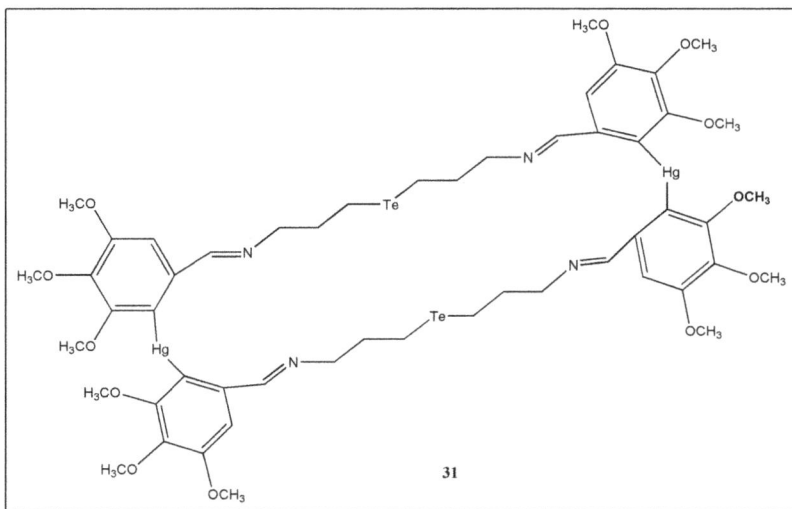

Template synthesis of octahedral 10- and 12-membered tellurium containing tetraazamacrocylic metal complexes (Te_2N_4M system) have been also described [53]. Also, a 2:2:1 condensation process on metal template has been reported to isolate 10 membered distorted octahedral ditellura tetraazamacrocyclic complexes (**32**) [54] (Scheme 2.11).

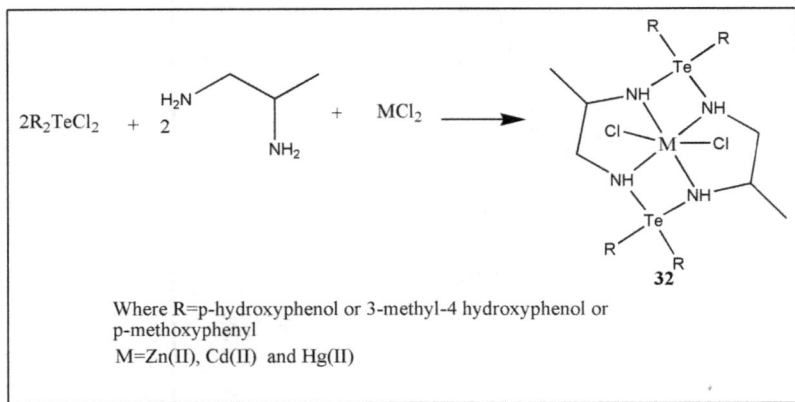

Where R=p-hydroxyphenol or 3-methyl-4 hydroxyphenol or
p-methoxyphenyl
M=Zn(II), Cd(II) and Hg(II)

Scheme 2.11: Scheme to synthesise ditellura tetraazamacrocyclic complexes by template reaction.

The corresponding 10-membered ditellura dithiadiazamacrocycles ($Te_2N_2S_2M$ system) and their metal complexes (**33**) are also obtained by reacting diaryltellurium dichlorides and 2-aminoethanethiol with metal dichlorides in 2:2:1 ratio on transition metal template (Scheme 2.12) by S.Kumari [55].

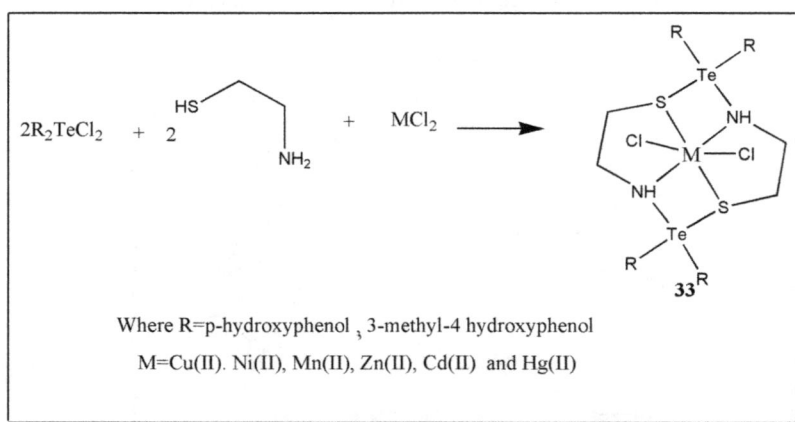

Where R=p-hydroxyphenol , 3-methyl-4 hydroxyphenol
M=Cu(II). Ni(II), Mn(II), Zn(II), Cd(II) and Hg(II)

Scheme 2.12: Synthesis of ditellura dithiadiazamacrocyclic complexes by template reaction.

Te/O macrocycles having Te—O—N bridges are also reported [56]. Levason and coworkers reported 18aneO$_4$Te$_2$ macrocycle (**34**) as main product and 9O$_2$Te as the minor product (**35**), synthesised from sodium telluride and 1,2-bis(2-chloroethoxy) ethane [57]. They are characterised by the formation of preparation of the Te(IV) derivatives 18O4Te2Me$_2$I$_2$, 18O4Te2Cl$_4$ [57].

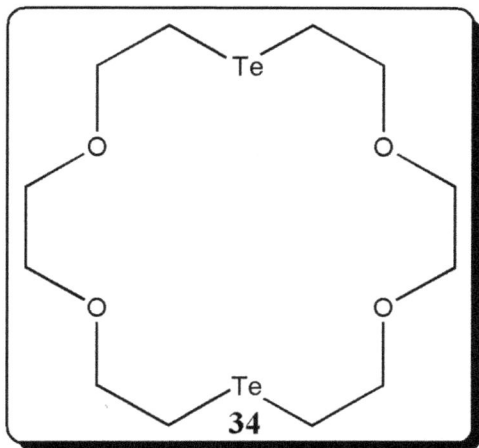

34

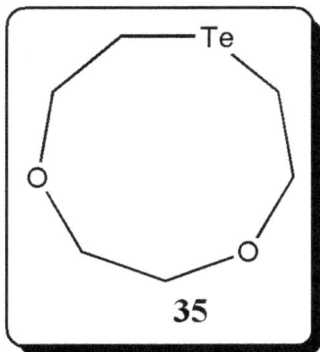

35

Kobayashi and coworkers [58] reported 21 and 18-membered (**36** and **37**), distorted trigonal bipyramidal multi-telluranes macrocycles with hypervalent Te—O apical linkages. This is the first such example of apical linkage in macrocycles and has been prepared by the [3 + 3] and [2 + 2] condensation of a phthalic acid disodium salt with cationic ditelluroxane respectively.

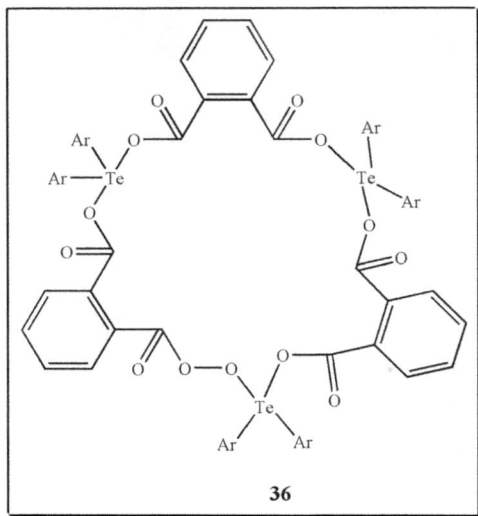

36

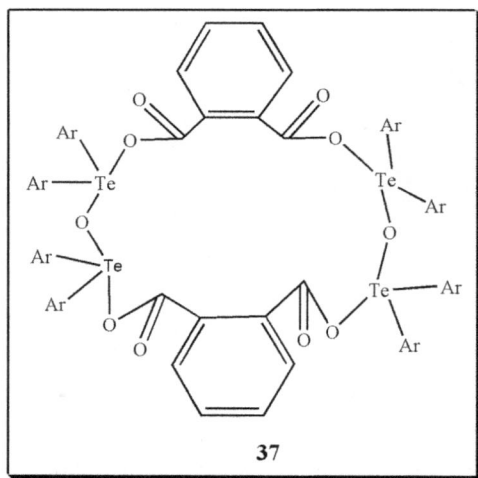

37

The reaction of the ditelluroxane (initiator) with the telluroxide (monomer) under mild conditions gives cationic oligotelluroxanes, a 14-membered pseudo-ring with hypervalent apical Te–O bonds. The distance between Te and oxygen is in 2.55 ± 2.79 A° range. The atoms in the ring are coplanar and the cavity of the macrocycle are accommodated by two counterions [59]. The aggregates of iso-tellurazole N-oxides comprises of cyclic tetramers, hexamers and a helical polymer, associated by Te .... O–N bonds. The Te–O bond is short and interaction between these atoms are strong. They show properties of actual macrocycles and act as host for small molecules and complexes with transition-metal ions. These macrocycles form addition product with fullerenes i.e., adduct [56]. Vargas along with his coworkers reported effects of chlorination on chalcogen (Te⋯O) bonds in aggregates of Iso-Tellurazole N-Oxides macrocycles [60].

9,11,12 and 14 membered macrocycles (**38–41**) with mixed tridendate (S_2Te) and tetradendate (S_3Te) donor atoms are isolated by reacting disodium telluride with various α-ω-dichlorothioalkane in [1 + 1] ratio [61–63] (Scheme 2.13). These macrocycles are light yellow coloured and poorly soluble. In $11S_2Te$ and $12S_2Te$, tellurium occupy corner and the Sulphur occupy position on edges with C–S–C angle (~100°) slightly larger than C–Te–C angles (~94°). Te (IV) diiodide species ($11S_2TeI_2$ and $12S_2TeI_2$) of $11S_2Te$ and $12S_2Te$ are also reported therein [64]. A geometry at Te (IV) in $12S_2TeI_2$ is distorted pseudo trigonal bipyramidal. Two atoms of iodine have trans position while the equatorial position is occupied by carbon and tellurium's lone pair electrons [63].

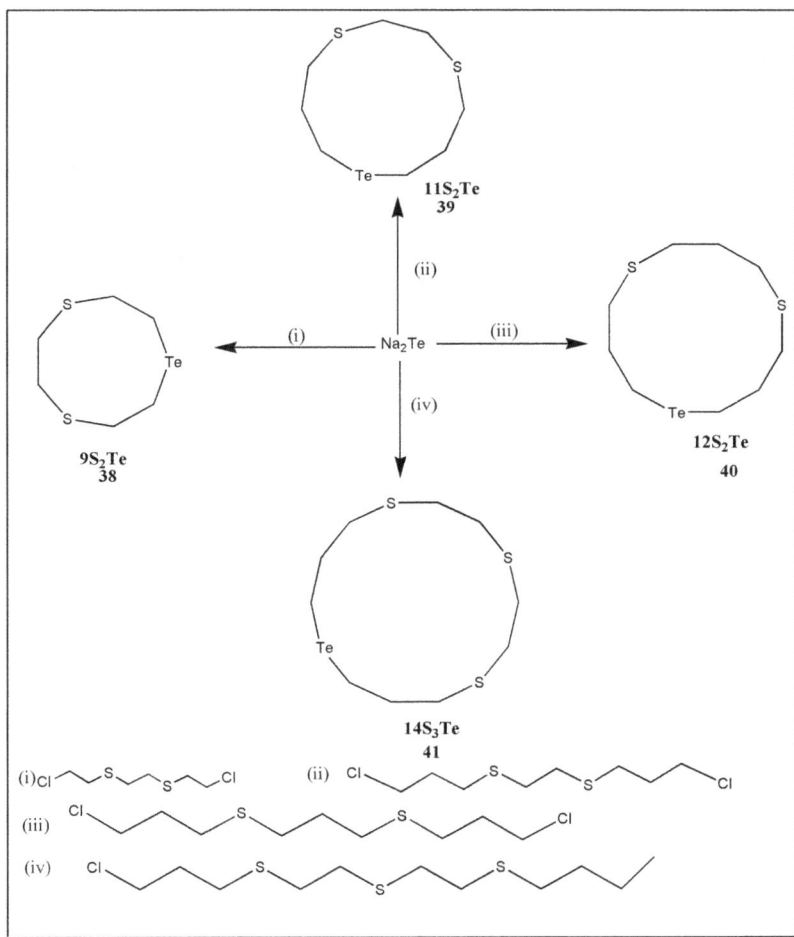

Scheme 2.13: Synthesis of macrocycles with tridendate (S_2Te) and tetradendate (S_3Te) donor atoms.

2.4 Conclusions

Tellurium act as sigma donors for transition metal centres with low oxidation state. Also, ^{125}Te-NMR help to monitor their structures in solutions. These structural characteristic of tellurium macrocycles make them excellent architect ligand that show some interesting coordinating bonding behaviour in these species. These factors are the driving force in creating curiosity in this field and chemistry of such tellurium macrocycles need to be explored. The novel synthetic routes for tellurium macrocycles via high dilution methods are being develop. The considerable work has been done on telluraporphyrins and telluracrown ethers, but its potential as efficient materials need to be explored further. Macrocycles involving Te donor's atom form complexes with variety of metal. Well-designed macrocycles containing mixed donor atoms can bind two different metal atoms with different nature within the same cavity and thereby such macrocycles play extensive and effective role as ligands for variety of transition-metal ions. In comparison to bonds C–S and C–Se, Te–C bond is highly reactive. This facile nature of Te–C bond will make the journey of chemists challenging in terms of synthesis and investigating the chemistry of tellurium macrocycles. Through this discussion, the interest in the tellurium macrocycles will be boosted up and the different aspects of their synthesis and practical application will be explored further.

References

1. Jain VK, Kedarnath G. Applications of metal selenium/tellurium compounds in materials science. Phys Sci Rev 2018;2017:1–38.
2. Jones PG, Ramírez de Arellano MC. Synthesis of aryl selenides using arylmercurials. Cyclopalladation of Se(R)Ph [R = C$_6$H$_3$(N=NC$_6$H$_4$Me-4′)-2,Me-5]. Crystal structures of Se$_2$R$_2$ and [Pd{C$_6$H$_3$[N=NC$_6$H$_3$(SePh)-2′,Me-4′]-2,Me-5}Cl]. J Chem Soc Dalton Trans 1996;2713–17.
3. Kienitz CO, Thone C, Jones PG. Coordination chemistry of 2,2′-dipyridyl diselenide: X-ray crystal structures of PySeSePy, [Zn(PySeSePy)Cl(2)], [(PySeSePy)Hg(C(6)F(5))(2)], [Mo(SePy)(2)(CO)(3)], [W(SePy)(2)(CO)(3)], and [Fe(SePy)(2)(CO)(2)] (PySeSePy = C(5)H(4)NSeSeC(5)H(4)N; SePy = [C(5)H(4)N(2-Se)-N,Se]). Inorg Chem 1996;35:3990–7.
4. Nishibayashi Y, Segawa K, Singh JD, Fukuzawa SI, Ohe K, Uemura S. Novel chiral ligands, diferrocenyl dichalcogenides and their derivatives, for the Rh(I)- and Ir(I)-catalyzed asymmetric hydrosilylation. Organometallics 1996;15:370–9.
5. Nishibayashi Y, Singh JD, Arikawa Y, Uemura S, Hidai M. Rhodium(I)-, iridium(I)-, and ruthenium(II)-catalyzed asymmetric transfer hydrogenation of ketones using diferrocenyl dichalcogenides as chiral ligands. J Organomet Chem 1997;531:13–8.
6. Chang Y, Emge TJ, Brennan JG. Pyridineselenolate complexes of tin and lead: Sn(2-SeNC5H4)2, Sn(2-SeNC5H4)4, Pb(2-SeNC5H4)2, and Pb(3-Me3Si-2-SeNC5H3)2. Volatile CVD precursors to group IV–Group VI semiconductors. Inorg Chem 1996;35:342–6.

7. Steigerwalk ML, Sprinkle CR. Application of phosphine tellurides to the preparation of Group II-VI (2-16) semiconductor materials. Organometallics 1988;7:245–6.

8. Hirpo W, Dhingra S, Sutorik AC, Kanatzidis MG. Synthesis of mixed copper-indium chalcogenolates. Single-source precursors for the photovoltaic materials CuInQ$_2$ (Q = S, Se). J Am Chem Soc 1993;115:1597–9.

9. Gange RR, Allison JL, Gall RS, Koval CA. Models for copper-containing proteins: structure and properties of novel five-coordinate copper(I) complexes. J Am Chem Soc 1977;99:7170–8.

10. Martin JWL, Johnston JH, Curtis NF. Complexes of 2,4,4-trimethyl-1,5,9-triazacyclododec-1-ene with cobalt(II), nickel(II), and copper(II); X-ray structure determination of di-isothiocyanato(2,4,4-trimethyl-1,5,9-triazacyclododec-1-ene)nickel(II). J Chem Soc Dalton Trans 1978;68:76.

11. Hughes MN. The inorganic chemistry of biological processes, 2nd ed. New York: Wiley; 1981.

12. Casella L, Gullotti M, Gioia LD, Monzani E, Chillemi F. Synthesis, ligand binding and biomimetic oxidations of deuterohaemin modified with an undecapeptide residue. J Chem Soc Dalton Trans 1991;2945–53. https://doi.org/10.1039/dt9910002945.

13. James SR, Margerum DW. Stability and kinetics of a macrocyclic tetrapeptide complex, tetradeprotonated(cyclo-(.beta.-alanylglycyl-.beta.-alanylglycyl))cuprate(II). Inorg Chem 1980;19:2784–90.

14. Panda A, Menon SC, Singh HB, Butcher RJ. Synthesis of some macrocycles/bicycles from bis(o-formylphenyl) selenide: X-ray crystal structure of bis(o-formylphenyl) selenide and the first 28-membered selenium containing macrocyclic ligand. J Organomet Chem 2001;623:87–94.

15. Pietraszkiewicz M. Synthetic methods in supramolecular chemistry. J Coord Chem 1992;27:151–99.

16. Fenton DE, Mathews RW, McPartlin M, Murphy BP, Scowen IJ, Tasker PA. Macrocyclic helicates: complexes of a 34-membered Schiff-base ligand. J Chem Soc Chem Commun 1994;1391–2. https://doi.org/10.1039/c39940001391.

17. Brunner H, Schiessling H. Dialdehyde + diamine—polymer or macrocycle? Angew Chem, Int Ed Engl 1994;33:125–6.

18. Nanda KK, Venkatsubramanian K, Majumdar D, Nag K. Synthesis of a novel octaamino tetraphenol macrocyclic ligand and structure of a tetranuclear nickel(II) complex. Inorg Chem 1994;33:1581–2.

19. Panda A, Menon SC, Singh HB, Morley CP, Bachman R, Cocker TM, et al. Synthesis, characterization and coordination chemistry of some selenium -containing macrocyclic Schiff bases. Eur J Inorg Chem 2005;2005:1114–26.

20. Alexander V. Design and synthesis of macrocyclic ligands and their complexes of lanthanides and actinides. Chem Rev 1995;95:273–342.

21. Beer PD, Graydon AR. New anion receptors based on cobalticinium-aza crown ether derivatives. J Organomet Chem 1994;466:241–7.

22. Van Veggel FCJM, Verboom W, Reinhoudt DN. Metallomacrocycles: supramolecular chemistry with hard and soft metal cations in action. Chem Rev 1994;94:279–99.

23. Alka A, Shetti VS, Ravikanth M. Telluraporphyrinoids: an interesting class of coremodified porphyrinoids. Dalton Trans 2019;48:4444–59.

24. Panda A. Developments in tellurium containing macrocycles. Coord Chem Rev 2009;253:1947–65.

25. Nogueira CW, Zeni G, Rocha JBT. Organoselenium and organotellurium compounds: toxicology and pharmacology. Chem Rev 2004;104:6255–86.

26. Ulman A, Manassen J, Frolow F, Rabinovich D. Synthesis of new tetraphenylporphyrin molecules containing heteroatoms other than nitrogen. III Tetraphenyl-21-tellura-23-thiaporphyrin: an internally-bridged porphyrin. Tetrahedron Lett 1978;19:1885–6.

27. Abe M, Hilmey DG, Stilts CE, Sukumaran DK, Detty MR. 21-Telluraporphyrins. 1. Impact of 21,23-heteroatom interactions on electrochemical redox potentials, 125Te NMR spectra, and absorption spectra. Organometallics 2002;21:2986–92.

28. Abe M, You Y, Detty MR. 21-Telluraporphyrins. 2. Catalysts for bromination reactions with hydrogen peroxide and sodium bromide. Organometallics 2002;21:4546–51.

29. Ulman A, Manassen J, Frolow F, Rabinovich D. Synthesis and properties of tetraphenylporphyrin molecules containing heteroatoms other than nitrogen. 5. High resolution nuclear magnetic resonance studies of inner and outer aromaticity. J Am Chem Soc 1979;101:7055–9.

30. Latos-Grażynski L, Pacholska E, Chmielewski PJ, Olmstead MM, Balch A. Alteration of the reactivity of a tellurophene within a core-modified porphyrin environment: synthesis and oxidation of 21-telluraporphyrin. Angew Chem Int Ed Engl 1995;34:2252–4.

31. Pacholska E, Latos-Grazynski L, Ciunik Z. A direct link between annulene and porphyrin chemistry —21-vacataporphyrin. Chem Eur J 2002;8:5403–5.

32. Pacholska-Dudziak E, Szterenberg L, Latos-Grażyński L. A flexible porphyrin–annulene hybrid: a nonporphyrin conformation for*meso*-tetraaryldivacataporphyrin. Chem Eur J 2011;17:3500–11.

33. Pacholska-Dudziak E, Vetter G, Góratowska A, Białońska A, Latos-Grazynsk L. Chemistry inside a porphyrin skeleton: Platinacyclopentadiene from tellurophene. Chem Eur J 2020;26:16011–8.

34. Abe M, Detty MR, Gerlits OO, Sukumaran DK. 21-Telluraporphyrins. 3. Synthesis, structure, and spectral properties of a 21,21-dihalo-21-telluraporphyrin. Organometallics 2004;23:4513–8.

35. Pacholska-Dudziak E, Szczepaniak M, Książek A, Latos-Grażyński L. A porphyrin skeleton containing a palladacyclopentadiene. Angew Chem Int Ed 2013;52:8898–903.

36. Von Zelewsky A. Stereochemistry of Coordination Compounds. Chichester: John Wiley; 1996.

37. Panda A, Menon SC, Singh HB, Butcher RJ. Synthesis of some macrocycles/bicycles from bis(o-formylphenyl) selenide: X-ray crystal structure of bis(o-formylphenyl) selenide and the first 28-membered selenium containing macrocyclic ligand. J Organomet Chem 2001;623:87–94.

38. Fujihara H, Ninoi T, Akaishi R, Erata T, Furukawa N. First example of tetraalkyl substituted ditelluride dication salt from 1,5-ditelluracyclooctane. Tetrahedron Lett 1991;32:4537–40.

39. Takaguchi Y, Horn E, Furukawa N. Preparation and X-ray structure analysis of 1,1,5,5,9,9-Hexachloro-1,5,9-tritelluracyclododecane (Cl6([12]aneTe3)) and its redox behavior. Organometallics 1996;15:5112–5.

40. AL-Salim N, Hamor TA, McWhinnie WR. Lewis acid and Lewis base behaviour of a tellurium(II) compound: a mercury(II) complex of a bis-telluride ligand with a 13-member macrocyclic chelate ring. J Chem Soc Chem Commun 1986;453–5. https://doi.org/10.1039/c39860000453.

41. Liaw WF, Lai CH, Chiou SJ, Horng YC, Chou CC, Liaw MC, et al. Synthesis and characterization of polymeric Ag(I)-telluroether and Cu(I)-diorganyl ditelluride complexes – crystal-structures of (Ag(MeTe(CH2)2TeMe)2]n[(BF4]n, ((μ2-MeTeTeMe)Cu(μ2-Cl)]n, and (Ag2(NCCH3)4(-(μ2-(p-C6H4F) TeTe(p-C6H4F))2[BF4]2. Inorg Chem 1995;34:3755–9.

42. Bali S, Singh AK, Sharma P, Drake JE, Hursthouse MB, Light ME. First example of bimetallic complex of platinum(II) with a hybrid organotellurium ligand [(4-MeOC6H4Te)CH2CH2OCH2CH2CH2 (2-C5H4N)] (L1) containing 20-membered metallomacrocycle ring: synthesis and crystal structure. Inorg Chem Commun 2003;6:1378–81.

43. Bali S, Singh AK, Drake JE, Light ME. Multidentate hybrid organotellurium ligands 1-(4-methoxyphenyltelluro)-2-[3-(6-methyl-2-pyridyl)propoxy]ethane (L1) 2-methyl-6-{3-[2-({2-[3-(6-methyl-2-pyridyl)propoxy]- ethyl}telluranyl)ethoxy]propyl}pyridine (L2) and their metal complexes: formation of 20-membered metallomacrocycle by L1. Polyhedron 2006;25:1033–42.

44. Guerriero P, Vigato PA, Fenton DE, Hellier PC. Synthesis and application of macrocyclic and macrocyclic schiff bases. Acta Chem Scand 1992;46:1025–46.

45. Nanda KK, Venkatsubramanian K, Majumdar D, Nag K. Synthesis of a novel octaamino tetraphenol macrocyclic ligand and structure of a tetranuclear nickel(II) complex. Inorg Chem 1994;33:1581–2.

46. Rissanen K, Breitcnbach J, Huuskonen J. An unusual copper(I) complex of a new macrocyclic ligand. J Chem Soc Chem Commun 1994;1265–6. https://doi.org/10.1039/c39940001265.

47. Menon SC, Singh HB, Patel RP, Kulshreshtha SK. Synthesis, structure and reactions of the first tellurium-containing macrocyclic Schiff base. J Chem Soc Dalton Trans 1996;1203–7. https://doi.org/10.1039/dt9960001203.

48. Kaur R, Menon SC, Singh HB. New aspects of intramolecular coordination in organochalcogen (Se/Te) chemistry. Proc Indian Acad Sci 1996;108:159–64.

49. Menon SC, Panda A, Singh HB, Butcher RJ. Synthesis and single crystal X-ray structure of the first cationic Pd(ii) complex of a tellurium-containing polyaza macrocycle: contrasting reactions of Pd(ii) and Pt(ii) with a 22-membered macrocyclic Schiff base. Chem Commun 2000;143–4. https://doi.org/10.1039/a908683h.

50. Menon SC, Panda A, Singh HB, Patel RP, Kulshreshtha SK, Darby WL, et al. Tellurium azamacrocycles: synthesis, characterization and coordination studies. J Organomet Chem 2004;689:1452–63.

51. Tripathi SK, Khandelwal BL, Gupta SK. A new family of chalcogen bearing macrocycles: synthesis and characterization of N4O2E2 (E = Se,Te) type compounds. Phosphorus, Sulfur Silicon Relat Elem 2002;177:2285–93.

52. Das S, Singh HB, Butcher RJ. Synthesis and characterization of 22-, 28- and 32-membered mercuraazamacrocycles: isolation of ring-Chain tautomer and Se/Te-containing macrocycles. Eur J Inorg Chem 2018;25:4702–10.

53. Srivastava S, Kalam A. Synthesis and characterization of manganese (II), cobalt (II), nickel (II), copper (II), zinc (II) complexes of a tellurium containing tetraazamacrocycle: a photoelectron spectroscopic study. J Indian Chem Soc 2006;83:563–7.

54. Kumari S, Verma KK, Garg S. Synthesis, spectral and antimicrobial studies of some d^{10} metal-ions ditellura tetraazamacrocyclic complexes. Chem Sci Trans 2017;6:77–86.

55. Kumari S, Verma KK, Garg S. Investigations on some divalent transition metal complexes of 10-membered tellurium containing N_2S_2 donor macrocycles. Int J Chem Sci 2017;15:207.

56. Ho PC, Szydlowski P, Sinclair J, Elder PJW, Kubel J, Gendy C, et al. Supramolecular macrocycles reversibly assembled by TeyO chalcogen bonding. Nat Commun 2016;7:11299.

57. Hesford MJ, Levason W, Matthews ML, Reid G. Synthesis and complexation of the mixed tellurium–oxygen macrocycles1-tellura-4,7-dioxacyclononane, [9]aneO2Te, and 1,10-ditellura-4,7,13,16-tetraoxac$_y$clooctadecane, [18]aneO4Te2 and their selenium analogues. Dalton Trans 2003:2852–8. https://doi.org/10.1039/b303365c.

58. Kobayashi K, Iizawa H, Yamaguchi K, Horn E, Furukawa N. Macrocyclic multi-telluranes with hypervalent Te–O apical linkages. Chem Commun 2001:1428–9. https://doi.org/10.1039/b103676a.

59. Kobayashi K, Deguchi N, Takahashi O, Tanaka K, Horn E, Kikuchi O, et al. Nucleophilic addition of telluroxides to a cationic ditelluroxane: oligotelluroxanes. Angew Chem Int Ed 1999;38:1638–40.

60. Ho PC, Lomax J, Tomassetti V, Britten JF, Vargas-Baca I. Competing effects of chlorination on the strength of Te···O chalcogen bonds select the structure of mixed supramolecular macrocyclic aggregates of iso-tellurazole N-oxides. Chem Eur J 2021;27:10849–53.

61. Evano G, Barde E. Four-membered rings with one selenium or tellurium atom. In: Comprehensive heterocyclic chemistry IV, 4th ed.; 2022, vol 2. pp. 327–34.

62. Levason W, Orchard SD, Reid G. Synthesis and properties of the first series of mixed thioether/telluroether macrocycles. Chem Commun 2001:427–8. https://doi.org/10.1039/b008370o.

63. Hesford MJ, Levason W, Matthews ML, Orchard SD, Reid G. Synthesis, characterisation and coordinating properties of the small ring S_2Te-donor macrocycles [9]aneS2Te, [11]aneS2Te and [12] aneS2Te. Dalton Trans 2003:2434–42. https://doi.org/10.1039/b302985a.
64. Barton AJ, Genge ARJ, Hill NJ, Levason W, Orchard SD, Patel B, et al. Recent developments in thio-, seleno-, and telluro-ether ligand chemistry. Heteroat Chem 2002;13:550–60.

Aparna Das* and Bimal Krishna Banik*

3 Semiconductor characteristics of tellurium and its implementations

Abstract: Tellurium (Te) gained worldwide attention because of its excellent properties, distinctive chained structures, and potential usages. Bulk Te is a p-type elemental helical semiconductor at room temperature and it also having a very limited band gap. Te presents fascinating characteristics such as nonlinear optical response, photoconductivity, good thermoelectric and piezoelectric properties. These charming characteristics induce Te a possible nominee for applications in field-effect transistors, IR acousto-optic deflectors, solar cells, self-developing holographic recording devices, photoconductors, gas sensors, radiative cooling devices, and topological insulators. The developments in these areas are incorporated in great detail. This study opens up the possibility of designing novel devices and considering modern applications of Tellurium.

Keywords: photoconductivity; semiconductor; sensors; solar cells; tellurium.

3.1 Introduction

Tellurium (Te) is a p-type low band gap semiconducting material. It is a semi-metallic and silvery-white element. Te is brittle and crystalline. Tellurium dissolve in nitric acid, at the same time it remains stable in hydrochloric acid or water. At room temperature, bulk tellurium has shown a band gap of 0.35 eV [1]. It is often doped with silver, gold, copper, or tin. It is found in the ores calaverite, sylvanite, and krennerite. From the refining and mining copper, Te can also be obtained as a byproduct. Tellurium (Te) has gained worldwide attention because of its distinctive chained structures, excellent characteristics, and possible usages. The important chemical, physical, mechanical, thermal, and electrical properties of Te are noted in Table 3.1.

Te is used for various applications. It is used as an additive to steel and alloyed to lead, tin, aluminum, or copper. In the case of lead, to raise the strength, durability, and corrosion resistance, Te is used as an additive to lead. Te is used for the fabrication of blasting caps, ceramics, cast iron, chalcogenide glasses, and solar panels. In the case of rubber, to accelerate the curing process, cut down the susceptibility to aging, and

---
*Corresponding authors: Aparna Das and Bimal Krishna Banik, Department of Mathematics and Natural Sciences, College of Sciences and Human Studies, Prince Mohammad Bin Fahd University, Al Khobar, 31952, Kingdom of Saudi Arabia, E-mail: aparnadasam@gmail.com (A. Das), bbanik@pmu.edu.sa, bimalbanik10@gmail.com (B.K. Banik). https://orcid.org/0000-0002-2502-9446 (A. Das)

As per De Gruyter's policy this article has previously been published in the journal Physical Sciences Reviews. Please cite as: A. Das and B. K. Banik "Semiconductor characteristics of tellurium and its implementations" *Physical Sciences Reviews* [Online] 2022. DOI: 10.1515/psr-2021-0108 | https://doi.org/10.1515/9783110735840-003

Table 3.1: Chemical, physical, mechanical, thermal and electrical properties of Te.

Chemical Properties	
CAS number	13494-80-9
Electrochemical equivalent	0.99 g/A/h
Cross section of thermal neutron	4.7 b per atom
Electronegativity	2.1
Ionic radius	0.56 Å
Absorption edge of X-ray	0.38 Å
Physical Properties	
Boiling point	1390 °C
Melting point	450 °C
Density	6.23 g/cm^3
Mechanical Properties	
Poisson's ratio	0.33
Tensile strength	11 MPa
Shear modulus	15.16 GPa
Modulus of elasticity	40 GPa
Hardness, Brinell	25
Material condition	Polycrystalline
Thermal Properties	
Thermal conductivity	1.97 W/mK
Thermal expansion co-efficient (@20–100 °C)	16.8 μm/m°C
Latent heat of fusion	138 J g^{-1}
Latent heat of evaporation	820 J g^{-1}
Specific heat @25C	201 J K^{-1} kg^{-1}
Electrical Properties	
Electrical resistivity @0 °C	1.6 × 105 μOhmcm

enhance the resistance to oil, Te is used as an additive to rubber. Tellurium is used for heterogenous catalysis. Compared to sulfur-vulcanized synthetic rubber, the synthetic rubber vulcanized with Te displays superior thermal and mechanical properties [2, 3].

For ceramics, Te-containing compounds are used as specialized pigments [4]. Studies have shown that tellurides and selenides significantly raise the light refraction of glass which is in a great degree employed in optical fiber applications such as in telecommunications [5, 6]. In the delay powder of electric blasting caps, mixtures of selenium and tellurium are applied as an oxidizer with barium peroxide [7]. One of the most common methods to make iodine-131 is the neutron bombardment of the tellurium technique. This iodine-131 can be applied for the treatment of thyroid disorders, and in hydraulic fracturing as a tracer material, along with different other practical usages.

For the epitaxial fabrication of II–VI compound semiconductors by metalorganic vapor phase epitaxy (MOVPE), organotellurium compounds are employed as precursors. The mainly used precursor compounds are diallyl telluride, diisopropyl telluride, diethyl telluride, dimethyl telluride and methyl allyl telluride [8]. For the low-temperature epitaxial growth of CdHgTe by MOVPE, the considered favored precursor is diisopropyl telluride [9]. In these procedures, the most pureness metalorganics of both tellurium and selenium are utilized. Adduct purification is mainly used to prepare the compounds for the semiconductor industry [10, 11]. Tellurium suboxide is employed in rewritable optical discs, especially in the media layer. For example in the media layer of digital video discs, compact discs, and blu-ray discs [12]. To make acousto-optic modulators for confocal microscopy, dioxide of tellurium is employed. In phase change memory chips also, tellurium compounds are used [13, 14]. Lead telluride and bismuth telluride is the mainly used for the preparation of thermoelectric instruments. To make far-IR detectors, tellurides of lead is also used.

Te has application in several photocathodes which are used for making solar-blind photomultiplier tubes [15]. For example, the Cs-Te photocathode has 3.5 electron volt photoemission threshold in poor vacuum environments. This photocathode presents the rare coordination of prominent durability and quantum efficiency. This has caused photoemission electron guns employed to drive the free-electron lasers. Even though several other Tellurium comprising photocathodes have been made employing different alkali metals like sodium, rubidium, and potassium, but they have not obtained the similar acceptance that Cs–Te photocathode has received. Te is also employed in high brightness photoinjectors which are used to drive modern particle accelerators.

Considering the biological applications of tellurium, in humans or other animals, it has some known biological function [16] as tellurium is identical in a chemical manner to various chalcogens like sulfur and selenium. However, Fungi can incorporate tellurium in place of selenium and sulfur into amino acids, for example, telluromethionine and telluro-cysteine [17, 18]. A highly varying tolerance to Te compounds has been shown by organisms. Several bacteria, say *Pseudomonas aeruginosa*, absorb tellurite and later reduce it to Te. The prepared elemental Te then gathers and induces a feature and sometimes cell darkening will occur [19]. Regarding yeast, the sulfate assimilation pathway mediates the reduction of tellurite [20] Accumulation of Te produces toxicity effects. To produce dimethyl telluride, several organisms metabolize tellurium partly. In hot springs, dimethyl telluride has been detected in low quantities [21, 22]. Tellurite agar technique is employed to detect the individuals of the genus corynebacterium, for example, *Corynebacterium diphtheriae*, which is the responsible microorganism for diphtheria [23]. In reality, some compounds of tellurium are toxic to human being in a moderate degree. However, the human body metabolizes it and makes dimethyl telluride as a byproduct [24]. Compounds of tellurium also have been investigated for their possible anti-cancer and anti-inflammatory activities [25, 26]. For example, it was reported that ammonium trichloro(dioxoethylene-O,O′)tellurate, or AS101 slowed down the cancer cell's growth and to be a potent immunomodulator [27].

Tellurium has p-type semiconductor properties. Owing to its low electronegativity Te interacts with a variety of materials and makes different materials with low band gaps. It is capable of being addressed by light with comparatively long-wavelength. Because of this special characteristic, it is applicable for several possible usages such as in solar cells, photoconductive materials, and IR sensors. The concerns about environmental impact and the low constancy of these materials are the major worries arresting the use in some applications. However, in some applications, tellurium compounds have shown the excellent property. For instance, in the case of cadmium telluride solar panels, these solar boards demonstrated some of the most outstanding efficiencies [28]. X-ray detectors based on tellurium, cadmium, and zinc have been reported [29]. For infrared radiation detection, mercury cadmium telluride is a good choice [30]. Detailed semiconductor properties and recent developments in these areas are discussed in this article.

3.2 Recent developments in Te- semiconductor research

Tellurium exhibits excellent semiconductor properties. For example, it had demonstrated excellent fascinating characteristics like nonlinear optical response, photoconductivity, and high thermoelectric and piezoelectric properties. The recent developments in this area are discussed in the following sections.

An elementary semiconductor, few-layer Tellurium (FL-Te), has shown and succeeded in some of the outstanding physical characteristics that black phosphorus provides. Few-layer tellurium could be practicably prepared using simple solution-based methods [31]. Few-layer tellurium is constituted of non-covalently bound parallel tellurium chains, within which covalent-like characteristic comes out. With the strength of intra-chain covalent bonding, the strength of this inter-chain covalent-like quasi-bonding was comparable, which heads to the closed stability of many tellurium allotropes. The electronic band arrangement of few-layer Te is demonstrated in Figure 3.1.

Few-layer tellurium also brings out a tunable bandgap (0.31–1.17 eV). It also introduces in the first Brillouin zone, a highly anisotropic, two (four) complex, and layer-dependent hole (electron) pockets. As the function of sample thickness, the variations in the indirect bandgaps and the locations of VBM (valence band maximum) and CBM (conduction band minimum) are shown in Figure 3.2.

It also has presented notable great hole mobility (around 10^5 cm^2/Vs). Together, a strong optical absorption along the non-covalently bound direction (Figure 3.3). The

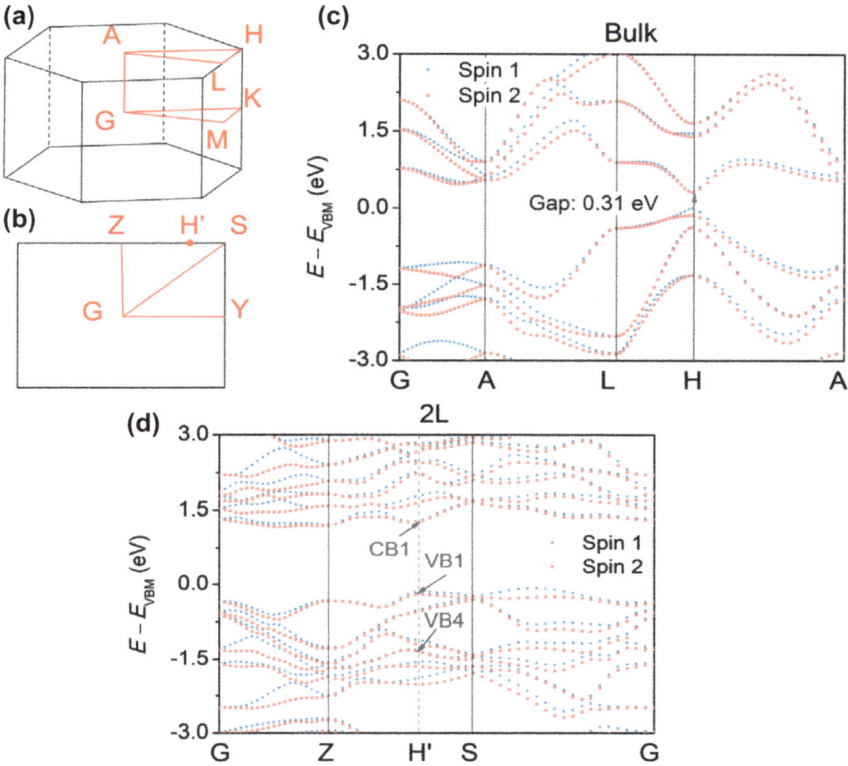

Figure 3.1: Electronic band structures of FL tellurium.
(b), (a) Brillouin zones of FL α-Te and bulk. (c) and (d) Band structures of bulk and bilayer α-Te. Adapted with permission from [31].

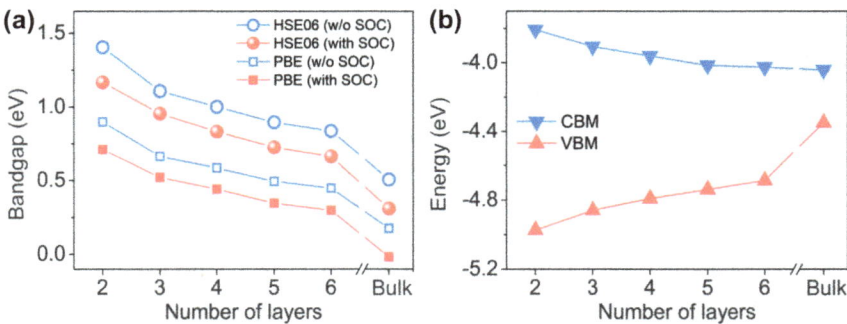

Figure 3.2: Changes in bandgap and energy.
(a) and (b) variations in the indirect bandgaps and the locations of conduction band minimum and valence band maximum as a function of the thickness of the sample. Adapted with permission from ref. [31].

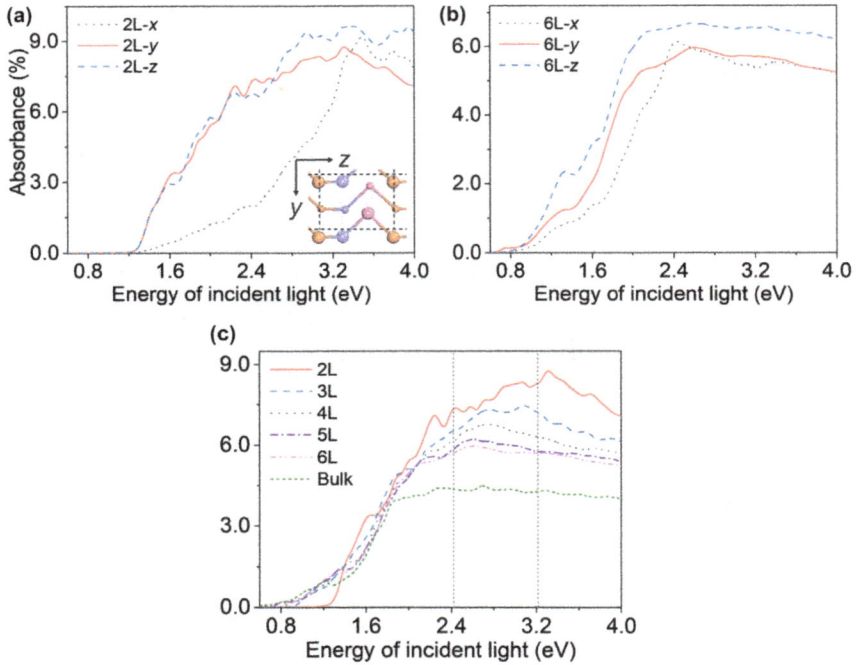

Figure 3.3: Optical absorption spectra of few-layer-α-Tellurium.
(a) Absorbance per layer of 2 L (b) absorbance per layer of 6 L-α-Te with the polarization direction of incident light along x, y, and z, respectively. (c) Absorbance per layer with the polarization direction of incident light along y for few-layer-α-Tellurium with the thickness changing from 2 L to 6 L and bulk. Adapted with permission from ref. [31].

optical properties were layer-dependent and nearly isotropic. A large ideal strength (above 20%), compared to black phosphorus, improved environmental stability, and unusual force constants crossover for breathing modes and interlayer shear were also observed. The study demonstrated that the FL tellurium is a remarkable semiconductor having great optical absorption, mobility, better environmental stability, tunable bandgap, intrinsic-anisotropy, nearly direct bandgap, and low-cost production is possible. Along with different geometrically alike layered materials, this few-layer α-tellurium may upgrade the outgrowth of a novel group of layered materials.

3.2.1 Te based Photoconductors

Using liquid-phase sedimentation together with the dry vibration milling method, fine powders containing Te grains (less than 10 nm) were raised, beginning from poly-crystalline powders having grain diameter of around ca. 30 μm [32]. By bonding these nanosized Te grains with poly (methyl methacrylate), nanocomposite films were prepared.

Raman spectrum obtained from the film is shown in Figure 3.4. The spectrum disclosed that the Te/PMMA films were grounded on the existence of the oxide of tellurium and crystalline phases of tellurium because of the partial oxidation of the grains in the air. The data obtained from optical analysis revealed that in the 310–2200 nm range the absorbance was nearly constant as shown in Figure 3.5. Along with that, a distinctive ultraviolet absorption peak from the Te nanostructures was observed roughly at 260 nm.

To get details about photoconductivity properties, an extensive characterization was actioned by irradiating the Te-poly(methyl methacrylate) films with rays selected from the ultraviolet to near-infrared region or white light (Figure 3.6). The obtained data

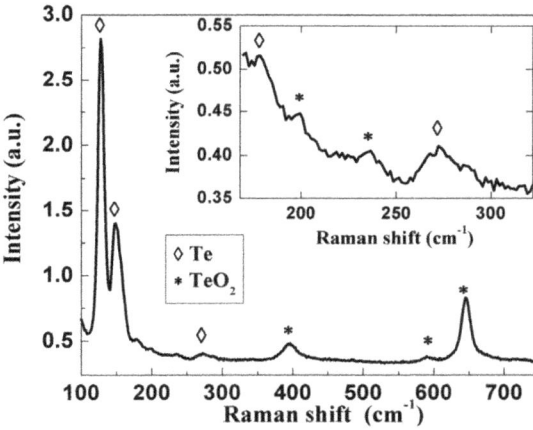

Figure 3.4: Raman spectrum obtained from Te/PMMA film. The magnified Raman spectrum is shown in the inset. Adapted with permission from ref. [32].

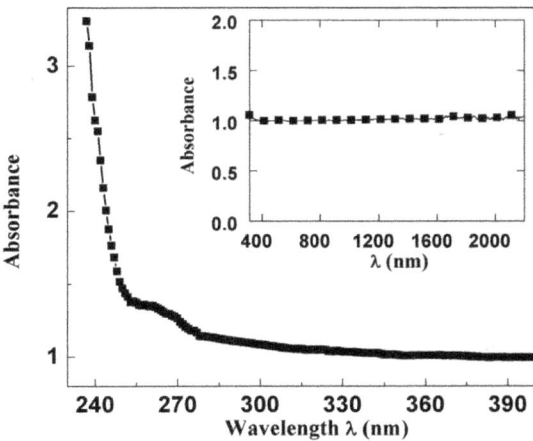

Figure 3.5: Absorbance spectrum in the ultraviolet region. The absorbance in the ultraviolet to near infrared region is shown in the inset. Adapted with permission from ref. [32].

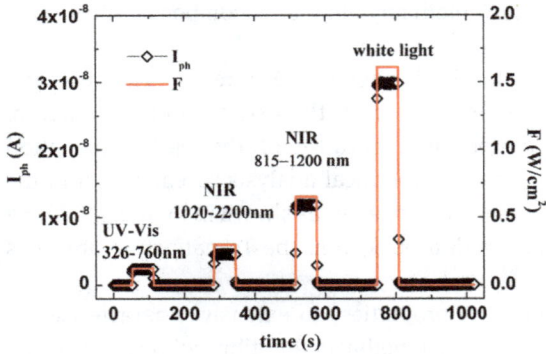

Figure 3.6: The changes in the photocurrent, for light-dark cycles. The sample was irradiated with radiations of the various spectral compositions having optical power density F. Adapted with permission from ref. [32].

demonstrated that the response to light is mainly associated with the absorption of light and is not related to the spectral constitution of the incident rays in the 310–2200 nm wavelength range. It was also observed that with the optical power density the photocurrent enhanced linearly over about three orders of magnitude. The reported photoconductivity characteristics of the Te-poly(methyl methacrylate) film may have potential applications in electronics and optoelectronics.

Pursuing a layer-by-layer fabrication method, tellurium nanowires prepared from stabilizer-depleted nanoparticles of CdTe were pieced in thin films [33]. Scanning electron microscopy imaging displayed that the tellurium nanowires were homogeneously distributed over the surface of the substrate (Figure 3.7a). To analyze the morphology of the surface of the film, microscopic analysis was used the image is shown in Figure 3.7b.

The photoresponses of the tellurium thin films were examined in detail. By chopping the incident red laser light, the time-resolved traces of the photocurrent at 1 mV were found (Figure 3.8a). In comparison with the initial dark current (0.16 nA), the

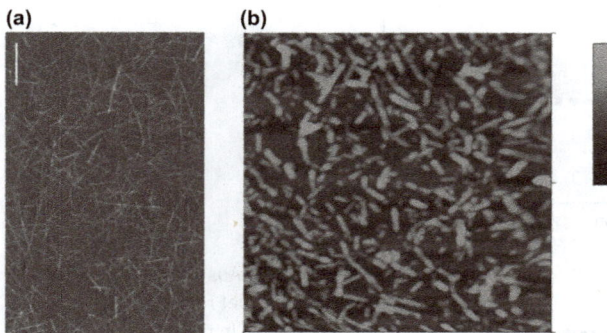

Figure 3.7: Microscopic analysis.
(a) Scanning electron microscopy image of (PDDA/NW). (b) Atomic force microscopy image of (PDDA/NW). Adapted with permission from ref. [33].

(a)

(b)

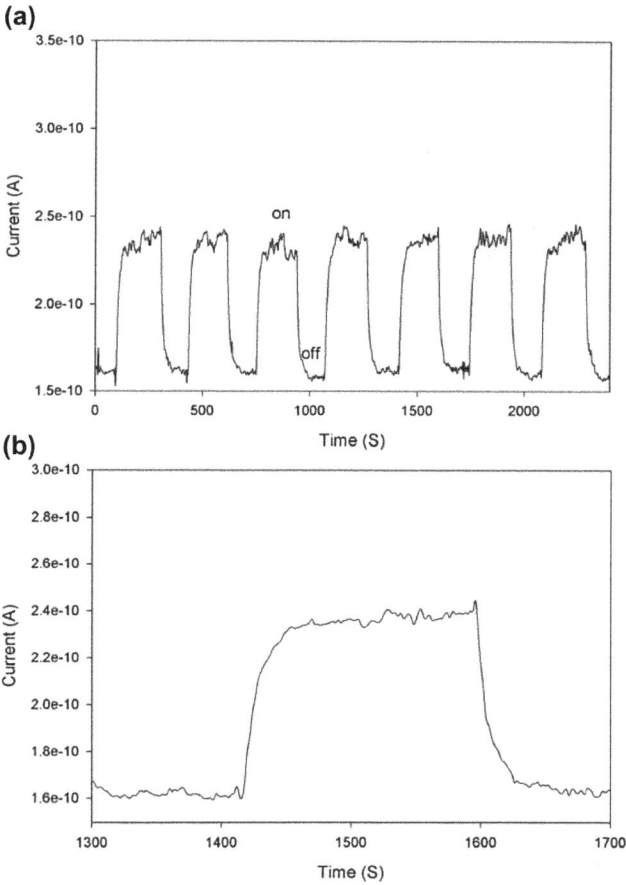

Figure 3.8: Photoresponses of the tellurium thin films.
(a) The response of the Te thin films to light for repetitious switching of the He:Ne laser light between "on" and "off" states. (b) Enlarged view of one on-off cycle. The bias between two contacts is 1 mV. Adapted with permission from ref. [33].

current under illumination raised to 0.24 nA. Figure 3.8b shows that the rise and decay times of the conductance in tellurium films were about 40 s.

The tellurium thin-film displayed a potent photoconductance result which is the feature of narrow-bandgap semiconductors. The films have an unusual metallic mirror-like visual aspect. Treatment with gold induced several variations in the characteristics of the film (Figure 3.9a). Reduction of Au, *in situ,* resulted in the creation of gold nanoparticles binding to tellurium nanowires. It heads to the loss of photo-conductivity of the tellurium thin film. Besides, the study was able to prove that the resistance can be changed by summing novel chemical constituents, for example,

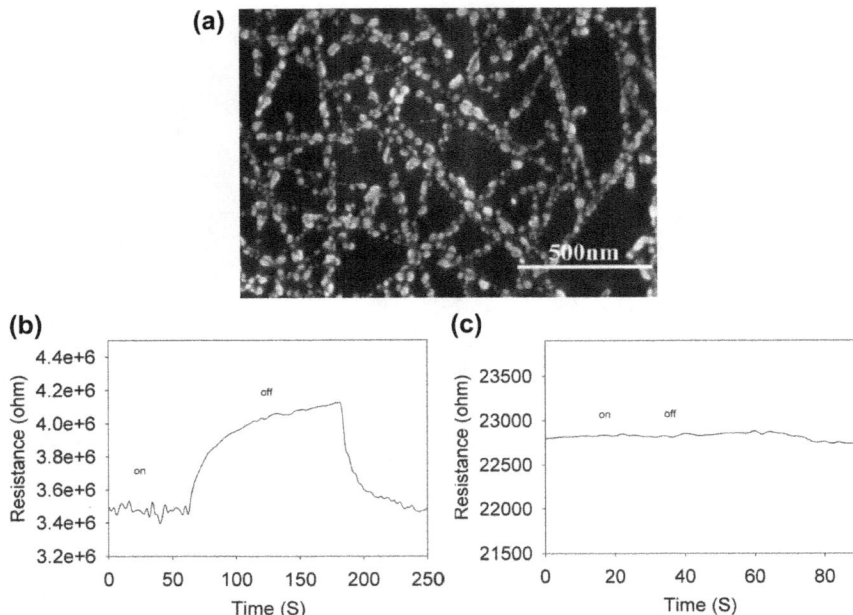

Figure 3.9: Characteristics of the film.
(a) SEM image of Te nanowires after the film was immersed in 1 mM $HAuCl_4$ for 3 min. The lateral resistance changes of Te nanowire thin film before (b) and after (c) immersing in 1 mM $HAuCl_4$ when the ambient light is turned on or off. Adapted with permission from ref. [33].

metallic Au, employing tellurium nanowires as a chemical template (Figures 3.9b and 3.9.c). The photocurrent measurements evidenced that following irradiation with the external light source, the tellurium thin films can be switched in a reversible manner among the higher- and lower-conductivity conditions. These types of photoconducting nanowires can serve as light detectors and switching devices for optoelectric purposes, as the binary conditions could be addressed in an optical manner by using the photoconducting nanowires.

Using the Langmuir–Blodgett technique, periodic chiselled mesostructures of hydrophilic ultrathin Te nanowires could be made [34]. Any extra functionalization hydrophobic or pretreatment was not used during this technique and the nanowires with aspect ratios of about or more than 10^4 were produced. Fabrication of nanomesh-like structures or advanced multilayered systems framed of ultrathin nanowires on a planar substrate was possible with packing the arrayed nanowire monolayers. The proper-arranged monolayer of tellurium nanowires having regular mesostructures can be promptly employed as a stamp to transport such nanopatterns with mesostructured to another substrate or can be implanted among a polymer matrix. When the light is on and off, the mesostructures of the sample displayed reversibly shifted photoelectric characteristics among the higher and lower conductivity conditions. It was also observed that the photocurrent was regulated by the count of mesostructured nanowire monolayer films and the intensity of the light. This technique can be applied for the inventing of other mesostructured assemblies of ultrathin nanotubes or nanowires.

3.2.2 Te-based Infrared acousto-optic deflectors

For laser wavelength purposes in the range of 5–20 μm, a single element 2-dimensional acousto-optic laser beam deflector by employing a Te crystal was projected and constructed [35]. The figure of merit values related with the two acousto-optic actions $(285,000 \times 10^{-15}\, s^3\, kg^{-1})$ were among two and three orders of magnitude greater than Ge, the immediate most effective acousto-optic material for 10.6 μm. The acoustic power densities of the crystal (nearly 0.1 Wmm^{-2}) were adequate to diffract all the light striking on it. The observational analysis supported the anticipated deflection slope of about 13°/25 MHz. The theoretical analysis regarding the amplitude and frequency modulation features of the 2-D acousto-optic deflector was also addressed.

3.2.3 Te-based Field effect transistors

Using an environmentally-friendly hydrothermal method, uniform and well-dispersed needle-like tellurium nanowires were fabricated in excellent yield [36]. The study demonstrated that the morphology of tellurium nanowires, reaction temperature and beta-cyclodextrin ligands play a critical role. To obtain uniform needle-like tellurium nanowires, an appropriate amount of beta-cyclodextrin and temperature for reaction are needed. An applicable formation mechanism of the needle-liked tellurium nanowires was also covered based on the experimental data. Using the photolithographic patterning technique, high-quality single tellurium nanowires field-effect transistors were fabricated. The demonstrated tellurium nanowires field-effect transistor presented carrier mobility of about 299 cm^2 V^{-1} s^{-1}. The performance of the device was influenced by surface species of nanowires, metal contacts of nanowire devices, crystallinity, and purity.

Using a catalyst-free one-step PVD (physical vapor deposition) technique, extremely arranged and well-arrayed 1-D tellurium nanostructures were fabricated [37]. The size, smooth structures, and density of nanowires were optimized consistently. To show remarkable dependence on nanostructure morphologies, field emission analysis was executed. The field emitter based on the ordered nanowire array presented a turn-on field as small as 3.27 V/μm and a more prominent field enhancement parameter of approximately 3270. The study demonstrated the potential of mastering the Te nanowire lays out growth and it opens the potential applications of tellurium in electronic and display devices.

For the development of new technologies grounded on 2-D materials, the authentic fabrication of 2-D crystals is necessary. But, most of the current synthesis technologies were subjected to a diversity of retracts, such as restrictions in crystal stability and size. Several synthetic methods were investigated to overcome the drawbacks in the synthesis. Using a substrate-free solution method, manufacturing of high-quality, large-area two-dimensional tellurene (tellurium) was reported [38]. Samples with the process-tunable thickness (monolayer to tens of nanometres) and with lateral dimensions of maximum 100 µm were possible with this approach. The chiral-chain van der Waals structure of tellurene produced influential thickness-dependent changes in Raman vibrational modes and greater in-plane anisotropic characteristics. These characteristics were not observed in other two-dimensional layered materials. Field-effect transistors based on tellurene were also demonstrated. At room temperature, the fabricated transistor showed air-stable functioning, field-effect mobilities of around 700 $cm^2\,V^{-1}\,s^{-1}$ and on/off ratios on the order of 106. In addition, by integrating with high-k dielectrics and scaling down the length of the channel, transistors with a considerable on-state current density of about 1 Amm^{-1} were achieved.

3.2.4 Te-based self-developing holographic recording devices

On Li-implanted Te thin films, the ablative holographic recording was executed employing a Nd:YAG laser [39]. Implantation dosages in the range of 2.9×10^{13} ions per cm^2 to 2.3×10^{15} ions per cm^2 inside the thin films were realized to make an increment in the writing threshold of about 20%. A decrease in the efficiency of the diffraction of the registered gratings was also observed at the elevated doses. A substantial rise in the constancy of the recording media has been detected, for dosages as small as 1.2×10^{14} ions per cm^2.

3.2.5 Te-based radiative cooling devices

The formulation of large area and homogenous thin layers of Te on thin polyethylene foils was reported [40]. The Te was prepared using room temperature breakdown of electrochemically created hydrogen telluride. To improve the tellurium homogeneity and adhesion, the polyethylene substrates were pre-treatment using $KMnO_4$ to produce a manganese oxide layer (Figure 3.10).

Applying spectroscopic technique, the optical characterization of the layers was executed. In the mid-IR region, the Te layers displayed eminent transmission and in the solar spectral region, blocking was observed (Figure 3.11). In a radiative cooling device, these characteristics are favorable for the role of solar radiation protective covering.

The greater solar reflectivity would amend the layers even more for the usage as a radiative cooling device. By control of deposit morphology, it can be accomplished.

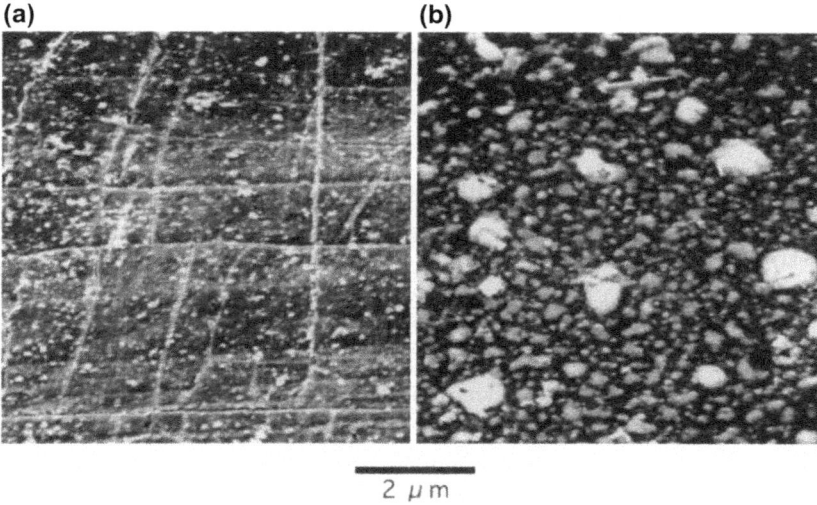

Figure 3.10: Scanning electron microscopic image of (a) manganese oxide coated polyethylene (b) Tellurium implant on the manganese oxide/polyethylene. Adapted with permission from ref. [40].

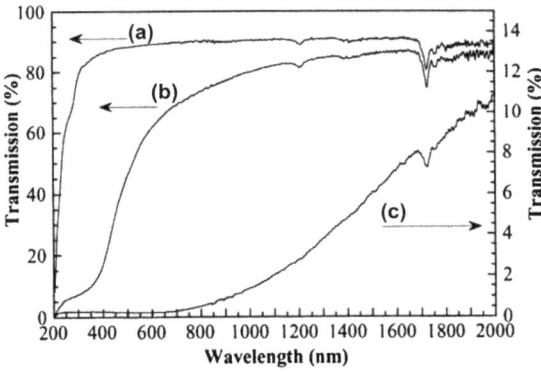

Figure 3.11: Visible-NIR total transmission spectra of polyethylene (a), manganese oxide covered polyethylene (b), tellurium film on manganese oxide covered polyethylene (c). Adapted with permission from ref. [40].

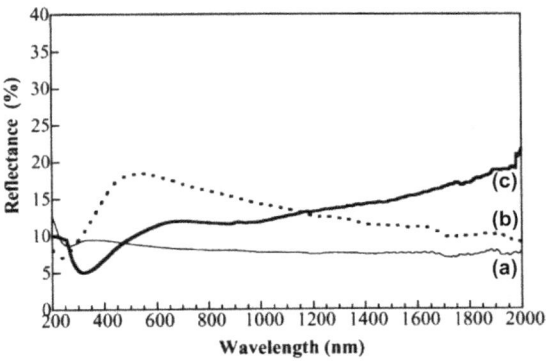

Figure 3.12: Total reflectance of the samples. Adapted with permission from ref. [40].

The total reflectance (diffuse reflectance and specular reflectance) of the samples is shown in Figure 3.12. The easiness of preparation of the very unstable hydrogen telluride was also used to evidence the establishment of size-quantized Cadmium-Tellurium nanocrystals.

3.2.6 Te-based Topological insulators

Electronic structure analysis using ab initio calculation demonstrated that under shear strain (uniaxial or hydrostatic), trigonal tellurium constituting feebly acting helical chains experiences a transition from less important insulator to potent topological insulator, for instance metal [41]. The changeover was presented by analyzing the concomitant band inversion, the band structure's strain evolution, and the topological Z2 constant. The proposed mechanism was the depopulation of the lone-pair orbitals linked with the valence band through suitable strain engineering. In that way, tellurium turns the prototype of a new group of chiral-based 3-D topological insulators with significant applications in thermoelectrics, magneto-optics and spintronics.

3.2.7 Te-based photodetectors

Because of the broad usages in the synthesis of ultrathin 2-D layered materials, Van der Waals epitaxy is of great concern. At the same time, the Van der Waals epitaxy of nonlayered useable substance was not very well investigated. Because of its chain-like structure, Te has a greater inclination to grow into 1-D nanoarchitecture, however, researchers successfully realized two-dimensional hexagonal tellurium nanoplates using Van der Waals epitaxy on flexible mica sheets [42]. For the lateral growth of hexagonal tellurium nanoplates, a chemically inert mica surface was observed to be essential. As it helps the tellurium adatom's migration through the surface of mica and grants a big lattice mismatch. Moreover, photodetectors based on two-dimensional Te hexagonal nanoplates were manufactured on flexible mica sheets *in situ*. Even afterwards deforming the device for hundred times, an efficient photoresponse was observed. The analysis indicated that the photodetectors, on mica sheets, based on two-dimensional Te hexagonal nanoplates possess outstanding purpose in wearable and flexible optoelectronic devices. The underlying knowledge of the consequence of Van der Waals epitaxy on the growth of two-dimensional tellurium hexagonal nanoplate can open a direction for leveraging Van der Waals epitaxy as a practicable method to obtain the two-dimensional geometry of other nonlayered materials.

The nonlinear optical response of the tellurium is also studied in detail. In single-crystal tellurium, on the coherent phonon formation the effect of chirped laser pulses was investigated [43]. It was observed that the amplitude of coherent phonons having

A_1 geometry was affected by the pulse chirp in the instance of strong irritation. By altering the chirp of a strong exciting pulse, the negatively chirped pulses were about twice greater efficient in the introduction of lattice coherence compared to positively chirped pulses.

3.2.8 Te-based thermoelectric materials

Tellurium has shown high thermoelectric properties. Zhang et al. presented a fabrication principle to make novel classes of thermoelectric nanowire heterostructures established on telluride via rational solution-phase interactions [44]. The catalyst-free production made Tellurium-Bi_2Te_3 "barbell" nanowire structures with a small length and diameter distribution together with a rough govern over the density of the hexagonal Bi_2Te_3 plates on the tellurium nanowires. As $PbTe$-Bi_2Te_3, it can be again changed to other telluride-based compositional-modulated nanowire heterostructures. The characterizations of the hot-pressed nanostructured bulk pellets of the Te-Bi_2Te_3 structure depicted greatly reduced thermal conductivity and a largely enhanced Seebeck coefficient. It resulted in a bettered thermoelectric figure of merit. This method pioneers novel platforms to look into energy filtering and phonon scattering.

The thermoelectric transport characteristics of elemental tellurium were studied using DFT together with the Boltzmann transport theory in the rigid band estimation [45]. It was reported that because of the isotropic (anisotropic) electron (hole) pockets of the Fermi surface, the thermoelectric transport characteristics perpendicular and parallel to the helical chains were extremely symmetric (asymmetric) for n-(p-) type doped Te. The electronic band structure indicated that the lone-pair deduced uppermost heavy-hole and highly light-hole lower valence bands allow the possibility to get both high electrical conductivity and Seebeck coefficient along the chains via bismuth or antimony doping. Moreover, relative to the Fermi energy, the stair-like density of states gives a great asymmetry for the transport distribution function which guides to large thermopower. The analysis showed that Te has the potency to be an efficient p-type thermoelectric material having an optimum figure of merit of around 0.31 at a hole concentration of about 1×10^{19} cm^{-3} at room temperature. Tapping the rich chemistry of lone pairs in chiral solids may have significant imports for the breakthrough of large-figure of merit thermoelectric materials based on polychalcogenide.

Utilizing ab initio calculations, the intrinsic lattice thermal conductivity of chiral Te was calculated [46]. The study showed that the interplay among the weak van der Waals interchain and strong covalent intrachain interactions produces the phonon band gap among the higher and lower optical phonon branches. The fundamental process of the great anisotropy of the thermal conductivity was the anisotropy of the anharmonic interatomic force constants and of the phonon group velocities. The large interchain anharmonic interatomic force constants were related to the lone electron pairs. It was anticipated that the low thermal conductivity of Te was because of the large three-phonon scattering phase space. Under implemented hydrostatic pressure, the thermal conductivity anisotropy decreases.

High conductivity is required for high-efficiency thermoelectric materials. A big count of degenerate band valleys provides lots of conducting channels to improve the conductivity without damaging impressions on the other attributes. Several semiconductors provide an inherent band nestification, along with the strategy of converging diverse bands, evenly allowing a big count of efficient band valley degeneracy. Tellurium displays a great thermoelectric figure of merit of unity, not only evidencing the concept but also filling the high-performance gap for elemental thermoelectrics in the range of 300–700 K [47]. A survey of the temperature-dependent figure of merit (zT) for elemental thermoelectrics is depicted in Figure 3.13. For comparison, both high-temperature Si/Si–Ge alloys and low-temperature Bi/Bi–Sb alloys are considered in the survey.

The Hall carrier concentration dependence in nested bands on the transport properties is shown in Figure 3.14.

Transport properties (temperature-dependent) are shown in Figure 3.15. In Figure 3.15a, for comparison, the ab initio computed Seebeck coefficient is admitted as solid curves. Many of the materials examined showed a dissipated p-type semiconducting behavior and a predominant phonon scattering, which is noted as a red curve in Figure 3.15d.

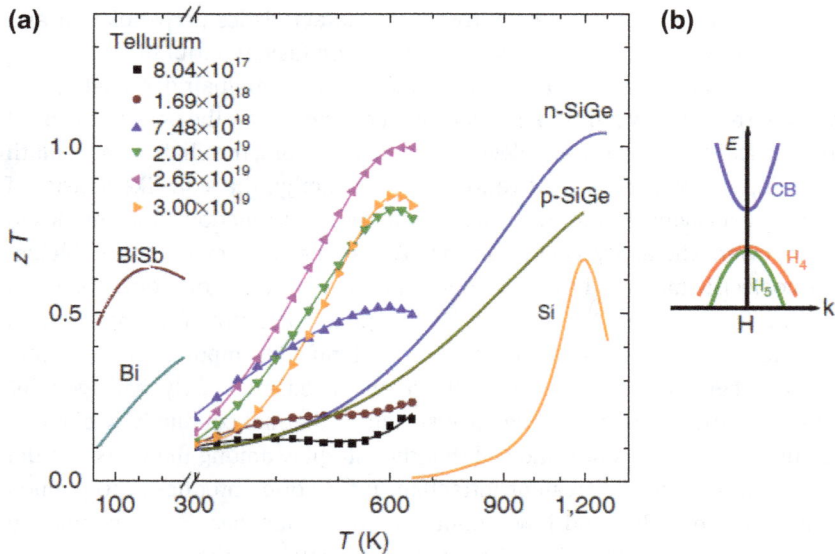

Figure 3.13: Survey of the figure of merit for elemental thermoelectrics.
(a) The temperature-dependent figure of merit for p-type polycrystalline Te with various carrier concentrations is shown in a unit of cm^{-3}. (b) p-type tellurium showed the highest zT in the temperature range of 300–700 K, largely relying on its inherently nested valence bands (H_4 and H_5). Adapted with permission from [47].

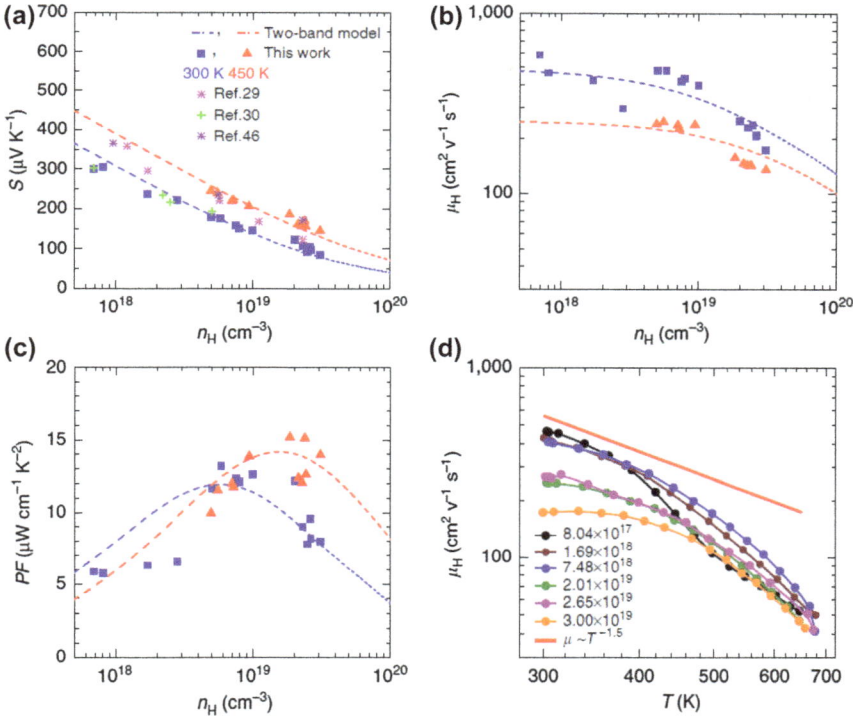

Figure 3.14: Changes in the parameters.
(a) Hall carrier concentration-dependent Seebeck coefficient, (b) Hall mobility, (c) power factor at 450 and 300 K and (d) temperature-dependent Hall mobility with a comparison to literature. Adapted with permission from ref. [47].

3.2.9 Te-based piezoelectric materials

Trigonal ultrathin tellurium nanowire was demonstrated as a novel group of piezoelectric nanomaterial bearing a radially asymmetric crystal geometry for a high-volume power-density nanogenerator [48]. Considering the crystal structure of trigonal tellurium nanowire, a perfect design for the trigonal tellurium nanowire-based nano-generator was proposed. Figure 3.16 shows the schematic representation of a most preferably projected trigonal tellurium nanowire-based nanogenerator.

To analyze the grade of the organization of the trigonal tellurium nanowires, the monolayer was detected employing different techniques (Figure 3.17). To assert the potency of the qualitative design, using a finite element method a theoretical analysis of two potential stretching modes of trigonal tellurium nanowires in a deforming functioning was carried out. As a result, it was confirmed that strain only in the radial way of the trigonal tellurium nanowires triggers its piezoelectricity. Using a monolayer

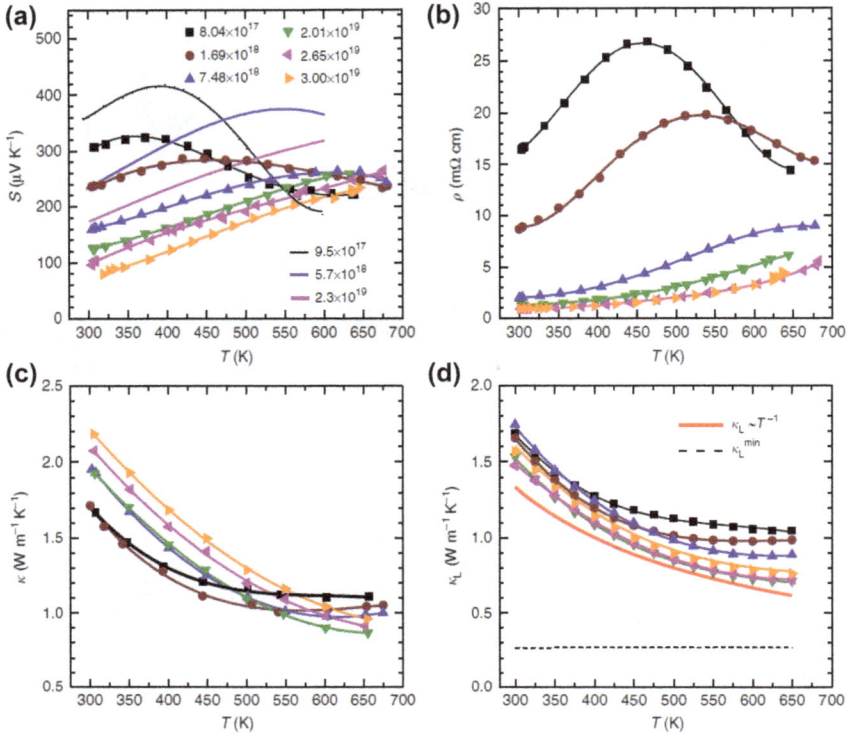

Figure 3.15: Temperature dependent transport properties. For p-tellurium, (a) Seebeck coefficient, (b) resistivity, (c) total thermal conductivity and (d) lattice thermal conductivity. Adapted with permission from ref. [47].

of well-aligned trigonal tellurium nanowires developed employing a fast-spreading process, a nanogenerator with a power density of 9 mWcm^{-3} was demonstrated (Figure 3.18). The nanogenerator showed distinct operation degrees concording to the ordinated way of the trigonal tellurium nanowires in reference to the deforming line, which was in accord with the theoretic calculations. From the view of bringing out the trigonal tellurium nanowire as a novel group of piezoelectric material, the bending frequency stability and long-term stability of the trigonal tellurium nanowire-based nanogenerator were evaluated, and its impressive stability in these positions was affirmed by experimentation. Accordingly, a project path in which the way of the chief piezoelectric strain in the trigonal tellurium nanowire-based nanogenerator has to be aligned to the direction of the asymmetry in its crystal structure was created. To manifest the higher quality piezoelectricity of the trigonal tellurium nanowire on the output power, the thickness of the dielectric layer in the nanogenerator was optimized. This process significantly promoted the functioning in comparison to that of a conventional nanogenerator. This investigation of the trigonal tellurium nanowire can be

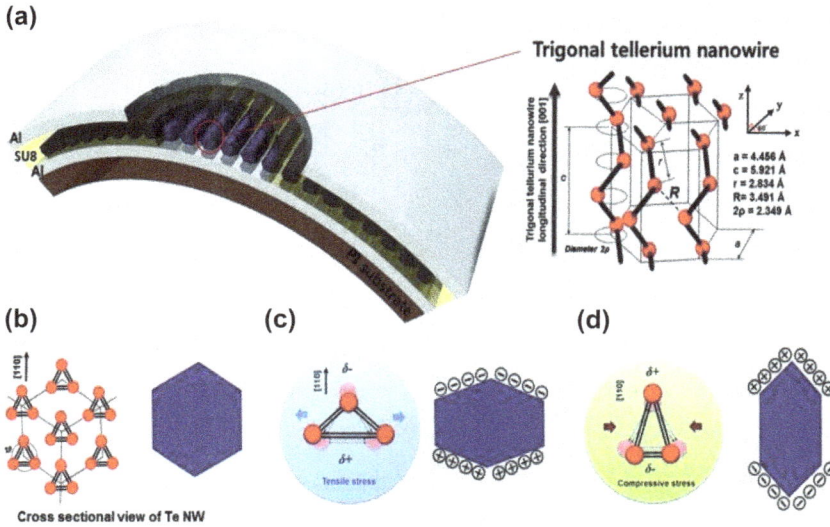

Figure 3.16: Schematic representation.
(a) A bent nanogenerator comprising of arrayed single-crystalline trigonal tellurium nanowires which are immersed in an SU8 layer bamong Al electrodes on a PI substrate. (b) A cross-sectional view of trigonal tellurium nanowires. (c) Under tensile stress, dipole moment arises along the [-1-10] direction. (d) Under compressive stress, the [110] directional dipole moment is generated. Adapted with permission from [48].

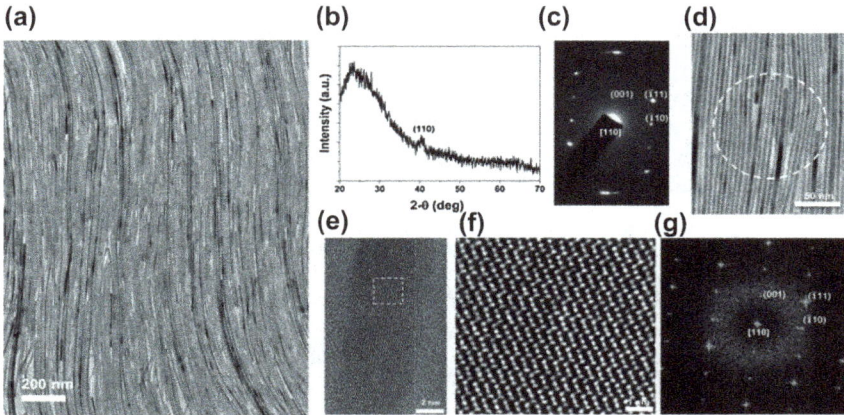

Figure 3.17: Microscopic images.
(a) The TEM image of the monolayer. (b) The XRD analysis of the monolayer on a glass substrate. (c) SAED pattern of the trigonal tellurium nanowire film. (d) The white dotted circle indicates the selected area. (e) High-resolution TEM analysis of a trigonal tellurium nanowire. The (110) orientation of the NW on the TEM grid definitely is confirmed from the magnification image (f) and the FFT image (g) of white dotted square. Adapted with permission from [48].

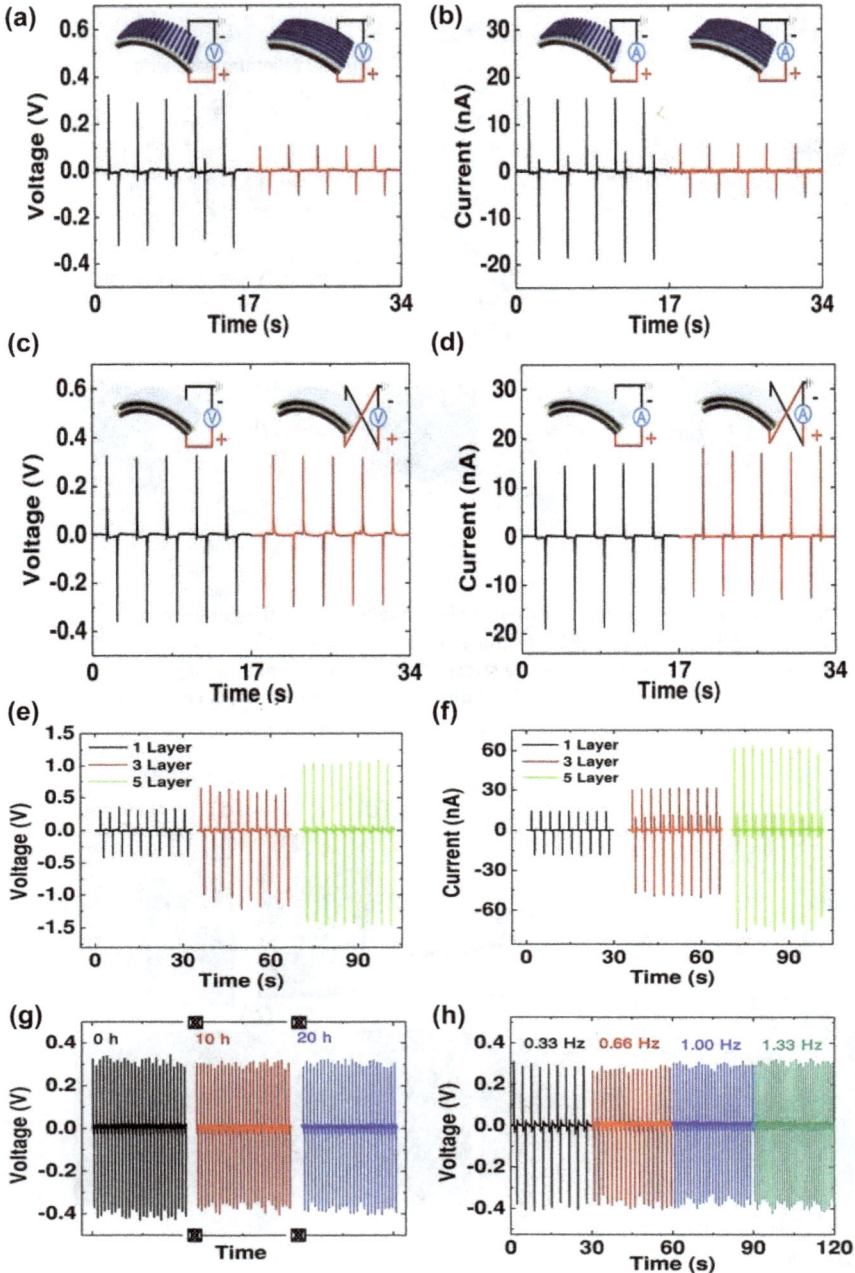

Figure 3.18: According to the aligned direction of the trigonal tellurium nanowires, the output voltage of nanogenerators (a) and the output current of nanogenerators (b) under tensile stress as induced by bending are shown. (c) The output voltage of the nanogenerator with forward and reverse connections. (d) The output current of the nanogenerator with forward and reverse connections. (e) The output voltage with 1, 3, and 5 layers. (f) The output current with 1, 3, and 5 layers. Stability depends on the (g) time and (h) frequency using a sample of a single layer. Adapted with permission from ref. [48].

elaborated to the different ultrathin chalcogenide nanowires. The trigonal tellurium nanowires can be utilized as a template to get several telluride materials posing practicable characteristics, exchanging different kind of ambient energy into electrical energy utilizing features like thermoelectricity, pyroelectricity, and piezoelectricity.

On a sample of Au-coated textile, the ultra-thin Te nanoflakes (NFs) were fabricated in a successful manner, which then was employed as an efficient piezoelectric [49]. The schematic drawing of the manufacturing method and characterization of the tellurium nanoflake nanogenerator device are shown in Figure 3.19. The schematic diagrams in Figure 3.19a show the preparation method for constructing a flexible tellurium nanoflake nanogenerator device. The photographs in Figure 3.19b represent three various bending conditions: flat, rolled and folded. The SEM images in Figure 3.19c and the insets present that 20 nm thick tellurium nanoflakes are well fabricated on textiles. The crystalline symmetry of the tellurium nanoflakes is analyzed by X-ray diffraction (Figure 3.19d). The crystalline nature of the tellurium nanoflakes was examined by employing HRTEM (high-resolution transmission electron microscopy) as shown in Figure 3.19e. The SAED (selected area electron diffraction) and high-resolution transmission electron microscopic images of individual tellurium NFs are demonstrated in Figure 3.19f.

The bending test with a driving frequency of around 10 Hz showed a current of 300 nA and an output voltage of around 4 V. To examine the practical applications, the tellurium NFs nanogenerator device was connected to the subject's arm. Employing periodic arm-bending motions, the electrical energy was created from mechanical energy. When the tellurium nanoflake nanogenerator device experienced a compression test with driving frequency of around 10 Hz and a compressive force of about 8 N, the optimized short-circuit current density and open-circuit voltage of approximately 17 $\mu A/cm^2$ and 125 V, respectively, were measured. The instant powering of 10 green LEDs (light-emitting diodes) was possible with this high-power generator, the diodes were glowed without any help from an external power supply.

Under identical strains, the tellurium nanoflake nanogenerator device reached an open-circuit voltage and a closed-circuit current of 3 V and 290 nA, respectively (Figure 3.20). The current (Figure 3.21a) and output voltage (Figure 3.21b) created by the continues straightening and bending activities of the human arm achieved up to ~650 nA and 2.5 V, respectively.

The experimental analysis of the piezotronic phenomenon in one dimensional van der Waals solid of p-type Te nanobelt was investigated [50]. The strain-gated charge carrier's transport properties were also studied. Through the volumetric action on the conducting medium and the interfacial effect on the Schottky contacts, the strain-stimulated polarization charges at the [10-10] surfaces of tellurium nanobelt can regulate electronic transport. The competing phenomenon among volumetric and interfacial effects has been examined. This study permits entry to a wide range of depiction and implementation of tellurium nanostructures for piezotronics. It could also lead the future study of piezotronic effect in different structures. The advance in

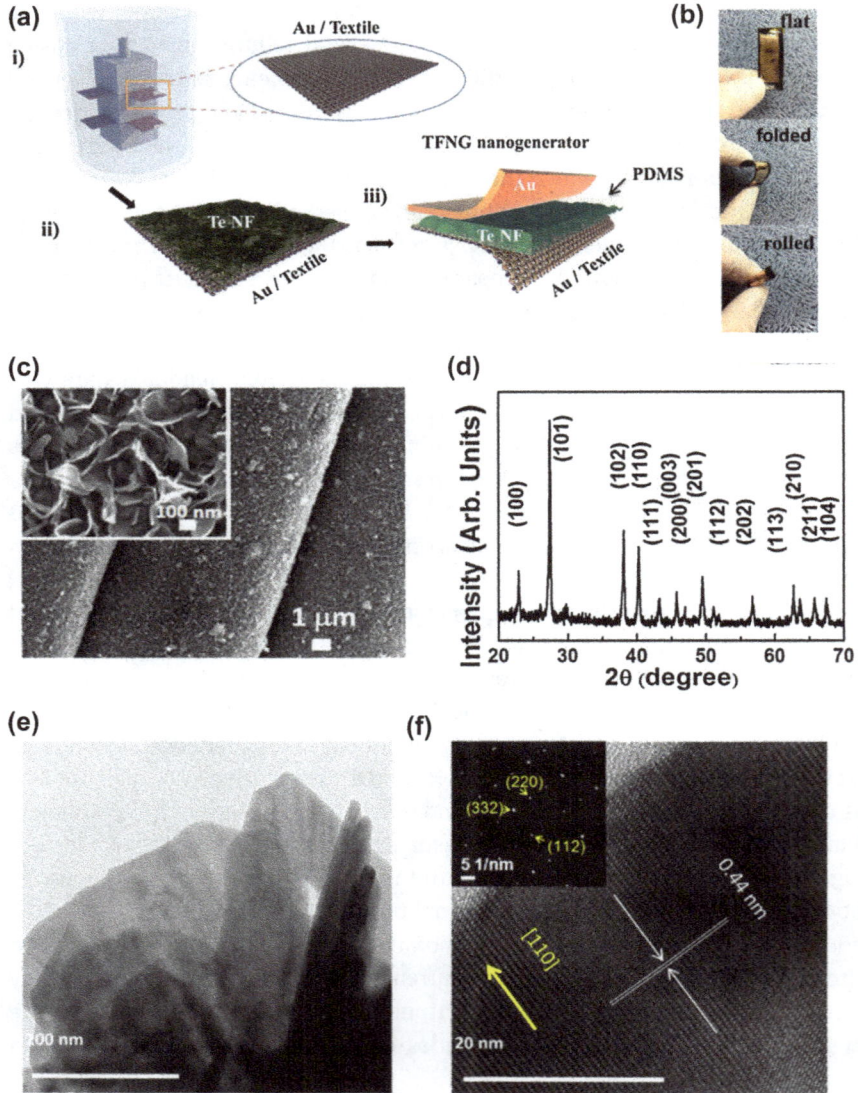

Figure 3.19: The schematic diagrams of the fabrication process and general characterization of the tellurium nanoflake nanogenerator device.
(a) Schematic of the process for fabricating the TFNG devices. (b) Optical images depicting the tellurium nanoflake nanogenerator devices in different states: Flat, folded, and rolled. (c) SEM image of the Te nanoflakes synthesized by the low hydrothermal method; the inset represents a magnified view. (d) XRD patterns of the Te nanoflakes as synthesized. (e) TEM image of Te nanoflakes and (f) HRTEM image showing individual nanoflake with [110] growth direction, Inset shows the SAED pattern of Te nanoflakes. Adapted with permission from ref. [49].

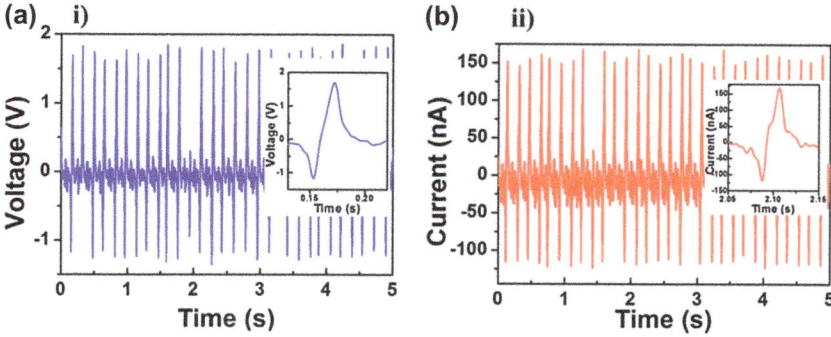

Figure 3.20: Electrical measurements of the TFNG device during bending test. Measured output voltage and current of the TFNG device during repeated bending and unbending motions. The device exhibited a maximum peak open-circuit voltage of 3.0 V (i) and a short-circuit current of 290 nA (ii), while the power was 0.87 mW. Adapted with permission from ref. [49].

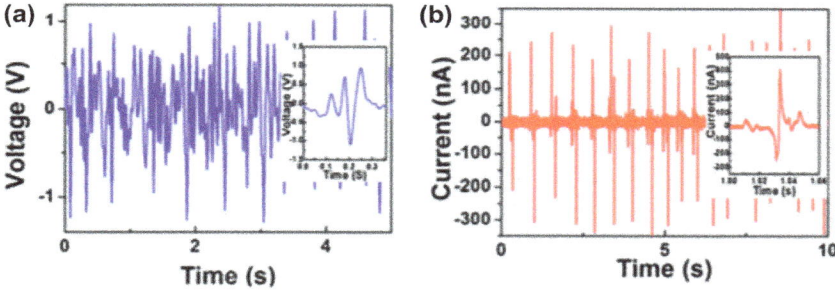

Figure 3.21: Electrical measurements of the TFNG device attached to a human arm.
(a) Open-circuit voltage during periodic bending and straightening of the human arm. (b) Closed-circuit currents are produced by the TFNG device during periodic bending and straightening of the human arm. Adapted with permission from ref. [49].

piezotronics, along with rising ways for settled manufacturing and gathering of nanostructures, heads to obligating possibilities for research from basic analysis of semiconductor properties and piezoelectricity in useable nanostructures to the exploitation of excellent optoelectronics and electronics.

3.2.10 Te-based Sensors

A single ultralong tellurium micrometer wire-based flexible strain sensors have been demonstrated [51]. The flexible strain sensor comprised of a flexible polystyrene substrate having a single Te micrometer wire was positioned on its surface and attached by silver paste. Two different kinds of strain sensors were demonstrated. Back-to-back Schottky barrier contacts were utilized in one sensor. Because of the alteration of the Schottky barrier height, this sensor had the diverse shape I–V behaviors, when there were no stretch, strain, and compressive strains. As the compressive strain raises the current also raises. A single Schottky barrier contact was employed in the second sensor. At a negative voltage, under compression strain, the current enhances more markedly. the second one was more appropriate for strain detection.

Using quaternary chalcogenides (As–Ge–S–Te), room temperature ac operating gas sensors were fabricated [52]. In order to check the sensing property at room temperature, the impedance spectra of quaternary As–Ge–S–Te based alloys were analyzed in both dry synthetic air and a mixture with NO_2. The quaternary compositions $As_2Te_{13}Ge_8S_3$ and $As_2Te_{130}Ge_8S_3$, with enhancing concentration of Te was counted along with pure Te films to analyze the effect of tellurium. At room temperature, the frequency-dependent sensitivity of the films against NO_2 is shown in Figure 3.22.

Sensors using CdTeS quantum dots (QDs) and silver nanoparticles (NPs) for the sensing of L-cysteine were demonstrated [53]. For the fabrication of the sensor, the ternary alloyed CdTeS QDs were fabricated in two steps, utilizing 3-mercaptopropionic acid as a stabilizer. The polyvinylpyrrolidone stabilized silver nanoparticles were made by the crystal-seed method. The CdTeS QDs/Ag NPs (QNs) emitting at around 580 nm were employed as detectors for L-cysteine. With the summation of L-Cysteine, the luminescence of QNs was raised. A linear relation was observed among the fluorescence intensity and the L-cysteine concentrations (20–400 µM) as shown in Figure 3.23. The sensing limitation was observed at 0.025 µM and the correlation coefficient was 0.99.

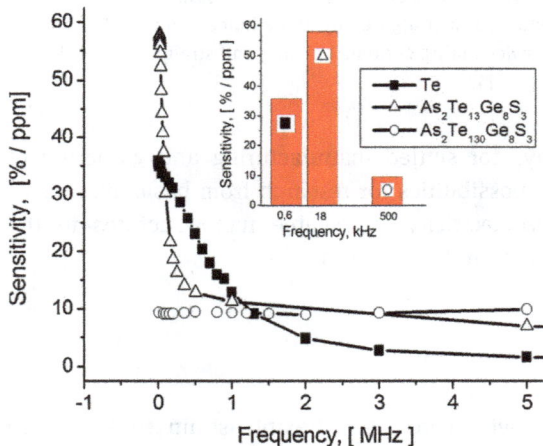

Figure 3.22: Sensitivity spectra of quaternary chalcogenide thin films to 1.5 ppm NO_2 in. dry air at room temperature. Inset shows the comparison diagram of the maximum sensitivities at respective frequencies for the materials in question. Adapted with permission from ref. [52].

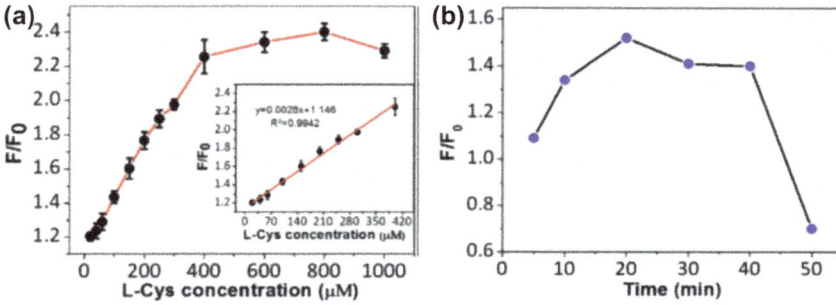

Figure 3.23: Changes in the fluorescence intensity.
(a) Relationship between fluorescence intensity F/F_0 and L-cysteine concentrations. Inset: Expanded linear region (20–400 µM) of the calibration curve. (b) The effect of the response time of CdTeS QDs/Ag NPs to L-cysteine on detection results. Adapted with permission from ref. [53].

3.2.11 Te-based Solar cells

Oladeji et al. demonstrated the potency of CdTe/CdS/Cd$_{1-x}$Zn$_x$S solar cell material [54]. The demonstrated solar cell structure was an option to CdTe/CdS material in photovoltaic usage. The solar cell materials fabricated on transparent conducting oxide-coated soda-lime glass having no antireflection covering showed a 10% effectiveness.

Recently, a glass/FTO/CdS/CdTe/Te/Al superstrate structured solar cell was demonstrated [55]. Tellurium was integrated among the CdTe layer and the Al electrode. The optical transmittance of the manufactured device was analyzed using UV–Visible-NIR-spectrometer in the wavelength range of 200–1100 nm. The optical transmittance spectrum of the demonstrated device is depicted in Figure 3.24. The device

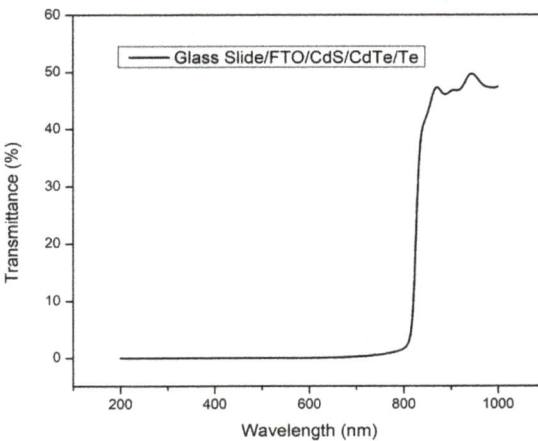

Figure 3.24: Transmittance spectrum in the wavelength range 200–1000 nm of the fabricated device (glass/FTO/CdS/CdTe/Te). Adapted with permission from ref. [55].

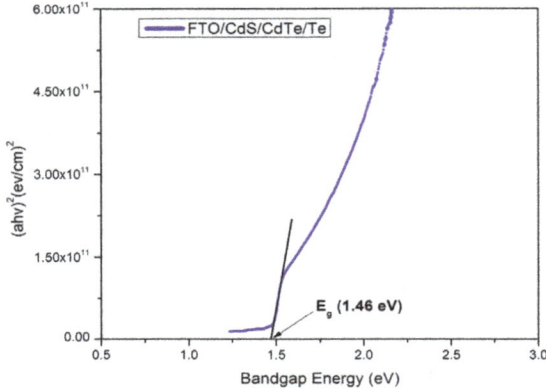

Figure 3.25: Plot of $(\alpha h\upsilon)^2$ versus photon energy $(h\upsilon)$ for the evaluation of energy band gap of. CdTe thin films. Adapted with permission from ref. [55].

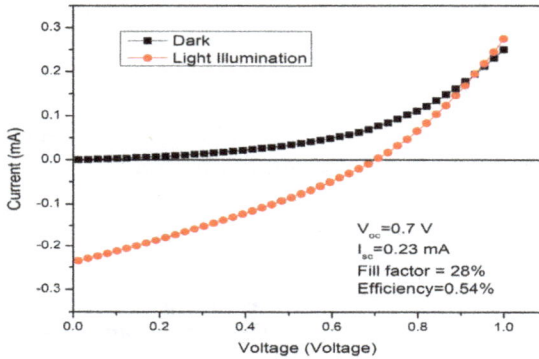

Figure 3.26: *I–V* response of the FTO/CdS/CdTe/Te/Al solar cell. Adapted with permission from ref. [55].

presented a transmittance of 48% in the infrared range (810–1000 nm) and poor transmittance in the range of 810–200 nm. The poor transmission was owing to the high absorption features of CdTe in that range.

To find the energy band gap of the film, the plot of $(\alpha h\upsilon)^2$ versus $h\upsilon$ was diagrammed (Figure 3.25). The calculated band gap was observed to be around 1.4 eV, the value was the optimized bandgap for CdTe absorber material. The I-V response of the solar cell is shown in Figure 3.26.

3.3 Conclusions

Tellurium is a promising and important functional material because of its unique physical and chemical properties. This article has discussed the semiconductor properties of tellurium and recent developments in these areas. This study opens up the possibility of designing novel devices and considering modern applications of Tellurium.

Acknowledgments: AD and BKB are grateful to Prince Mohammad Bin Fahd University for support.

References

1. Anzin VB, Eremets MI, Kosichkin YV, Nadezhdinskii AI, Shirokov AM. Measurement of the energy gap in tellurium under pressure. Phys Status Solidi 1977;42:385–90.
2. Morton M. Rubber technology. Heidelberg, Germany: Springer Netherlands; 1987.
3. Knockaert G. Tellurium and tellurium compounds In: Ullmann's Encyclopedia of industrial chemistry [Internet]. John Wiley & Sons; 2000. Available from: https://onlinelibrary.wiley.com/doi/abs/10.1002/14356007.a26_177 [Accessed 19 Nov 2021].
4. Lide DR. CRC handbook of chemistry and physics. Boca Raton, Florida, United States: CRC Press; 2004.
5. Nishii J, Morimoto S, Inagawa I, Iizuka R, Yamashita T, Yamagishi T. Recent advances and trends in chalcogenide glass fiber technology: a review. J Non-Cryst Solids 1992;140:199–208.
6. El-Mallawany RAH. Tellurite glasses handbook: physical properties and data. Boca Raton, Florida, United States: CRC Press; 2014.
7. Johnson LB. Correspondence. Representing delay powder data. Ind Eng Chem 1960;52:868.
8. Capper P, Elliott CT. Infrared detectors and emitters: materials and devices: materials and devices. Berlin/Heidelberg, Germany: Springer Science & Business Media; 2001.
9. Shenai-Khatkhate DV, Webb P, Cole-Hamilton DJ, Blackmore GW, Brian Mullin J. Ultra-pure organotellurium precursors for the low temperature MOVPE growth of II/VI compound semiconductors. J Cryst Growth 1988;93:744–9.
10. Mullin JB, Cole-Hamilton DJ, Shenai-Khatkhate DV, Webb P. Method for purification of tellurium and selenium alkyls. United States US5117021A, filed December 2, 1988, issued May 26, 1992. https://patents.google.com/patent/US5117021A/en.
11. Shenai-Khatkhate DV, Parker MB, McQueen AED, Mullin JB, Cole-Hamilton DJ, Day P, et al. Organometallic molecules for semiconductor fabrication. Phil Trans Roy Soc Lond Math Phys Sci 1990;330:173–82.
12. Nishiuchi K, Kitaura H, Yamada N, Akahira N. Dual-layer optical disk with Te–O–Pd phase-change film. Jpn J Appl Phys 1998;37:2163.
13. Hudgens S, Johnson B. Overview of phase-change chalcogenide nonvolatile memory technology. MRS Bull 2004;29:829–32.
14. Geppert L. The new indelible memories. IEEE Spectrum 2003;40:48–54.
15. Taft E, Apker L. Photoemission from cesium and rubidium tellurides. J Opt Soc Am 1953;43:81–3.
16. Cunha RLOR, Gouvea IE, Juliano L. A glimpse on biological activities of tellurium compounds. An Acad Bras Cienc 2009;81:393–407.
17. Ramadan SE, Razak AA, Ragab AM, El-Meleigy M. Incorporation of tellurium into amino acids and proteins in a tellurium-tolerant fungi. Biol Trace Elem Res 1989;20:225.
18. Rahman A. Studies in natural products chemistry. Amsterdam, Netherlands: Elsevier; 2008.
19. Chua SL, Sivakumar K, Rybtke M, Yuan M, Andersen JB, Nielsen TE, et al. C-di-GMP regulates Pseudomonas aeruginosa stress response to tellurite during both planktonic and biofilm modes of growth. Sci Rep 2015;5:10052.
20. Ottosson L-G, Logg K, Ibstedt S, Sunnerhagen P, Käll M, Blomberg A, et al. Sulfate assimilation mediates tellurite reduction and toxicity in Saccharomyces cerevisiae. Eukaryot Cell 2010;9:1635–47.

21. Chasteen TG, Bentley R. Biomethylation of selenium and tellurium: microorganisms and plants. Chem Rev 2003;103:1–26.
22. Taylor A. Biochemistry of tellurium. Biol Trace Elem Res 1996;55:231–9.
23. Kwantes W. Diphtheria in Europe. In: Epidemiology & infection. Cambridge University Press; 1984. p. 433–7, vol 93.
24. Chasteen TG, Fuentes DE, Tantaleán JC, Vásquez CC. Tellurite: history, oxidative stress, and molecular mechanisms of resistance. FEMS (Fed Eur Microbiol Soc) Microbiol Rev 2009;33: 820–32.
25. Sredni B. Immunomodulating tellurium compounds as anti-cancer agents. Semin Cancer Biol 2012;22:60–9.
26. Zare B, Nami M, Shahverdi A-R. Tracing tellurium and its nanostructures in biology. Biol Trace Elem Res 2017;180:171–81.
27. Seng H-L, Tiekink ERT. Anti-cancer potential of selenium- and tellurium-containing species: opportunities abound. Appl Organomet Chem 2012;26:655–62.
28. Zweibel K. The impact of tellurium supply on cadmium telluride photovoltaics. Science 2010;328: 699–701.
29. Saha GB. Physics and radiobiology of nuclear medicine. Berlin/Heidelberg, Germany: Springer Science & Business Media; 2001.
30. Kasap S, Willoughby A. In: Capper P, Garland J, editors. Mercury cadmium telluride: growth, properties and applications, 1st ed. Hoboken, NJ: Wiley; 2010.
31. Qiao J, Pan Y, Yang F, Wang C, Chai Y, Ji W. Few-layer Tellurium: one-dimensional-like layered elementary semiconductor with striking physical properties. Sci Bull 2018;63:159–68.
32. Coscia U, Ambrosone G, Palomba M, Binetti S, Le Donne A, Siliqi D, et al. Photoconductivity of tellurium-poly(methyl methacrylate) in the ultraviolet–visible-near infrared range. Appl Surf Sci 2018;457:229–34.
33. Wang Y, Tang Z, Podsiadlo P, Elkasabi Y, Lahann J, Kotov NA. Mirror-like photoconductive layer-by-layer thin films of Te nanowires: the fusion of semiconductor, metal, and insulator properties. Adv Mater 2006;18:518–22.
34. Liu J-W, Zhu J-H, Zhang C-L, Liang H-W, Yu S-H. Mesostructured assemblies of ultrathin superlong tellurium nanowires and their photoconductivity. J Am Chem Soc 2010;132:8945–52.
35. Souilhac D, Billerey D, Gundjian A. Infrared two-dimensional acoustooptic deflector using a tellurium crystal. Appl Opt 1990;29:1798–804.
36. Tao H, Liu H, Qin D, Chan K, Chen J, Cao Y. High mobility field effect transistor from solution-processed needle-like tellurium nanowires. J Nanosci Nanotechnol 2010;10:7997–8003.
37. Safdar M, Zhan X, Niu M, Mirza M, Zhao Q, Wang Z, et al. Site-specific nucleation and controlled growth of a vertical tellurium nanowire array for high performance field emitters. Nanotechnology 2013;24:185705.
38. Wang Y, Qiu G, Wang R, Huang S, Wang Q, Liu Y, et al. Field-effect transistors made from solution-grown two-dimensional tellurene. Nat Electron 2018;1:228–36.
39. Beauvais J, Lessard RA, Galarneau P, Knystautas EJ. Self-developing holographic recording in Li-implanted Te thin films. Appl Phys Lett 1990;57:1354–6.
40. Engelhard T, Jones ED, Viney I, Mastai Y, Hodes G. Deposition of tellurium films by decomposition of electrochemically-generated H2Te: application to radiative cooling devices. Thin Solid Films 2000;370:101–5.
41. Agapito LA, Kioussis N, Goddard WA, Ong NP. Novel family of chiral-based topological insulators: elemental tellurium under strain. Phys Rev Lett 2013;110:176401.
42. Wang Q, Safdar M, Xu K, Mirza M, Wang Z, He J. Van der Waals epitaxy and photoresponse of hexagonal tellurium nanoplates on flexible mica sheets. ACS Nano 2014;8:7497–505.

43. Misochko OV, Dekorsy T, Andreev SV, Kompanets VO, Matveets YA, Stepanov AG, et al. Effect of intense chirped pulses on the coherent phonon generation in Te. Appl Phys Lett 2007;90: 071901.
44. Zhang G, Fang H, Yang H, Jauregui LA, Chen YP, Wu Y. Design principle of telluride-based nanowire heterostructures for potential thermoelectric applications. Nano Lett 2012;12:3627–33.
45. Peng H, Kioussis N, Snyder GJ. Elemental tellurium as a chiral p-type thermoelectric material. Phys Rev B 2014;89:195206.
46. Peng H, Kioussis N, Stewart DA. Anisotropic lattice thermal conductivity in chiral tellurium from first principles. Appl Phys Lett 2015;107:251904.
47. Lin S, Li W, Chen Z, Shen J, Ge B, Pei Y. Tellurium as a high-performance elemental thermoelectric. Nat Commun 2016;7. https://doi.org/10.1038/ncomms10287.
48. Lee TI, Lee S, Lee E, Sohn S, Lee Y, Lee S, et al. High-power density piezoelectric energy harvesting using radially strained ultrathin trigonal tellurium nanowire assembly. Adv Mater 2013;25:2920–5.
49. He W, Van Ngoc H, Qian YT, Hwang JS, Yan YP, Choi H, et al. Synthesis of ultra-thin tellurium nanoflakes on textiles for high-performance flexible and wearable nanogenerators. Appl Surf Sci 2017;392:1055–61.
50. Gao S, Wang Y, Wang R, Wu W. Piezotronic effect in 1D van der Waals solid of elemental tellurium nanobelt for smart adaptive electronics. Semicond Sci Technol 2017;32:104004.
51. Liang T, Zha J-W, Wang D, Dang Z-M. Remarkable piezoresistance effect on the flexible strain sensor based on a single ultralong tellurium micrometre wire. J Phys D Appl Phys 2014;47:505103.
52. Tsiulyanu D, Ciobanu M. Room temperature a.c. operating gas sensors based on quaternary chalcogenides. Sensor Actuator B Chem 2016;223:95–100.
53. Yang M, Yan Y, Shi H, Wang C, Liu E, Hu X, et al. Efficient inner filter effect sensors based on CdTeS quantum dots and Ag nanoparticles for sensitive detection of l-cysteine. J Alloys Compd 2019;781: 1021–7.
54. Oladeji IO, Chow L, Ferekides CS, Viswanathan V, Zhao Z. Metal/CdTe/CdS/Cd1−xZnxS/TCO/ glass: a new CdTe thin film solar cell structure. Sol Energy Mater Sol Cell 2000;61:203–11.
55. Venkatesh R, Banapurmath NR, Ramesh K, Venkatesh A, Khandake SA, Kurade PR, et al. Enhancement of open circuit voltage of CdTe solar cell. Mater Today Proc 2020;27:117–9.

Aparna Das*, Devalina Ray and Bimal Krishna Banik*

4 Tellurium in carbohydrate synthesis

Abstract: In this article, we discuss about the influence of tellurium in carbohydrate synthesis. Mainly the chapter focuses on the importance of the tellurium during the synthesis of glycosides and during the oxidation of glucose.

Keywords: catalyst; gluconic acid; glucose; glycosides; oxidation; tellurium.

4.1 Introduction

Carbohydrates are accountable for various cell functioning, together with metabolism of energy [1, 2]. For example, carbohydrates are observed in extra- and intra-cellular media adhered to lipids and proteins [3–5]. In intercellular space and on cell surfaces, carbohydrates also pose as free polysaccharides and it is utilized by virus and bacteria as attachment locations [6]. Likewise, carbohydrates are also observed as units in various natural products such as macrolactin O, salicin, oleandrin, tunicamycin, dapagliflozin, empagliflozin, aureonuclemycin and catalpol. These compounds have showed a broad range of medicinal activities. Thus, it is important to note various routes for the carbohydrates synthesis.

In this chapter, we focus on the different methods for the synthesis of glycosides and the oxidation of glucose. Mainly, there are four glycosides: *O*-glycoside, *N*-glycoside, *S*-glycoside and *C*-glycoside. The major constituent of nucleic acids and nucleotides, adenosine, is an *N*-glycoside. To a great degree, *S*-glycosides were observed in the Brassica family. Examples of *C*-glycosides isolated from different plant genus include aloin, carminic acid, saponarin and scoparin. *O*-glycosides are widely found in nature in the higher plants, including rhubarb, frangula, and senna. In this article we discuss only about the synthesis of *O*- and *C*-glycosides.

Tellurium (Te) is a metalloid and it is a low bandgap semiconductor. Te is useful for various applications. For example, it is used for photovoltaic and thermoelectric applications. Te is used as a coloring agent in glass and ceramics and also in copying machines. Te is also utilized as a vulcanizing agent to make durable products in chemical industry. Te is also used in the semiconductor industry such as in integrated circuits, laser diodes and sensors, in medical instrumentation, and automobile industry. Tellurium compounds are also used in organic synthesis. For instance, it uses

---
*Corresponding authors: Aparna Das and Bimal Krishna Banik, Department of Mathematics and Natural Sciences, College of Sciences and Human Studies, Prince Mohammad Bin Fahd University, Al Khobar 31952, Kingdom of Saudi Arabia, E-mail: aparnadasam@gmail.com (A. Das), bimalbanik10@gmail.com (B.K. Banik). https://orcid.org/0000-0002-2502-9446 (A. Das)
Devalina Ray, Amity Institute of Biotechnology, Amity University, Noida 201313, UP, India

As per De Gruyter's policy this article has previously been published in the journal Physical Sciences Reviews. Please cite as: A. Das, D. Ray and B. K. Banik "Tellurium in carbohydrate synthesis" *Physical Sciences Reviews* [Online] 2022. DOI: 10.1515/psr-2021-0109 | https://doi.org/10.1515/9783110735840-004

for carbohydrate synthesis. In organic synthesis, in addition to utilize as reagents for oxidation, tellurium compounds are also described as versatile electrophiles practicable in diverse organic transformations. Tellurium-induced cyclization of alkenyl compounds, as well as alkynyl compounds, is explored very well considered for the study. In the following sections, some remarkable and recent, breakthrough in this field is highlighted. The importance of the tellurium during carbohydrate synthesis is presented in this chapter.

4.2 Synthesis of glycosides

4.2.1 *O*-Glycosides

In carbohydrate chemistry, one of the main interests is the exploitation of synthetic methods for the stereoselective and efficient synthesis of glycosides. Among several glycosides, the synthetic method for 2,3-unsaturated-*O*-glycosides *O* has gained much interest due to its grandness as intermediates in the synthesis of various crucial molecules [7–11]. Figure 4.1 shows examples of natural products containing an O-glycosidic bond.

Ferrier rearrangement is one of the significant *O*-glycosylation procedure for synthesizing 2,3-unsaturated-*O*-glycosides. This method requires an allylic shift in a glycal together with a nucleophilic substitution reaction and is originally promoted by $BF_3 \cdot Et_2O$ [12].

Other Lewis acids can also be used in the reaction, for example montmorillonite-K10 [13], $InCl_3$ [14], $SnCl_4$ [15], $FeCl_3$ [16], $BiCl_3$ [17], $Sc(OTf)_3$ [18], $LiBF_4$ [19], $ZnCl_2$ [20], $AuCl_3$ [21], $Dy(OTf)_3$ [22], $ZrCl_4$ [23] and $CeCl_3$ [24]. Besides, some oxidizing agents can also promote the reaction, for example DDQ [25], iodine [26], NIS [27], CAN [28], $I(Coll)_2ClO_4$ [29] and $HClO_4$ on silica gel [30]. However, majority of these process need a big quantity of alcohol, some have drawbacks in yields, selectivity and generality.

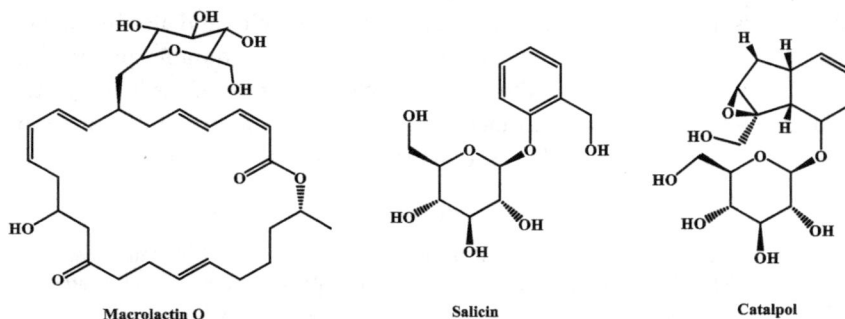

Macrolactin O   Salicin   Catalpol

Figure 4.1: Natural products bearing an *O*-glycosidic bond.

Due to the structural versatility and possible applications as synthons in several chemical interactions, tellurium tetrahalides (TeI$_4$, TeCl$_4$ and TeBr$_4$) have attracted considerable attention [31]. With both Lewis acids and Lewis bases, tellurium tetra-halides can interact [32]. The Te–X bonding which is partially ionic might be the reason for the amphoteric behaviour of tellurium tetrahalides.

Freitas et al. reported the usage of a catalytic quantity of tellurium(IV) tetra-bromide to elevate the glycal's O-glycosylation to produce 2,3-unsaturated-O-glyco-sides [33]. Figure 4.2 shows the synthetic route. At room temperature, a solution of glycal 2 and alcohol in CH$_2$Cl$_2$ was processed with catalytic quantity of TeBr$_4$ to produce 2,3-unsaturated-O-glycosides 3. The products were found in high α-selectivity and good yields. Nine selected products are shown in Table 4.1.

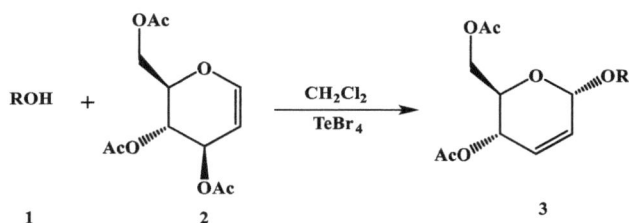

Figure 4.2: Preparation of 2,3-unsaturated-O-glycosides promoted by TeBr$_4$.

Table 4.1: Selected products.

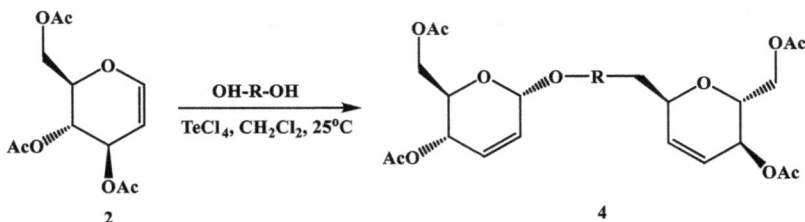

Figure 4.3: TeCl$_4$ promoted preparation of 2,3-unsaturated-O-glycosides.

The same research team demonstrated the O-glycosides dimers synthesis catalyzed by tellurium(IV) tetrachloride [34]. The synthesis of glycosides promoted by TeCl$_4$ is shown in Figure 4.3.

At room temperature, glucal 2 and diol in dichloromethane were processed with a tellurium catalyst to produce 2,3-unsaturated-O-glycosides 4. The interaction produced the desired products in impressive quantity and with well α-anomeric selectivity. The obtained products are shown in Table 4.2.

The pseudoglycoside synthesis catalyzed by tellurium (IV) tetrachloride under mild conditions was also reported and the reaction was stereoselective [35]. To boost the 3,4,6-tri-O-acetyl-D-glucal-O-acetyl-D-'s O-glycosylation to produce the representing 2,3-unsaturated-O-glycosides, catalytic amounts of TeCl$_4$ were used. Using only 2 mol% of the catalyst the hoped compounds were found in very good anomeric selectivity and good yields with simple alcohols, in a short reaction time. The interaction functioned substantially for diverse alcohols (Figure 4.4). The interaction with alkynols also produced the desired compounds with an excellent selectivity after short reaction times. Nine selected products are shown in Table 4.3.

Table 4.2: Selected products.

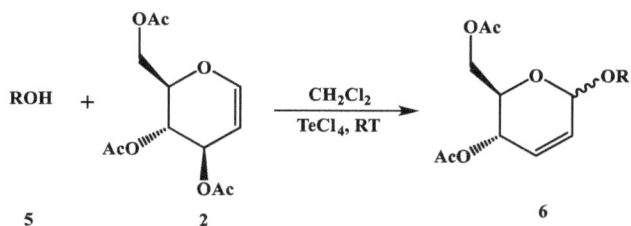

Figure 4.4: Preparation of 2,3-unsaturated-*O*-glycosides 6 promoted by TeCl₄.

Table 4.3: Selected products.

Using catalytic amounts of TeCl₄, the interaction of 3,4,6-tri-*O*-acetyl-D-glucal 2 with diverse alkynyl diols as nucleophiles produced the corresponding glycosides 8 in well selectivities, in short reaction times, and in good quantity (Figure 4.5). The obtained products are shown in Table 4.4.

The interaction of 3,4,6-tri-*O*-acetyl-D-glucal 2 with glycol ethers produced the desired products 10 in reasonable selectivities and yields (Figure 4.6). The obtained products are shown in Table 4.5.

Enediyne motifs and enyne are structural components observed in many natural products, which present a range of medicinal activities. For example, neocarzinostatin

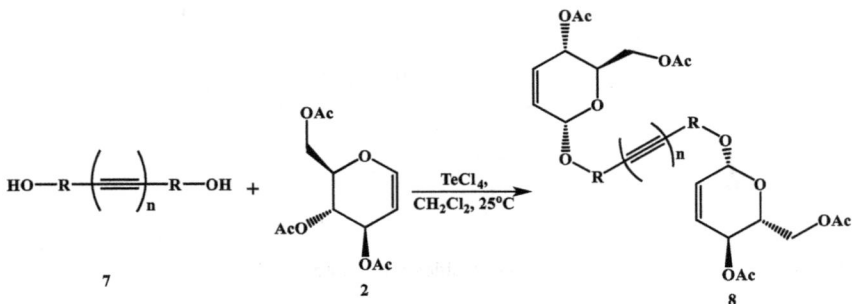

Figure 4.5: Preparation of alkynyl 2,3-unsaturated *O*-glycosides **8** promoted by TeCl$_4$.

Table 4.4: Selected products.

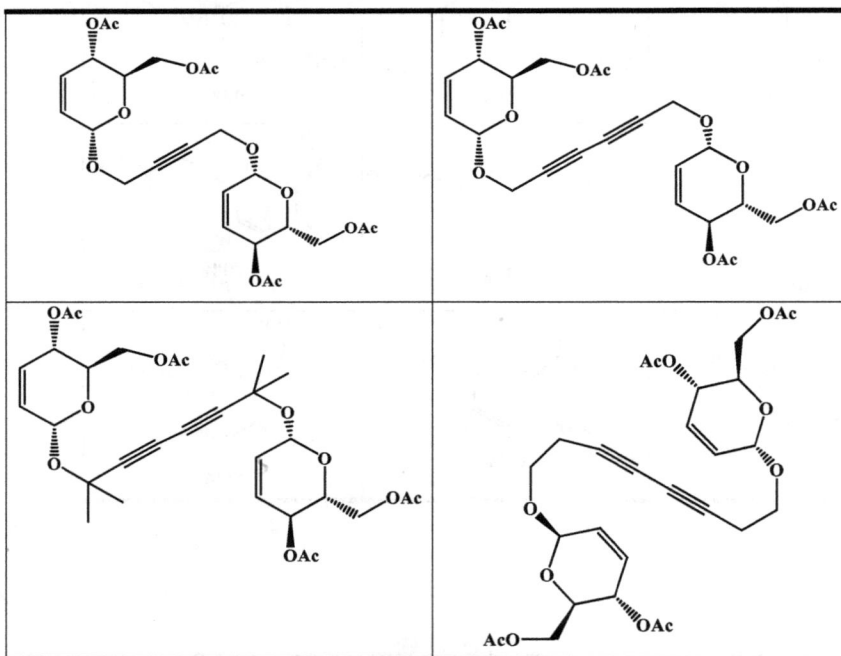

chromophore [36, 37] and phorbaside are biologically active molecules (Figure 4.7). However, because of the scarcity from natural resources and toxicity, the practical application of the mentioned compounds is confined. Hence, the synthesis of correspondent compounds with a less structural complication that could imitate the mode of

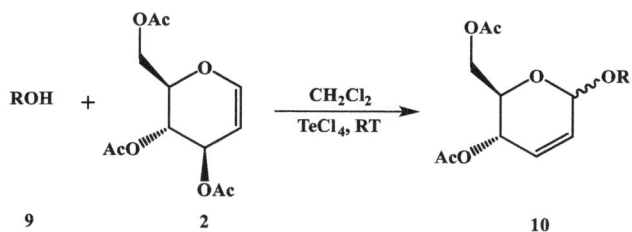

Figure 4.6: Interaction of compound 2 with glycol ethers promoted by TeCl₄.

Table 4.5: Products obtained.

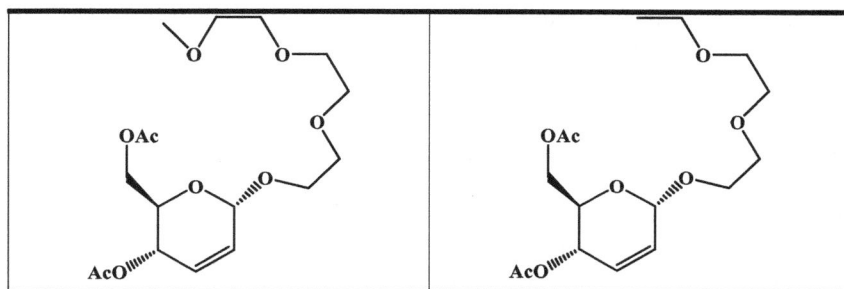

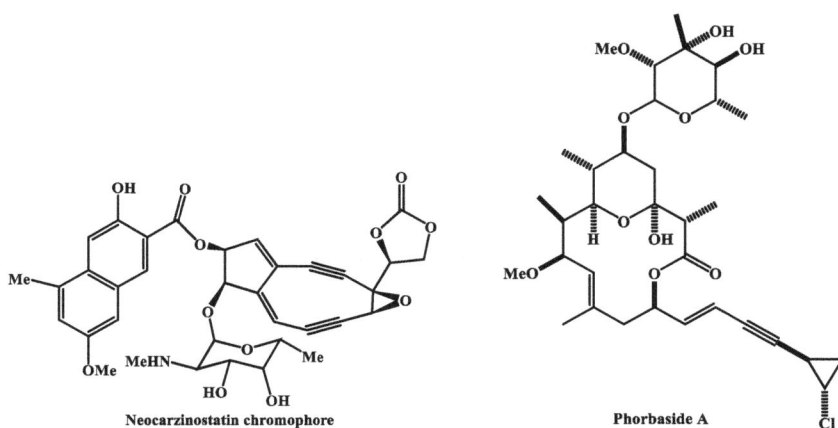

Figure 4.7: Molecular structure of phorbaside A and the neocarzinostatin chromophore.

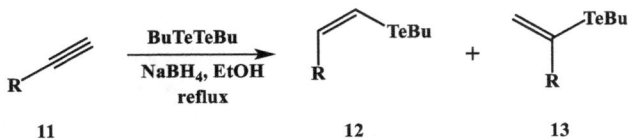

Figure 4.8: Preparation of stereoselective vinyl tellurides.

function of these compounds is important. For the preparation of stereodefined enynes, several methods can be employed [38–47].

Stereoselective synthesis of Z-enyne pseudoglycosides was investigated by Dantas et al. [48]. The preparation of Z-1,3-enynes was based on the coupling interaction of Z-vinyl tellurides and alkynes comprising a pseudoglycoside moiety. The desired compounds were received through a stereoselective pathway in good yields.

The stereoselective synthesis of vinyl tellurides is shown in Figure 4.8. Alkynes **11** were experienced to hydrotelluration circumstances to produce the desired Z-vinyl tellurides **12** in well quantity in a very regio- and stereo-selective path. The obtained vinyl tellurides are shown in Table 4.6.

As a glycosyl donor 3,4,6-tri-O-acetyl-D-glucal, **2**, was utilized in a reaction using various alcohols catalyzed by TeCl$_4$ (Figure 4.9). The corresponding pseudoglycosides **15** were produced in anomeric selectivities and good yields. The obtained products are depicted in Table 4.7.

The cross-coupling reaction of vinyl tellurides **16** with α-pseudoglycosides **17** catalyzed by palladium is shown in Figure 4.10. The corresponding enyne compounds are shown in Table 4.8.

In all cases good yields were obtained without isomerization of the Z-double bond, so the method was robust and the reaction seems not to be sensitive to the type of functional group present in the starting telluride. The reaction also does not seem to be

Table 4.6: Corresponding vinyl tellurides.

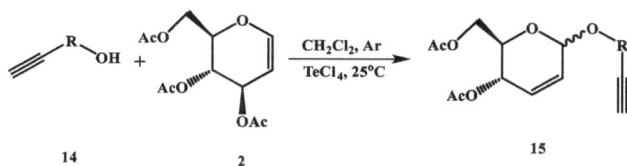

Figure 4.9: TeCl$_4$ catalyzed synthesis of pseudoglycosides.

Table 4.7: Corresponding pseudoglycosides.

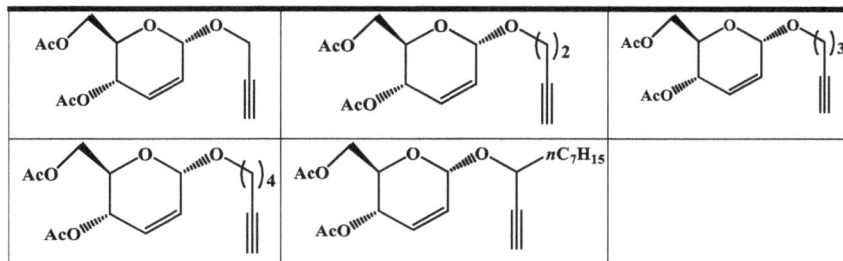

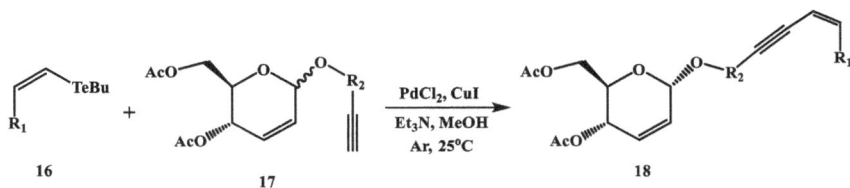

Figure 4.10: The coupling interaction of α-pseudoglycosides and Z-vinyl tellurides.

affected by the type of substituent present on the aromatic ring. The increment in the distance of the glycosyl moiety from the triple bond did not affect the yield considerably.

The synthesized compounds were tested against three cancer cell lines (NCI-H292, HL-60 and MCF-7) and the data showed that the desired products are hopeful intermediates for the synthesis of more biologically active compounds.

4.2.2 *C*-Glycosides

C-Glycosides are central parts of a large count of bioactive compounds, drugs and natural products. Some important *C*-Glycosides are shown in Figure 4.11. Structurally,

Table 4.8: The obtained enyne compounds.

89%	86%
87%	85%
85%	85%
84%	

they are similar to *O*-glycosides, in which the C–C bond in place of C–O acetal linkage. The changes increased importantly the stability of *C*-glycosides for hydrolytic enzymes *in vivo*. These have great potential as biological probes and therapeutic agents.

Various methods were described for *C*-glycosides synthesis. α-Alkoxyacyl telluride-associated interactions, reactions catalyzed by transition-metals and photo-mediated reactions are a few among the methods.

Using radical-based coupling reactions of sugar derivatives employing two-and three-component, the single-step building of different carboskeletons which are densely oxygenated was demonstrated [49]. During the process, an Et radical generated from Et_3B/O_2 would induce the C–Te homolysis of 20, leading to the acyl radical, the decarbonylation of which would give rise to -alkoxy carbon radicals. The nucleophilic radical would then react intermolecularly with the electrophilic C=N bond and the C=C bond to generate the corresponding N- and C-radicals, respectively. Et_3B in

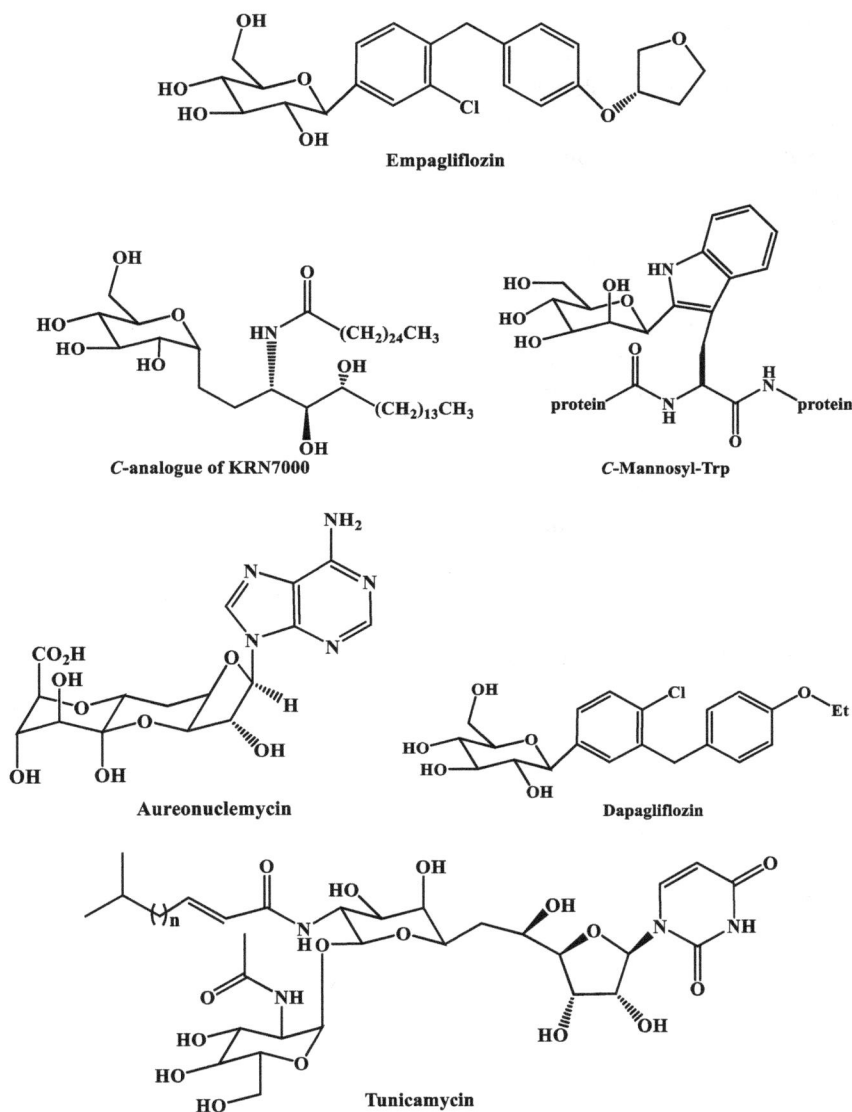

Figure 4.11: Selected *C*-glycosides.

turn would transform these radicals into the polar intermediates through ejection of an Et radical, and subsequent protonation would afford the two-component adducts 22 and 21, respectively. The three-component adducts were made using an intermolecular aldol reaction among the boron enolates and the aldehyde yielded by the capture of the radical intermediates (two-component) by Et_3B.

The synthesis of α-alkoxyacyl tellurides α- and their interactions with enones and glyoxylic oxime ether is shown in Figure 4.12. Acid derivative of β-D-Arabino-2-hexulopyranosonic β-D, β-D-Ribofuranosiduronic β-D-, and 2-Keto-L-gulonic were used as carboxylic acids with four neighbouring oxygen-bonded stereocenters. The alkoxyacyl tellurides 20 were then prepared in one pot from these sugar-derived acids. The two-component radical interactions among the three radical acceptors and three radical donors 20 produced nine complex adducts (Table 4.9).

The nucleoside antibiotics, polyoxins J and L, have showed significant activities. Synthesis of polyoxins and their artificial analogues with fluorouracil and trifluorothymine structures was reported [50]. The radical coupling reaction which is decarbonylative among a chiral glyoxylic oxime ether and α-alkoxyacyl tellurides head to stereo- and chemo-selective building of the ribonucleoside α-amino acid symmetry of polyoxins without altering the preinstalled nucleobases.

The synthesis of α-alkoxyacyl tellurides is shown in Figure 4.13. The C4″-radical precursors 24 and C3″-radical precursor 26 were made in three steps from commercially obtainable ribonucleosides 23 and (+)-2,3-O-isopropylidene-1-threitol 25, respectively.

The radical coupling reactions of 24 and 26 are shown in Figure 4.14. By the action of O_2 and Et_3B, the α-alkoxyacyl tellurides 24 and 26 were changed into the desired α-alkoxy radicals. The total synthesis of polyoxins is shown in Figure 4.15.

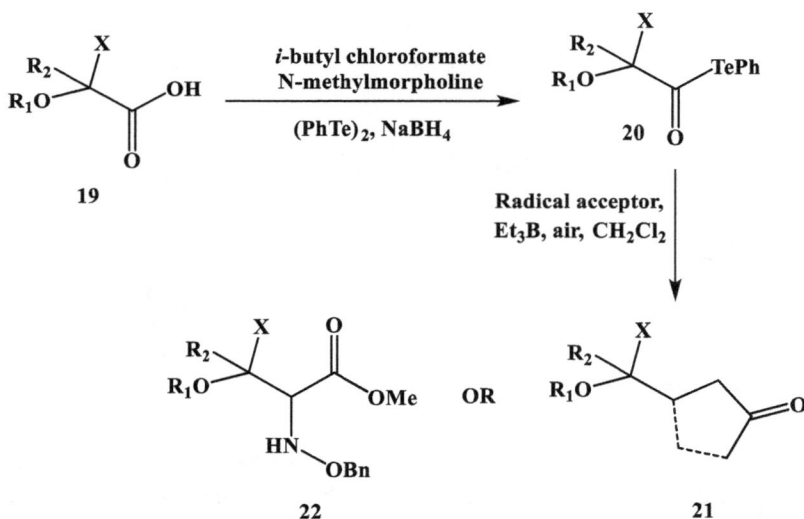

Figure 4.12: Synthesis of α-alkoxyacyl tellurides and their interactions with enones and glyoxylic oxime ether.

Table 4.9: The obtained compounds.

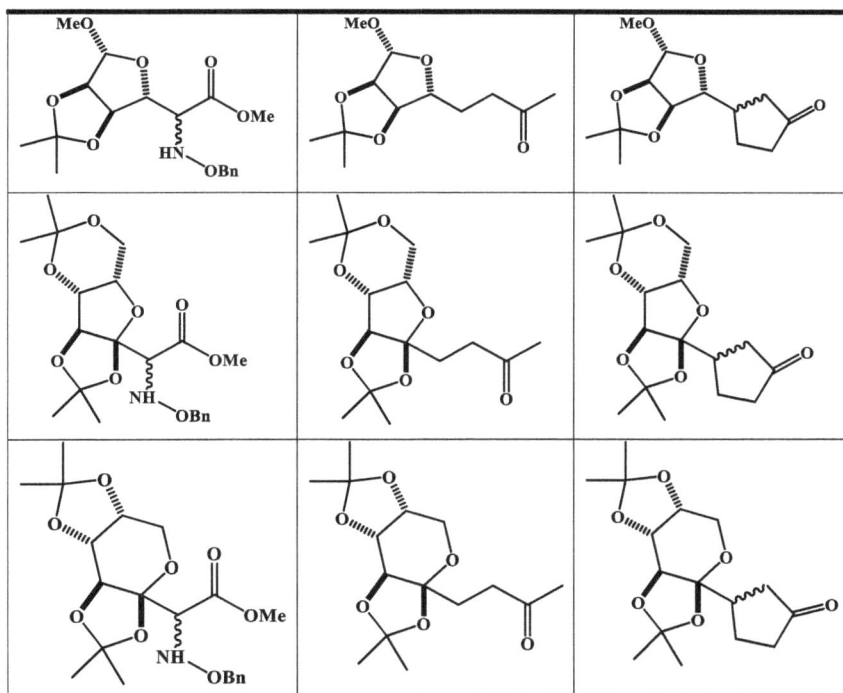

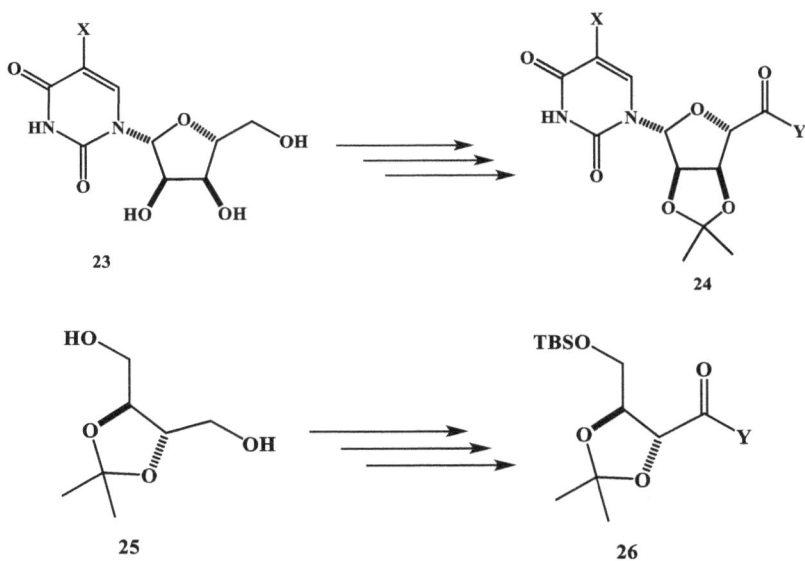

Figure 4.13: Synthesis of α-alkoxyacyl tellurides.

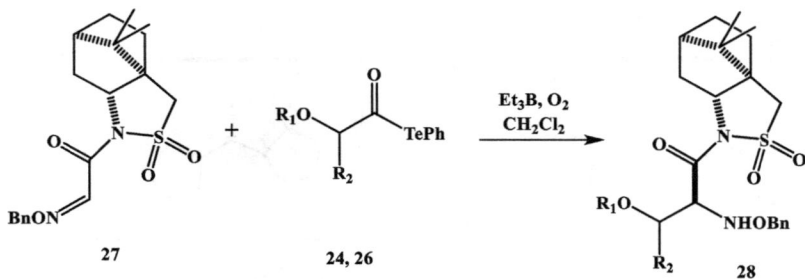

Figure 4.14: Decarbonylative radical reactions of 24 and 26.

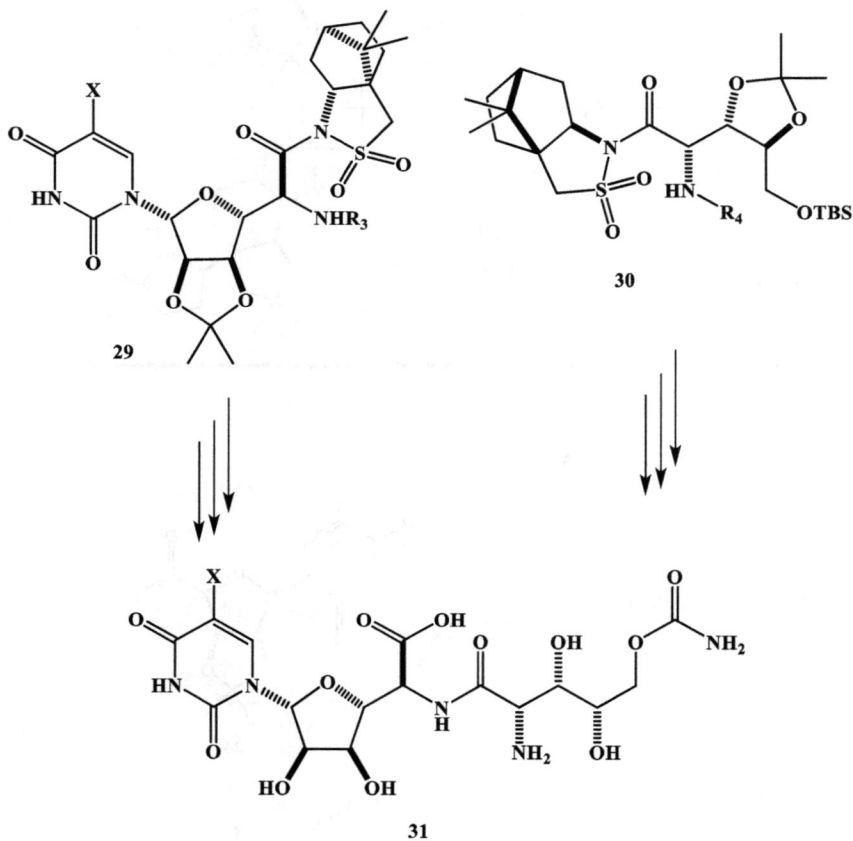

Figure 4.15: Unified total preparation of the compound 31.

Asimicin is a biologically active compound, it can extracted from the seeds and bark of the tree pawpaw [51, 52]. Asimicin displays cytotoxicity towards several tumour cell lines and powerful antileishmanial and pesticidal activities, also it is an assuring initiating material for the exploitation of novel agrochemicals and pharmaceuticals [53]. A convergent total synthesis of asimicin from D-gulose derivative in seventeen steps was reported [54]. Decarbonylative radical–radial homo-coupling of α-alkoxyacyl telluride in an efficient manner created the C_2-symmetric core, which was then converted into asimicin via the affixation of the two side-chains stepwise and functional group manipulations (Figure 4.16).

4.3 Oxidation of glucose to gluconic acid

Carbohydrates such as cellulose, lactose or glucose form the main part of the biomass. The chemical transformations of the carbohydrate lead to the creation of many products with interesting properties. The oxidation of glucose is an example and one of the main compounds obtained from the glucose oxidation is gluconic acid. Gluconic acid has a broad spectrum of usages, for example in medical industries, cosmetic and food industries [55–58].

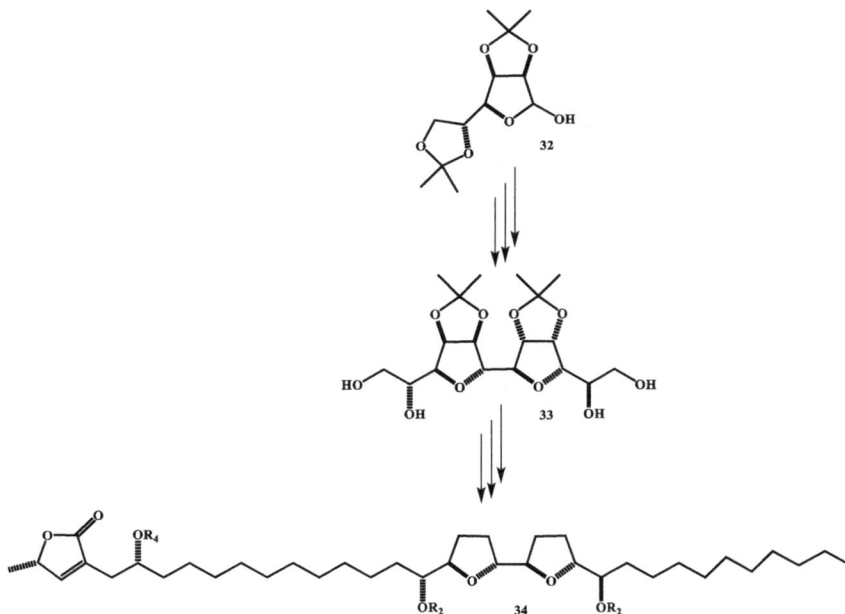

Figure 4.16: Total preparation of asimicin.

Catalytic glucose oxidation on monometallic systems, Pt/support and Pd/support produces gluconic acid. However, selectivity was poor in this process. Besides, as the oxygen dissolves on the surface layers of metals, these systems easily undergo deactivation [59–61]. The application of bimetallic and polymetallic catalysts can improve the glucose oxidation process.

It is reported that the insertion of metals like Bi, Pb, Sn, or Tl to Pt and Pd supported catalysts can alter their selectivity and activity during the oxidation of carbohydrates in the liquid phase [62–70]. Another important metal of choice for the improvement of the glucose oxidation reaction is the tellurium. Studies have shown that in the reaction of oxidation tellurium may modify the properties of metallic catalysts, especially the selectivity [71–82].

Frajtak et al. described in the glucose oxidation to gluconic acid the effect of tellurium addition on catalytic behaviour of Pd catalysts, supported by silica [83]. The Pd catalysts modified with Te showed improved selectivity, stability and activity than monometallic supported Pd catalysts. ToF-SIMS (Time of flight secondary ion mass spectrometry) and XRD (X-ray powder diffraction) techniques were used to characterize the modified catalysts. The characterization analysis evidenced the existence of the intermetallic compound PdTe. The intermetallic compound might be the reason for the increased selectivity and activity of Pd-Te/SiO$_2$ catalysts.

The oxidation of glucose by using the electrocatalyst, nanoparticles of Nd(OH)$_3$-NiTe$_2$, was also reported [84]. Chronoamperometry, cyclic voltammetry, and electrochemical impedance spectroscopy were used to analyze the functioning of the electrocatalyst for oxidation of glucose in alkaline electrolytes. The morphology of the nanoparticle was observed by SEM.

Cyclic voltammetry (CV) was used to analyze the electrocatalytic activity of the prepared materials for the oxidation of glucose (Figure 4.17). Figure 4.17A shows the CV of Nd(OH)$_3$/GCE, NiTe$_2$/GCE, and Nd(OH)$_3$-NiTe$_2$/GCE in NaOH with glucose (0.1 M) and without glucose. The analysis displayed that Nd(OH)$_3$ is an effective substance and it encourages the redox reaction of NiTe$_2$ [85]. Besides, the current density of Nd(OH)$_3$-NiTe$_2$/GCE was the highest, which points that the synergistic effect of Nd(OH)$_3$ and NiTe$_2$ plays a critical role in the non-noble metal catalysts for glucose electrocatalysis. Thus, for the electrocatalytic oxidation of glucose Nd(OH)$_3$-NiTe$_2$ is a promising material [86].

The changes in the CV curves of the Nd(OH)$_3$-NiTe$_2$/GCE in NaOH electrolyte with diverse concentrations of glucose is shown in Figure 4.17B. As the glucose concentration increases (0 to 0.1 M), the peak current densities increase gradually. The strong independence of current density on glucose concentration pointed to a good electrocatalytic activity of the Nd(OH)$_3$-NiTe$_2$ for glucose [87].

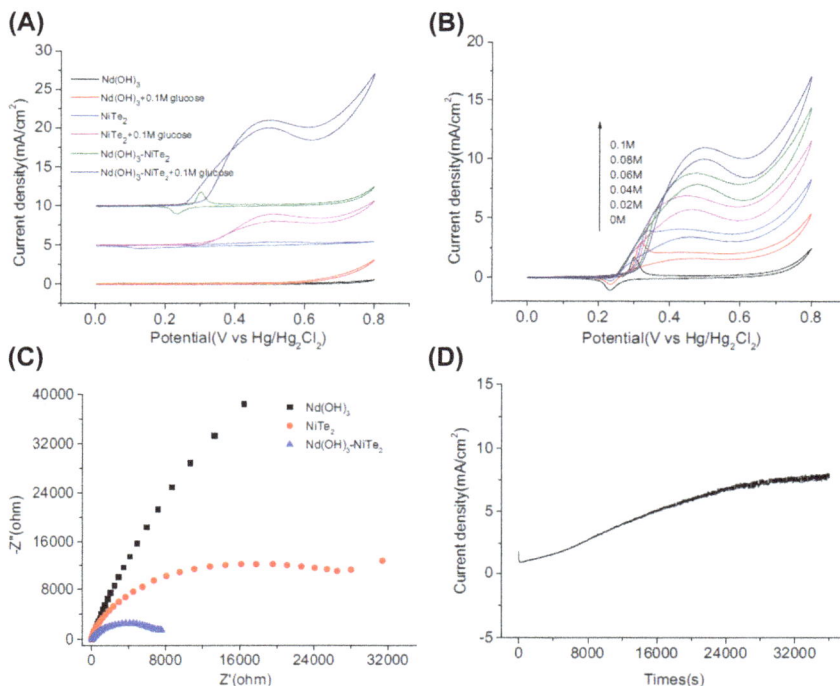

Figure 4.17: (A) The CV curves of Nd(OH)$_3$/GCE, NiTe$_2$/GCE, and Nd(OH)$_3$-NiTe$_2$/GCE in 1 M NaOH without and with 0.1 M glucose at the scan rate of 0.1 V/s. (B) CVs of Nd(OH)$_3$-NiTe$_2$/GCE at the scan rate of 0.1 V/s in 1 M NaOH with 0, 0.02, 0.04, 0.06, 0.08, 0.1 M glucose. (C) Impedance spectra of Nd(OH)$_3$/GCE, NiTe$_2$/GCE, and Nd(OH)$_3$-NiTe$_2$/GCE in 1 M NaOH with 0.1 M glucose. (D) The chronoamperometric curve for Nd(OH)$_3$-NiTe$_2$/GCE in 1 M NaOH with 0.1 M glucose at an oxidation potential 0.5 V. Adapted with permission from [84].

4.4 Conclusions

Te is a promising and important functional material due to its unique physical and chemical properties. For instance, Te is brittle and crystalline. Tellurium dissolves in nitric acid, at the same time it remains stable in hydrochloric acid or water. Tellurium has gained worldwide attention because of its distinct chained structures, excellent properties, and potential applications. This article has discussed the importance of Te during carbohydrate synthesis. Te-based synthesis methods have shown promising results. A great improvement in the carbohydrate synthesis field [88] can be contribute by the unique features and advantages of these synthesis methods.

Acknowledgements: AD and BKB are grateful to Prince Mohammad Bin Fahd University for support.

References

1. Horlacher T, Seeberger PH. The utility of carbohydrate microarrays in glycomics. OMICS A J Integr Biol 2006;10:490–8.
2. Roseman S. Reflections on glycobiology * 1. J Biol Chem 2001;276:41527–42.
3. Gama CI, Hsieh-Wilson LC. Chemical approaches to deciphering the glycosaminoglycan code. Curr Opin Chem Biol 2005;9:609–19.
4. Becker DJ, Lowe JB. Fucose: biosynthesis and biological function in mammals. Glycobiology 2003; 13:41R–53.
5. Murrey HE, Hsieh-Wilson LC. The chemical neurobiology of carbohydrates. Chem Rev 2008;108: 1708–31.
6. McReynolds KD, Gervay-Hague J. Chemotherapeutic interventions targeting HIV interactions with host-associated carbohydrates. Chem Rev 2007;107:1533–52.
7. Durham TB, Miller MJ. Conversion of glucuronic acid glycosides to novel bicyclic β-lactams. Org Lett 2002;4:135–8.
8. Reddy BG, Vankar YD. A convenient synthesis of methyl N-acetyl-α-d-lividosaminide from d-glucal. Tetrahedron Lett 2003;44:4765–7.
9. Krohn K, Gehle D, Kamp O, van Ree T. Highly deoxygenated sugars. II. Synthesis of chiral cyclopentenes via novel carbocyclization of C-4 branched deoxysugars. J Carbohydr Chem 2003; 22:377–83.
10. Chambers DJ, Evans GR, Fairbanks AJ. Synthesis of C-glycosyl amino acids: scope and limitations of the tandem Tebbe/Claisen approach. Tetrahedron Asymmetry 2005;16:45–55.
11. Domon D, Fujiwara K, Ohtaniuchi Y, Takezawa A, Takeda S, Kawasaki H, et al. Synthesis of the C42–C52 part of ciguatoxin CTX3C. Tetrahedron Lett 2005;46:8279–83.
12. Ferrier RJ, Prasad N. Unsaturated carbohydrates. Part IX. Synthesis of 2,3-dideoxy-α-D-erythro-hex-2-enopyranosides from tri-O-acetyl-D-glucal. J Chem Soc C Org 1969;570–5. https://doi.org/10.1039/j39690000570.
13. de Freitas Filho JR, Srivastava RM, da Silva WJP, Cottier L, Sinou D. Synthesis of new branched-chain amino sugars. Carbohydr Res 2003;338:673–80.
14. Nagaraj P, Ramesh NG. Direct Ferrier rearrangement on unactivated glycals catalyzed by indium(III) chloride. Tetrahedron Lett 2009;50:3970–3.
15. Grynkiewicz G, Priebe W, Zamojski A. Synthesis of alkyl 4,6-di-o-acetyl-2,3-dideoxy-α-d-threo-hex-2- enopyranosides from 3,4,6-tri-o-acetyl-1,5-anhydro-2-deoxy- d-lyxo-hex-1-enitol (3,4,6-tri-o-acetyl-d-galactal). Carbohydr Res 1979;68:33–41.
16. Masson C, Soto J, Bessodes M. Ferric chloride: a new and very efficient catalyst for the ferrier glycosylation reaction. Synlett 2000;2000:1281–2.
17. Swamy NR, Venkateswarlu Y. An efficient method for the synthesis of 2,3-unsaturated glycopyranosides catalyzed by bismuth trichloride in ferrier rearrangement. Synthesis 2002; 2002:0598–600.
18. Yadav JS, Reddy BVS, Murthy CVSR, Kumar GM. Scandium triflate catalyzed ferrier rearrangement: an efficient synthesis of 2,3-unsaturated glycopyranosides. Synlett 2000;2000:1450–1.
19. Babu BS, Balasubramanian KK. Lithium tetrafluoborate catalyzed ferrier rearrangement — facile synthesis of alkyl 2,3-unsaturated glycopyranosides. Synth Commun 1999:29:4299–305.
20. Bettadaiah BK, Srinivas P. ZnCl2-catalyzed Ferrier reaction; synthesis of 2,3-unsaturated 1-O-glucopyranosides of allylic, benzylic and tertiary alcohols. Tetrahedron Lett 2003;44:7257–9.
21. Balamurugan R, Koppolu SR. Scope of AuCl3 in the activation of per-O-acetylglycals. Tetrahedron 2009;65:8139–42.

22. Yadav JS, Reddy BVS, Reddy JSS. Dy(OTf)3-immobilized in ionic liquids: a novel and recyclable reaction media for the synthesis of 2,3-unsaturated glycopyranosides. J Chem Soc Perkin 2002;1: 2390–4.

23. Smitha G, Reddy CS. ZrCl4-Catalyzed efficient ferrier glycosylation: a facile synthesis of pseudoglycals Synthesis 2004:2004:834–6.

24. Yadav JS, Reddy BVS, Reddy KB, Satyanarayana M. CeCl3·7H2O: a novel reagent for the synthesis of 2-deoxysugars from d-glycals. Tetrahedron Lett 2002;43:7009–12.

25. Toshima K, Ishizuka T, Matsuo G, Nakata M, Kinoshita M. Glycosidation of glycals by 2,3-dichloro-5,6-dicyano-p-benzoquinone (DDQ) as a catalytic promoter. J Chem Soc Chem Commun 1993: 704–6. https://doi.org/10.1039/c39930000704.

26. Koreeda M, Houston TA, Shull BK, Klemke E, Tuinman RJ. Iodine-catalyzed ferrier reaction 1. A mild and highly versatile glycosylation of hydroxyl and phenolic Groups1. Synlett 1995;1995:90–2.

27. Cristobal Lopez J, Gomez AM, Valverde S, Fraser-Reid B. Ferrier rearrangement under nonacidic conditions based on iodonium-induced rearrangements of allylic n-pentenyl esters, n-pentenyl glycosides, and phenyl thioglycosides. J Org Chem 1995;60:3851–8.

28. Yadav JS, Reddy BVS, Pandey SK. Ceric(IV) ammonium nitrate-catalyzed glycosidation of glycals: a facile synthesis of 2,3-unsaturated glycosides. New J Chem 2001;25:538–40.

29. López JC, Fraser-Reid B. n-Pentenyl esters facilitate an oxidative alternative to the Ferrier rearrangement. an expeditious route to sucrose. J Chem Soc Chem Commun 1992:94–6.

30. Agarwal A, Rani S, Vankar YD. Protic acid (HClO4 supported on silica gel)-mediated synthesis of 2,3-unsaturated-O-glucosides and a chiral furan diol from 2,3-glycals. J Org Chem 2004;69: 6137–40.

31. Zeni G, Lüdtke DS, Panatieri RB, Braga AL. Vinylic tellurides: from preparation to their applicability in organic synthesis. Chem Rev 2006;106:1032–76.

32. Närhi SM, Oilunkaniemi R, Laitinen RS, Ahlgrén M. The reactions of tellurium tetrahalides with triphenylphosphine under ambient conditions. Inorg Chem 2004;43:3742–50.

33. Freitas JCR, de Freitas JR, Menezes PH. Stereoselective synthesis of 2,3-unsaturated-O-Glycosides promoted by TeBr4. J Braz Chem Soc 2010;21:2169–72.

34. Freitas JCR, Couto TR, Filho JRF, Menezes PH. Synthesis of O-glycosides dimers catalysed by tellurium (IV) tetrachloride. In: Blucher Chem Proc [Internet] 2013. p. 218. Available from: https://www.proceedings.blucher.com.br/article-details/synthesis-of-o-glycosides-dimers-catalysed-by-tellurium-iv-tetrachloride-8072.

35. Freitas JCR, Couto TR, Paulino AAS, de Freitas Filho JR, Malvestiti I, Oliveira RA, et al. Stereoselective synthesis of pseudoglycosides catalyzed by TeCl4 under mild conditions. Tetrahedron 2012;68:8645–54.

36. Kobayashi S, Hori M, Wang GX, Hirama M. Formal total synthesis of neocarzinostatin chromophore. J Org Chem 2006;71:636–44.

37. Hirama M, Gomibuchi T, Fujiwara K, Sugiura Y, Uesugi M. Synthesis and DNA cleaving abilities of functional neocarzinostatin chromophore analogs. Base discrimination by a simple alcohol. J Am Chem Soc 1991;113:9851–3.

38. Tikad A, Hamze A, Provot O, Brion J-D, Alami M. Suzuki Coupling Reactions of (E)- and (Z)-Chloroenynes with Boronic Acids: Versatile Access to Functionalized 1,3-Enynes. Weinheim, Germany: Wiley-VCH Verlag; 2010, 2010:725–31 pp.

39. Silveira CC, Braga AL, Vieira AS, Zeni G. Stereoselective synthesis of enynes by nickel-catalyzed cross-coupling of divinylic chalcogenides with alkynes. J Org Chem 2003;68:662–5.

40. Jahier C, Zatolochnaya OV, Zvyagintsev NV, Ananikov VP, Gevorgyan V. General and selective head-to-head dimerization of terminal alkynes proceeding via hydropalladation pathway. Org Lett 2012; 14:2846–9.

41. Alami M, Ferri F, Linstrumelle G. An Efficient palladium-catalysed reaction of vinyl and aryl halides or triflates with terminal alkynes. Tetrahedron Lett 1993;34:6403–6.
42. Hatakeyama T, Yoshimoto Y, Gabriel T, Nakamura M. Iron-catalyzed enyne cross-coupling reaction. Org Lett 2008;10:5341–4.
43. Wu M, Mao J, Guo J, Ji S. The use of a bifunctional copper catalyst in the cross-coupling reactions of aryl and heteroaryl halides with terminal alkynes. Eur J Org Chem 2008;2008:4050–4.
44. Bates CG, Saejueng P, Venkataraman D. Copper-catalyzed synthesis of 1,3-enynes. Org Lett 2004; 6:1441–4.
45. Saejueng P, Bates CG, Venkataraman D. Copper(I)-Catalyzed coupling of terminal acetylenes with aryl or vinyl halides. Synthesis 2005;2005:1706–12.
46. Wang L, Li P, Zhang Y. The Sonogashira coupling reaction catalyzed by ultrafine nickel(0) powder. Chem Commun 2004;514–5. https://doi.org/10.1039/b314246a.
47. Beletskaya IP, Latyshev GV, Tsvetkov AV, Lukashev NV. The nickel-catalyzed Sonogashira–Hagihara reaction. Tetrahedron Lett 2003;44:5011–3.
48. Dantas CR, de Freitas JJR, Barbosa QPS, Militão GCG, Silva TDS, da Silva TG, et al. Stereoselective synthesis and antitumoral activity of Z -enyne pseudoglycosides. Org Biomol Chem 2016;14: 6786–94.
49. Nagatomo M, Kamimura D, Matsui Y, Masuda K, Inoue M. Et 3 B-mediated two- and three-component coupling reactions via radical decarbonylation of α-alkoxyacyl tellurides: single-step construction of densely oxygenated carboskeletons. Chem Sci 2015;6:2765–9.
50. Fujino H, Nagatomo M, Paudel A, Panthee S, Hamamoto H, Sekimizu K, et al. Unified total synthesis of polyoxins J, L, and fluorinated analogues on the basis of decarbonylative radical coupling reactions. Angew Chem Int Ed 2017;56:11865–9.
51. Zhao G-X, Chao J-F, Zeng L, Rieser MJ, McLaughlin JL. The absolute configuration of adjacent bis-THF acetogenins and asiminocin, a novel highly potent asimicin isomer from Asimina triloba. Bioorg Med Chem 1996;4:25–32.
52. McLaughlin J, Rupprecht K, Chang C, Cassady J, Mikolkajczak K, Weisleder D. Asimicin, a new cytotoxic and pesticidal acetogenin from the pawpaw, asimina triloba (annonaceae). Heterocycles 1986;24:1197.
53. Newman DJ, Cragg GM. Natural products as sources of new drugs from 1981 to 2014. J Nat Prod 2016;79:629–61.
54. Kawamata T, Yamaguchi A, Nagatomo M, Inoue M. Convergent total synthesis of asimicin via decarbonylative radical dimerization. Chem Eur J 2018;24:18907–12.
55. Ramachandran S, Fontanille P, Pandey A, Larroche C. Gluconic acid: properties, applications and microbial production. Food Technol Biotechnol 2006;44:185–95.
56. Comotti M, Pina CD, Rossi M. Mono- and bimetallic catalysts for glucose oxidation. J Mol Catal Chem 2006;251:89–92.
57. Baatz C, Prüße U. Preparation of gold catalysts for glucose oxidation. Catal Today 2007;122:325–9.
58. Singh OV, Kumar R. Biotechnological production of gluconic acid: future implications. Appl Microbiol Biotechnol 2007;75:713–22.
59. Légaré P, Hilaire L, Maire G. Interaction of polycrystalline platinum and a platinum-silicon alloy with oxygen: an XPS study. Surf Sci 1984;141:604–16.
60. Jacobs JWM, Schryvers D. A high-resolution electron microscopy study of photodeposited Pd particles on TiO2 and their oxidation in air. J Catal 1987;103:436–49.
61. Nikov I, Paev K. Palladium on alumina catalyst for glucose oxidation: reaction kinetics and catalyst deactivation. Catal Today 1995;24:41–7.
62. Wenkin M, Touillaux R, Ruiz P, Delmon B, Devillers M. Influence of metallic precursors on the properties of carbon-supported bismuth-promoted palladium catalysts for the selective oxidation of glucose to gluconic acid. Appl Catal Gen 1996;148:181–99.

63. Abbadi A, van Bekkum H. Effect of pH in the Pt-catalyzed oxidation of d-glucose to d-gluconic acid. J Mol Catal Chem 1995;97:111–8.
64. Dirkx JMH, van der Baan HS. The oxidation of glucose with platinum on carbon as catalyst. J Catal 1981;67:1–13.
65. Besson M, Lahmer F, Gallezot P, Fuertes P, Fleche G. Catalytic oxidation of glucose on bismuth-promoted palladium catalysts. J Catal 1995;152:116–21.
66. Gallezot P. Selective oxidation with air on metal catalysts. Catal Today 1997;37:405–18.
67. Wenkin M, Ruiz P, Delmon B, Devillers M. The role of bismuth as promoter in Pd–Bi catalysts for the selective oxidation of glucose to gluconate. J Mol Catal Chem 2002;180:141–59.
68. Karski S. Activity and selectivity of Pd–Bi/SiO2 catalysts in the light of mutual interaction between Pd and Bi. J Mol Catal Chem 2006;253:147–54.
69. Karski S, Witońska I, Gołuchowska J. Catalytic properties of Pd–Tl/SiO2 systems in the reaction of liquid phase oxidation of aldoses. J Mol Catal Chem 2006;245:225–30.
70. Karski S, Witońska I. Bismuth as an additive modifying the selectivity of palladium catalysts. J Mol Catal Chem 2003;191:87–92.
71. Ischenko EV, Andrushkevich TV, GYa P, Bondareva VM, Chesalov YA, TYu K, et al. Formation of active component of MoVTeNb oxide catalyst for selective oxidation and ammoxidation of propane and ethane. In: Gaigneaux EM, Devillers M, Hermans S, Jacobs PA, Martens JA, Ruiz P, editors. In: Stud Surf Sci Catal [Internet]. Elsevier; 2010. p. 479–82. Available from: https://www.sciencedirect.com/science/article/pii/S0167299110750898.
72. Naraschewski FN, Praveen Kumar C, Jentys A, Lercher JA. Phase formation and selective oxidation of propane over MoVTeNbOx catalysts with varying compositions. Appl Catal Gen 2011;391:63–9.
73. Huynh Q, Schuurman Y, Delichere P, Loridant S, Millet JMM. Study of Te and V as counter-cations in Keggin type phosphomolybdic polyoxometalate catalysts for isobutane oxidation. J Catal 2009; 261:166–76.
74. López Nieto JM, Botella P, Solsona B, Oliver JM. The selective oxidation of propane on Mo-V-Te-Nb-O catalysts: the influence of Te-precursor. Catal Today 2003;81:87–94.
75. Botella P, López Nieto JM, Solsona B. Selective oxidation of propene to acrolein on Mo-Te mixed oxides catalysts prepared from ammonium telluromolybdates. J Mol Catal Chem 2002;184: 335–47.
76. Millet JMM, Roussel H, Pigamo A, Dubois JL, Jumas JC. Characterization of tellurium in MoVTeNbO catalysts for propane oxidation or ammoxidation. Appl Catal Gen 2002;232:77–92.
77. Allen MD, Hutchings GJ, Bowker M. Iron antimony oxide catalysts for the ammoxidation of propene to acrylonitrile: comments on the method of preparation of tellurium promoted catalysts. Appl Catal Gen 2001;217:33–9.
78. Florea M, Mamede A-S, Eloy P, Parvulescu VI, Gaigneaux EM. High surface area Mo–V–Te–Nb–O catalysts: preparation, characterization and catalytic behaviour in ammoxidation of propane. Catal Today 2006;112:139–42.
79. Zhonghua D, Hongxin W, Wenling C, Weishen Y. Influence of the reducing atmosphere on the structure and activity of Mo-V-Te-Nb-O catalysts for propane selective oxidation. Chin J Catal 2008; 29:1032–6.
80. Holmberg J, Häggblad R, Andersson A. A study of propane ammoxidation on Mo–V–Nb–Te-oxide catalysts diluted with Al2O3, SiO2, and TiO2. J Catal 2006;243:350–9.
81. Kum SS, Jo BY, Moon SH. Performance of Pd-promoted Mo–V–Te–Nb–O catalysts in the partial oxidation of propane to acrylic acid. Appl Catal Gen 2009;365:79–87.
82. López Nieto JM, Botella P, Concepción P, Dejoz A, Vázquez MI. Oxidative dehydrogenation of ethane on Te-containing MoVNbO catalysts. Catal Today 2004;91–92:241–5.

83. Frajtak M, Witońska I, Krolak A, Krawczyk N, Karski S. The influence of tellurium addition to supported palladium catalysts on their catalytic properties in glucose oxidation. Rev Roum Chem 2011;56:631–5.
84. Yu B, Jia D, Miao Y. Nanoparticles of Nd(OH)3-NiTe2 for electrocatalytic oxidation of glucose. J Mater Sci Mater Electron 2020;31:2360–9.
85. Yu Z, Li H, Zhang X, Liu N, Tan W, Zhang X, et al. Facile synthesis of NiCo2O4@Polyaniline core–shell nanocomposite for sensitive determination of glucose. Biosens Bioelectron 2016;75:161–5.
86. Wang S, Zhang S, Liu M, Song H, Gao J, Qian Y. MoS2 as connector inspired high electrocatalytic performance of NiCo2O4 nanoplates towards glucose. Sensor Actuator B Chem 2018;254:1101–9.
87. Li R, Liu X, Wang H, Wu Y, Chan KC, Lu Z. Sandwich nanoporous framework decorated with vertical CuO nanowire arrays for electrochemical glucose sensing. Electrochim Acta 2019;299:470–8.
88. Nicotra F, Airoldi C, Cardona F. Synthesis of C-, S-Glycosides. In: Comprehensive Glycoscience. Amsterdam, Netherlands: Elsevier; 2007, 1:647–83 pp.

Aparna Das* and Bimal Krishna Banik*

5 Tellurium-based solar cells

Abstract: In this article, we discuss about various Tellurium-based solar cells. Mainly this analysis focuses on the CdTe solar cells. The latest development in this area is incorporated in great detail. Te doping in various other solar cells is also discussed in the last part of the article.

Keywords: CdTe; luminescence; nanostructures; semiconductors; solar cells; tellurium.

5.1 Introduction

Tellurium (Te) is a metalloid and is available rarely on Earth. Te is used for manufacturing films which is essential to solar cells. It is reported that when Te is alloyed with other elements, for example with cadmium (Cd), it creats a compound that presents raised electrical conductivity. Thus, it can take in sunlight in an efficient manner and change it into electric current. Working like an additive to steel, copper, and lead alloys, in thermoelectric cooling applications Te improves machine efficiency. In that way, ductility and tensile strength improve, and also it aids to prevent sulfuric acid corrosion.

More than two-thirds of global Te consumption was account for photovoltaic and thermoelectric applications. Te is used as a coloring agent in glass and ceramics and also in copying machines. In the chemical industry, it is also used as a vulcanizing agent to make durable products. Te is also used in the semiconductor industry such as in integrated circuits, laser diodes, and sensors, in medical instrumentation, and automobile industry. In this article, Tellurium-based solar cells are discussed in great detail. Mainly this study focuses on the CdTe based solar cells. Future trends and the latest development in this area are considered. Te doping in various solar cells is also discussed briefly.

5.2 CdTe solar cell

Cadmium telluride (CdTe) is considered as a potent candidate for thin-film based solar cell usages [1–3]. The bandgap energy of CdTe (1.45 eV) is perfect for photovoltaic energy changeover. All the incident radiation with energy above 1.45 eV is took in

*Corresponding authors: Aparna Das and Bimal Krishna Banik**, Department of Mathematics and Natural Sciences, College of Sciences and Human Studies, Prince Mohammad Bin Fahd University, Al Khobar 31952, Kingdom of Saudi Arabia, E-mail: aparnadasam@gmail.com (A. Das), bimalbanik10@gmail.com (B.K. Banik). https://orcid.org/0000-0002-2502-9446 (A. Das)

As per De Gruyter's policy this article has previously been published in the journal Physical Sciences Reviews. Please cite as: A. Das and B. K. Banik "Tellurium-based solar cells" *Physical Sciences Reviews* [Online] 2022. DOI: 10.1515/psr-2021-0110 | https://doi.org/10.1515/9783110735840-005

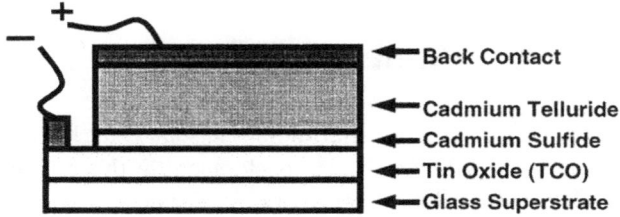

Figure 5.1: The typical CdTe/CdS superstrate configuration. Adapted with permission from Ref. [4].

inside 1–2 μm from the top layer due to its great coefficient of optical absorption. The Cadmium telluride-based cells are mainly heterojunctions, accompanying CdS (cadmium sulfide) as the n-type junction.

Figure 5.1 shows the superstrate configuration of the thin-film CdTe/CdS solar cell. Ferekides et al. reported about the usage of the close-spaced sublimation process for the fabrication of CdTe/CdS solar cells [4]. The close-spaced sublimation technique has appealing features such as efficient material utilization and high deposition rates which are useful for large-area applications.

The optical and structural properties of close-spaced sublimation CdS and CdTe films and junctions were reported. Figure 5.2 shows the SEM images of close-spaced sublimation CdTe films synthesized at two different temperatures. The XRD analysis of a set of close-spaced sublimation CdTe films prepared at various temperatures (500–600 °C) are shown in Figure 5.3.

The possibility of CdTe/CdS/Cd$_{1-x}$Zn$_x$S solar cell system was demonstrated by Oladeji et al. [5]. This solar cell structure was an option to CdTe/CdS system in photovoltaic usage. The presented solar cell system prepared on transparent

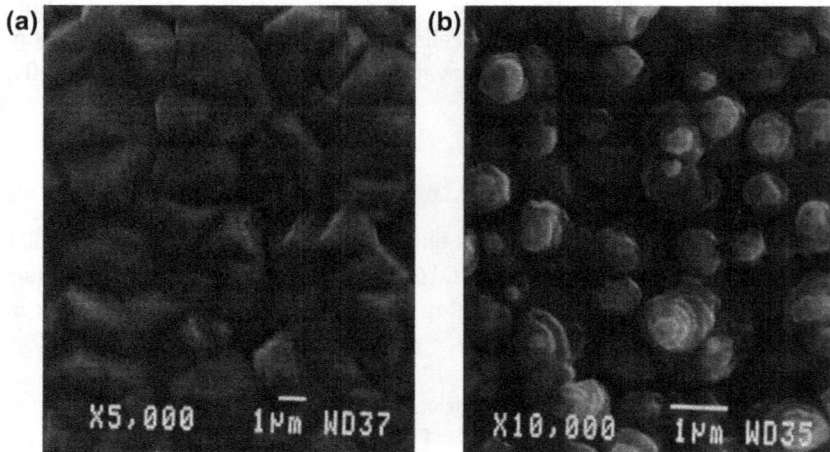

Figure 5.2: SEM images of CSS-CdTe films deposited at (a) 600 and (b) 500 °C. Adapted with permission from Ref. [4].

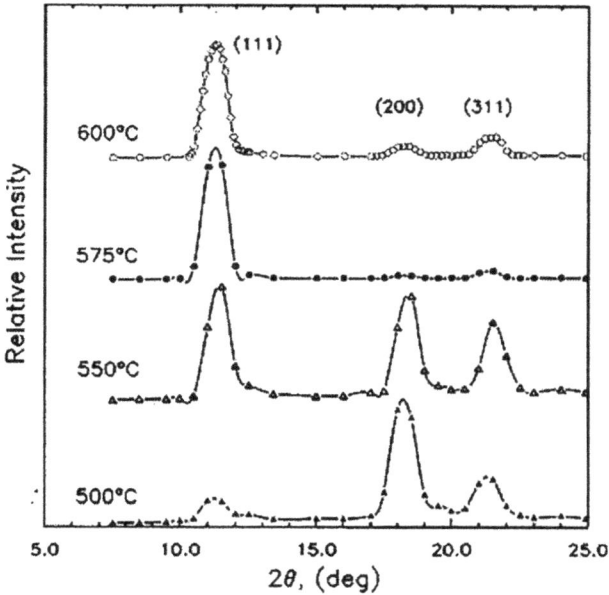

Figure 5.3: XRD spectra of CSS-CdTe films deposited at temperatures in the range 500–600 °C. Adapted with permission from Ref. [4].

conducting soda-lime glass coated with oxide having no antireflection coating showed an efficiency of 10%. The J–V features of thin-film CdTe solar cells are presented in Figure 5.4. The data showed a higher short circuit current (J_{sc}) value for type 1 cell and that of CdS/CdTe is the least. Figure 5.5 shows the CdTe thin-film solar cell's spectral response curves. Figure 5.6 presents the SIMS depth profiles of solar cells. Zinc seemed to have penetrated via to the CdS/CdTe interface and much into the CdTe layer in the cell type 2 compared to the cell type 1 was also observed.

Figure 5.7 compares the diagrammatic representation of the CdTe solar cells in conventional 'substrate' as well as 'superstrate' forms. In the 'superstrate' configuration, the layers are prepared onto the polymer substrates or transparent conducting oxide-coated glass substrate. At the same time, metal-coated glass or metal foils or polymers are employed for 'substrate' configuration which is less efficient.

Solar cells based on polycrystalline thin-film CdTe/CdS were reported. In that configuration, on p-CdTe a transparent conducting layer of ITO (indium tin oxide) was employed as a back electrical contact [6]. Solar cells were prepared on SnO_x:F-coated glass substrates, with an efficiency of around 8%. In the cell, both the back and front contacts were conducting and transparent. In this way, as a 'bi-facial cell, the solar cell can be excited from both or either side at the same time. Another important feature is this solar cell can be employed in tandem solar cells. Under accelerated test conditions, the solar cells having transparent conducting oxide back contact showed long-term

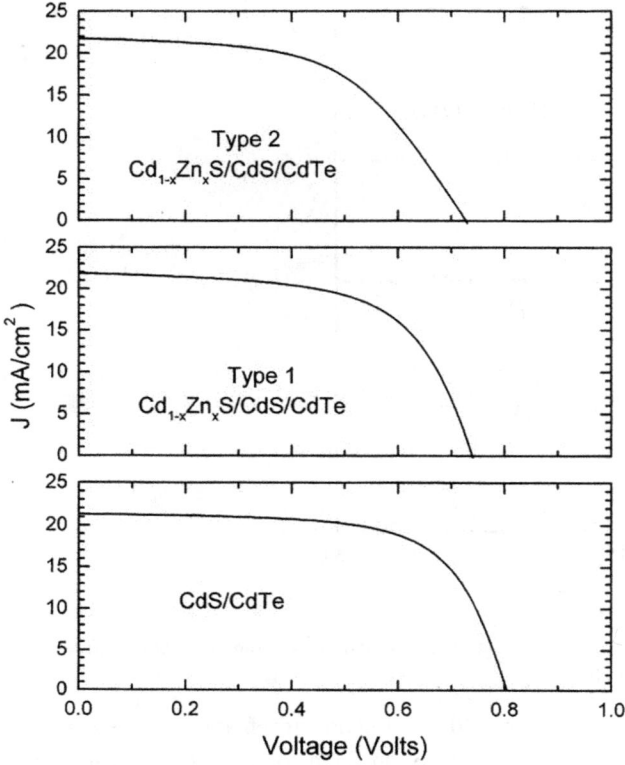

Figure 5.4: J–V characteristics of thin-film CdTe solar cells. Adapted with permission from Ref. [5].

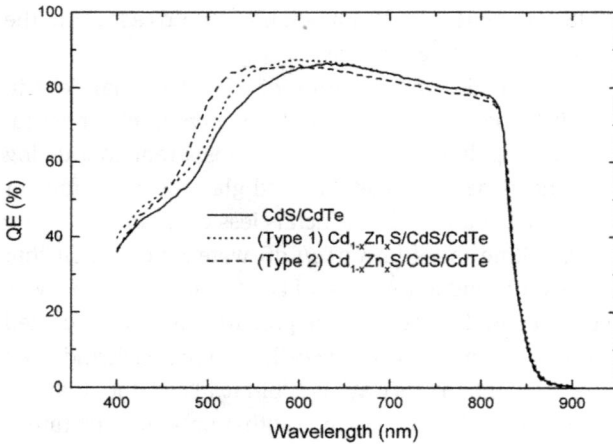

Figure 5.5: Spectral response curves of CdTe thin-film solar cells. Adapted with permission from Ref. [5].

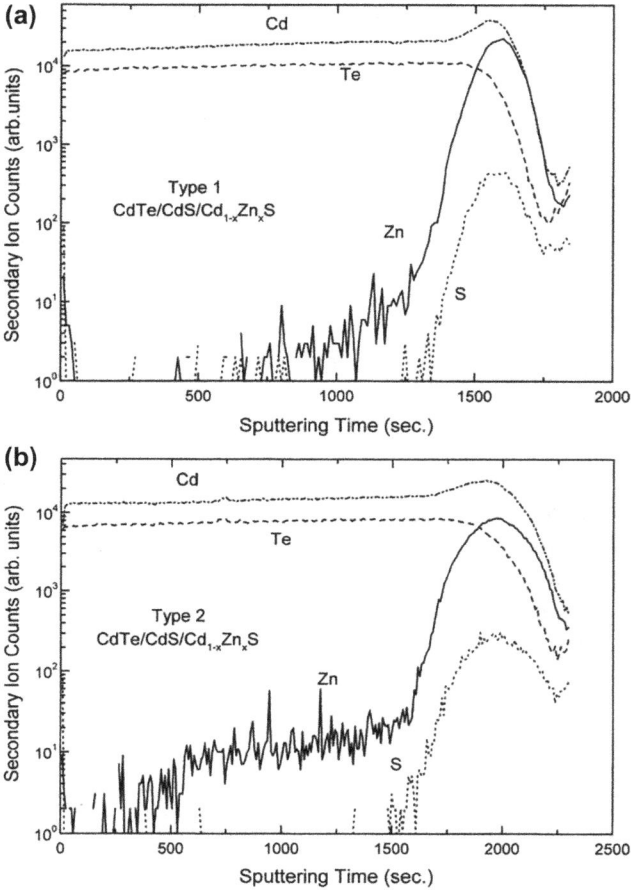

Figure 5.6: SIMS depth profiles of (a) type 1 and (b) type 2 CSS-CdTe/CBD-CdS/Cd$_{1-x}$Zn$_x$S solar cells. Adapted with permission from Ref. [5].

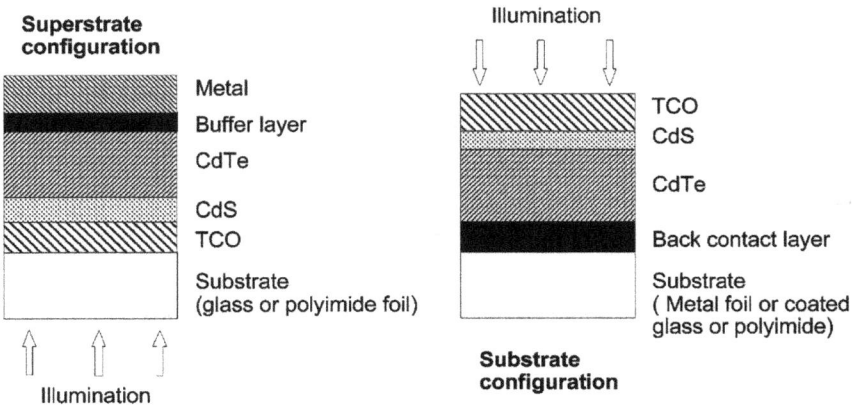

Figure 5.7: Schematics of the CdTe solar cells in conventional 'superstrate' and 'substrate' configurations. Adapted with permission from Ref. [6].

unchanging functioning. The current-voltage plot of the highest efficiency solar cell is shown in Figure 5.8 and the cell was activated from the FTO/glass position.

Lin et al. prepared the freestanding high-quality boron-doped graphene (BG) and pristine graphene (PG) [7]. Boron-doped graphene is more conductive than free-standing pristine graphene because of a larger density of state-produced close the Fermi level. Bettered photovoltaic efficiency and hole-collecting capacity for CdTe solar cells were achieved using boron-doped graphene as a back electrode.

The Nyquist plots present a smaller semicircle for freestanding pristine graphene than for reduced graphene oxide (r-GO), pointing an improved conductivity for free-standing pristine graphene; the conductivity of boron-doped graphene was even higher (Figure 5.9(a)). Solar cells are demonstrated in Figure 5.9(b) were prepared to demonstrate the suitability of boron-doped graphene and freestanding pristine gra-phene as back electrode substance for CdTe based solar cells. Figure 5.9(c) shows the current-voltage plot of the cell.

Wolden et al. reported the purposes of ZnTe buffer layers on functioning of CdTe based solar cell [8]. To analyze the development of the back contact region during RTP (rapid thermal processing) of layer, APT (atom probe tomography) and HR-TEM (high-resolution transmission electron microscopy) were used. After activation, the ZnTe layer transformed into a bilayer system comprising of a disordered part in touch with CdTe. The Copper, co-evaporated within ZnTe in a uniform manner, was observed to be dramatically combine

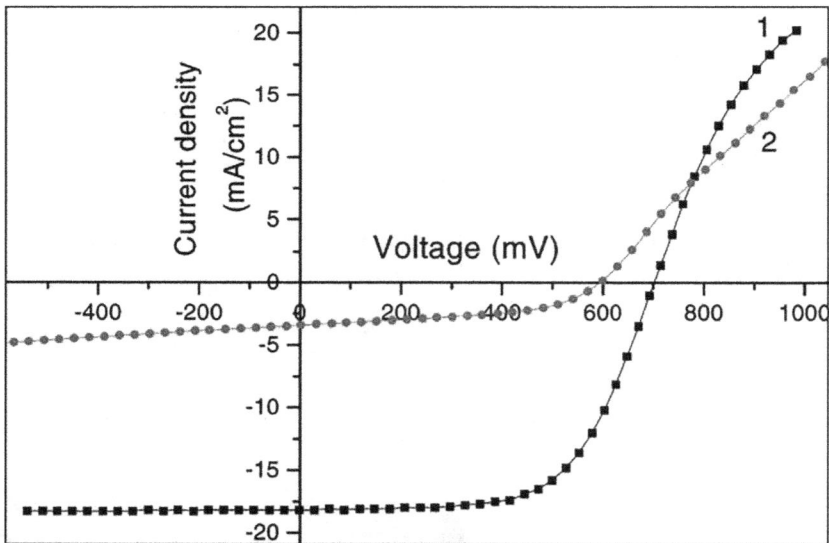

Figure 5.8: I–V curves of the FTO/CdS/CdTe/ITO solar cell illuminated from: (1) the FTO side (Voc = 702 mV, Jsc = 18.2 mA/cm^2, FF = 0.62, η = 7.9%); and (2) the ITO side (Voc = 591 mV, Jsc = 3.4 mA/cm^2, FF = 0.50, η = 1.0%) under simulated AM1.5 illumination. Adapted with permission from Ref. [6].

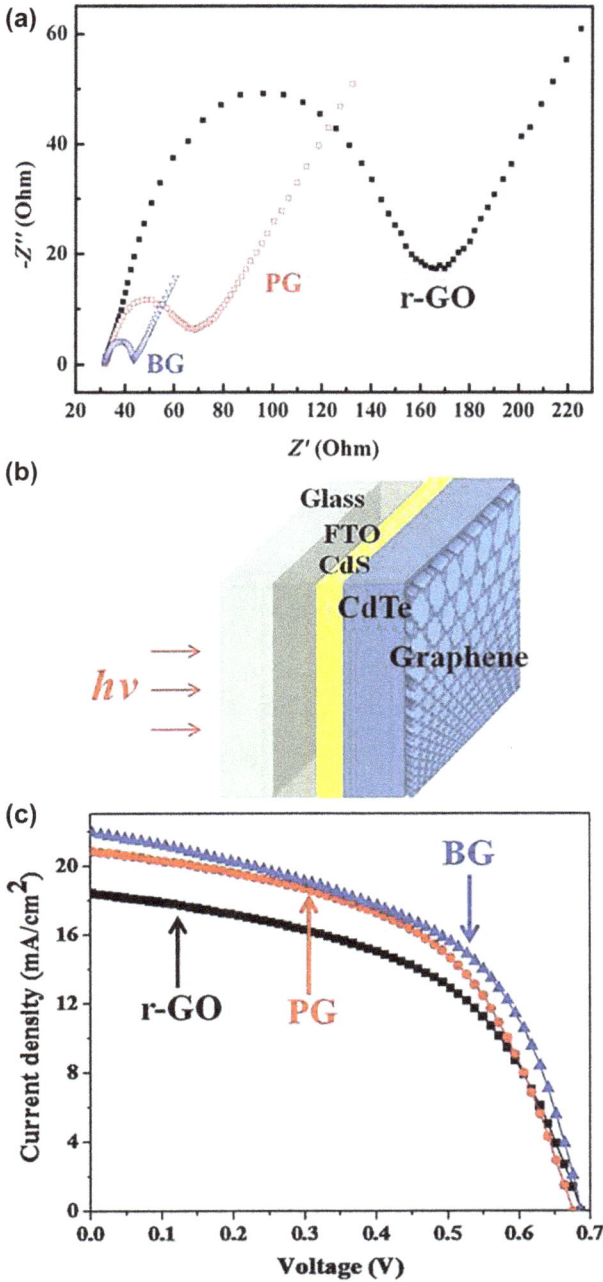

Figure 5.9: Characteristics of solar cell.
(a) The typical electrochemical Nyquist plots of the r-GO-modified, PG-modified, and BG-modified electrodes, (b) the schematic of the CdTe solar cell with a graphene back electrode, and (c) I–V characteristics of CdTe solar cells with different back electrodes.

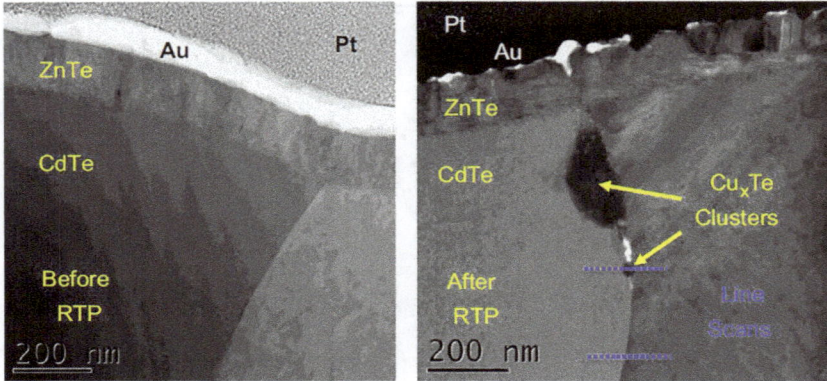

Figure 5.10: HAADF image (left) and a bright-field TEM image (right) of the back contact region before (left) and after (right) RTP treatment. Adapted with permission from Ref. [8].

and segregate after rapid thermal processing (Figure 5.10). The TEM analysis disclosed that Zn gather together at the border of the clusters. Figure 5.11 shows the high-resolution transmission electron microscopy of the CdTe|ZnTe interfacial area.

Instead of CdS, the use of $Zn_xCd_{1-x}S$ in heterojunction solar cells can head to an increment in photocurrent. It was provided by the compatibility in the electron affinities of the absorber material and window [9, 10]. In addition, the open-circuit voltage and the short circuit current can be bettered in heterojunction solar cells with the

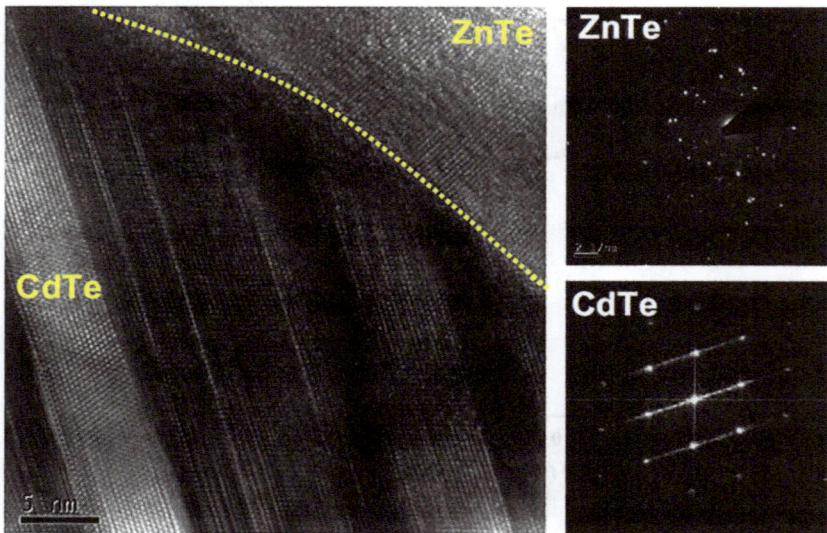

Figure 5.11: High-resolution TEM of the CdTe|ZnTe interfacial region after RTP activation and selected area diffraction images taken from both layers. Adapted with permission from Ref. [8].

internalization of Zn to CdS. This is because of the decrement of losses due to absorption in the window region [11, 12].

The effect of the layer dimension of radio frequency sputtered CdTe thin film in $Zn_xCd_{1-x}S/CdTe$ system at smaller concentrations of Zn was reported [13]. To investigate the surface morphology of CdTe layers (0.58–4.5 μm thick layers), FESEM images were taken (Figure 5.12). The Schematic representation of the solar cell device is

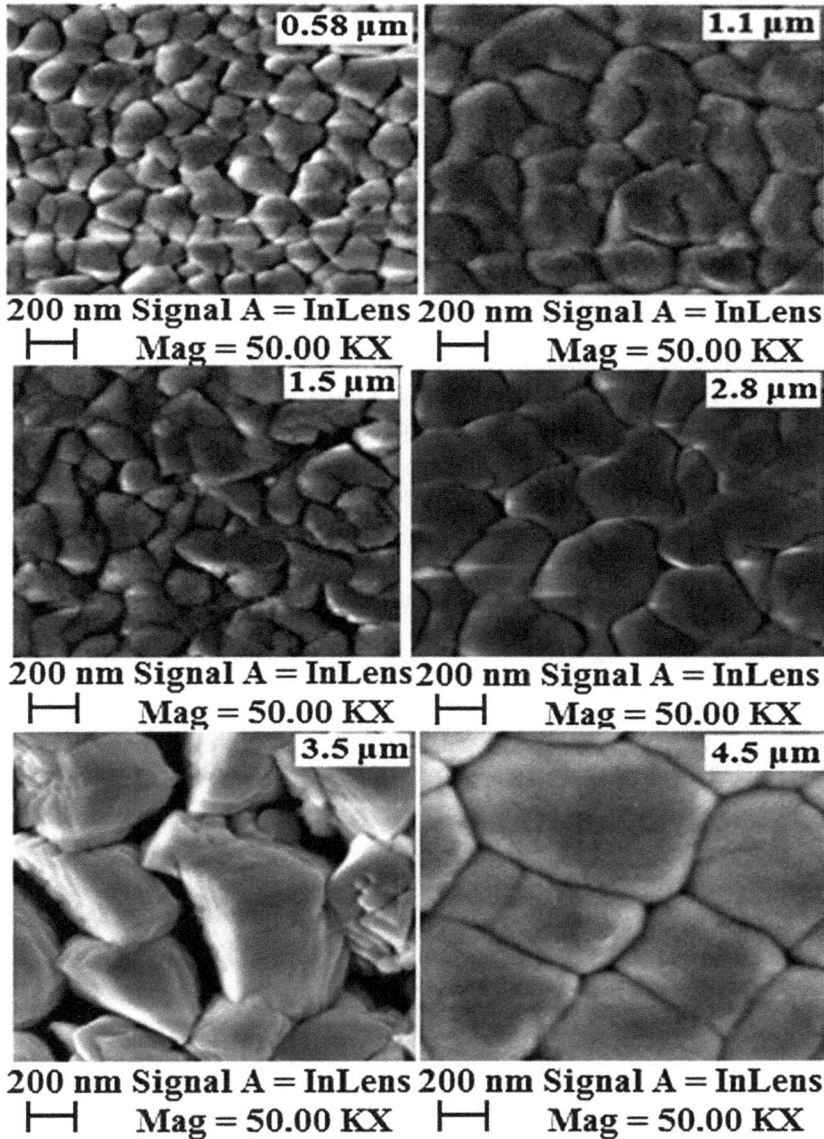

Figure 5.12: FESEM images of CdTe thin film with thickness variation. Adapted with permission from Ref. [13].

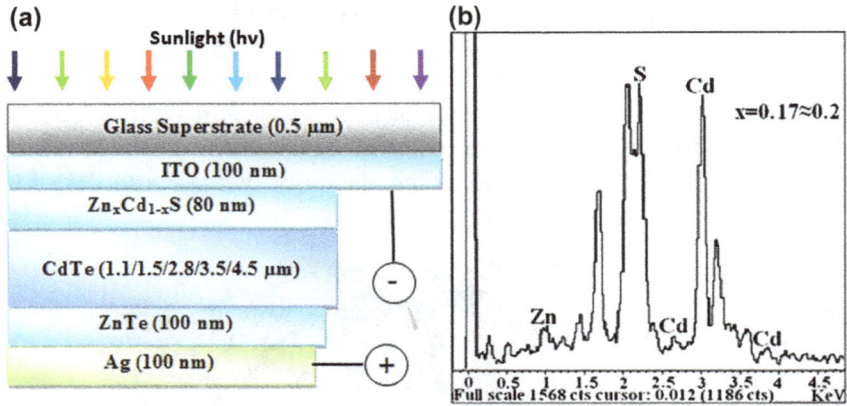

Figure 5.13: $Zn_xCd_{1-x}S/CdTe$ solar cell device.
(a) Schematic diagram of $Zn_xCd_{1-x}S/CdTe$ solar cell device and (b) EDX spectrum $Zn_xCd_{1-x}S$ thin film for $x \approx 0.2$. Adapted with permission from Ref. [13].

demonstrated in Figure 5.13(a). The concentrations of elements were analyzed by EDX phenomena (Figure 5.13(b)). The study reported that an additional reduction of CdTe thickness (0.58 µm) heads to inferior functioning of the solar cell. The J–V plot for the calibrated CdTe thickness (3.5×10^{-6} m) for better efficiency of cell is shown in Figure 5.14.

Many issues such as materials, chemical etching of the film, post-growth treatment, and choice of back contacts hinder the functioning of the CdTe based solar cell.

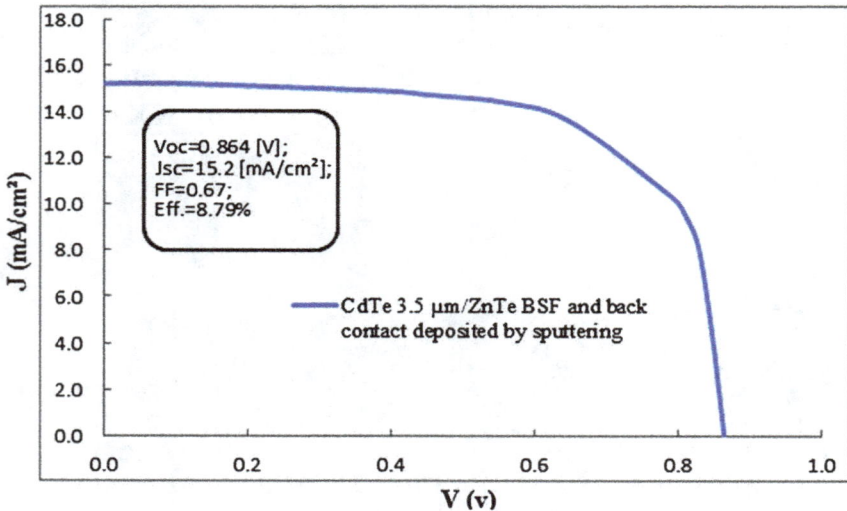

Figure 5.14: J–V characteristic of $Zn_{0.2}Cd_{0.8}S/CdTe$ solar cells. Adapted with permission from Ref. [13].

To improve efficiency more studies are needed in these areas. Apart from these, to get better efficiency, many studies are centering on various kinds 'of preparation techniques, materials for electrode, and anti-reflection covering on the glass substrate. To make an extremely effective CdTe solar cell, the most important and common step is treatment with $CdCl_2$. This procedure aids in the growth of grain, recrystallization, and also it reduces the lattice mismatch among CdTe and CdS [14, 15].

The Fermi energy pinning is one of the major elements that cuts down the open-circuit voltage. It develops because of the potential energy barrier among the upper metal electrode and the absorber layer. CdTe has a higher work function (5 eV) and it is important to choose a substance that equalizes the band alignment to both the top aluminum metal electrode and CdTe layer. Among different materials, tellurium is found to be the best material to cut down the potential energy barrier among the CdTe layer and top Al electrode [16].

Recently, Venkatesh et al. demonstrated a glass/FTO/CdS/CdTe/Te/Al superstrate integrated solar cell, and tellurium was integrated among the CdTe layer and Al electrode [17]. The optical transmittance of the prepared device was analyzed utilizing Ultra Violet–Visible-Near Infrared -spectrometer in the wavelength range of 200–1100 nm. The optical transmittance analysis of the fabricated system is depicted in Figure 5.15. The system showed a transmittance of around 50% in the Infrared range

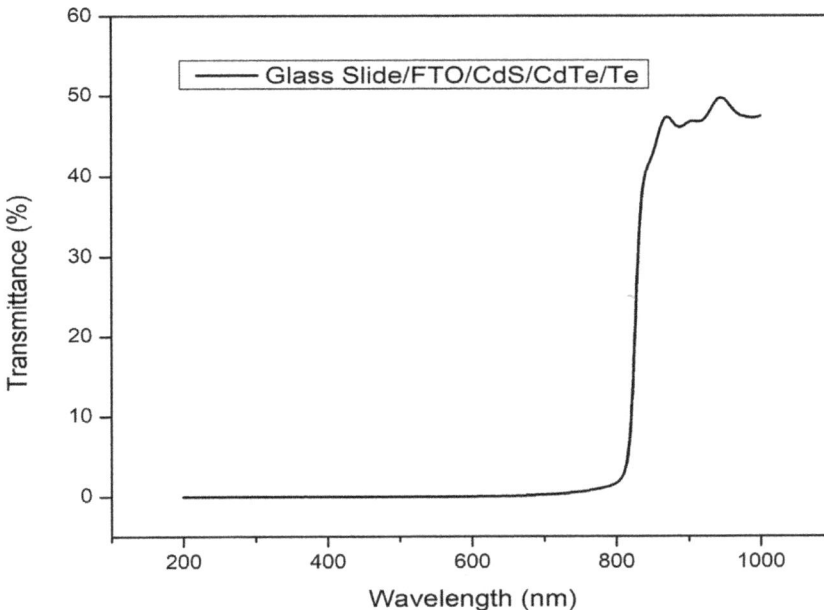

Figure 5.15: Transmittance spectrum in the wavelength range 200–1000 nm of the fabricated device (Glass/FTO/CdS/CdTe/Te). Adapted with permission from Ref. [17].

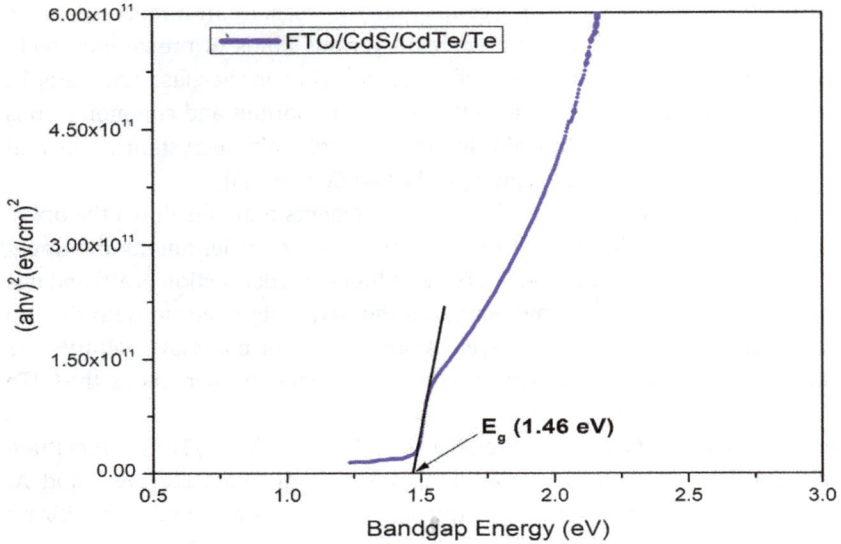

Figure 5.16: Plot of $(ahu)^2$ versus photon energy (hu) for the evaluation of energy band gap of. CdTe thin films. Adapted with permission from Ref. [17].

(nearly 800–1000 nm) and poor transmittance in the region of around 200–800 nm. The eminent absorption characteristics of CdTe was the reason for the poor transmission in that range.

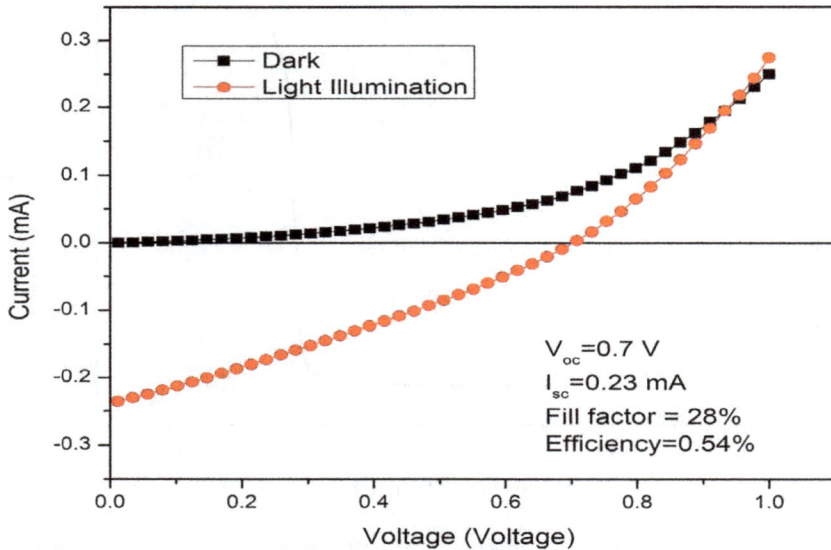

Figure 5.17: I–V response of the FTO/CdS/CdTe/Te/Al solar cell. Adapted with permission from Ref. [17].

The plot of $(\alpha h\upsilon)^2$ versus $h\upsilon$ was diagrammed to find the bandgap energy of the film, (Figure 5.16). The calculated bandgap was found to be around 1.4 eV, the value was the optimized bandgap for CdTe absorber material. The current-voltage response of the solar cell is presented in Figure 5.17.

It was reported that making an ohmic contact on the back interface of CdTe solar cells for hole extraction is hard due to the deep work function of CdTe. Copper/gold is mainly employed as back contact material, however, the existence of a Schottky potential barrier limits the device performance [18, 19]. Usage of a back contact buffer layer with a suitable work function is the one common process to amend the open-circuit voltage. To reduce the height of back-barrier between the standard back contact and CdTe, several studies are reported based on Copper bearing interfacial layers [20–22]. Another important approach was the tellurium layer deposition on the back interface, it also cuts down the height of the back-barrier [23–25]. For better device performance, the back contact interface layer helps transport of hole in the direction of the electrode and by producing an upward band bending, it also suppresses the transport of the electron towards the back contact [26, 27].

Subedi et al. synthesized p-type semi-transparent sulfide of barium copper (α-BaCu$_4$S$_3$, BCS) as a back contact interface layer for solar cells based on CdTe [28]. The study showed that the barium copper sulfide deposition procedure creates an enhancing tellurium rich surface on CdTe, it was occurred in a selective manner taking out cadmium from the surface. The barium copper sulfide interface layer acts dual operations for CdTe photovoltaic systems as a hole transport substance and as an engrave, heightening the performance of the device. With the barium copper sulfide buffer layer, a substantial increment in the open-circuit voltage of CdTe solar cells was also observed. The scanning electron microscope images of CdTe layers without and with the barium copper sulfide layer are shown in Figure 5.18(a) and (b) respectively.

Figure 5.19(a) shows a diagrammatic representation as well as cross-sectional micrograph of the solar cells and Figure 5.19(b) presents the J–V curve of the solar cells. To approximate the concentration of doping for CdTe devices with and without barium copper

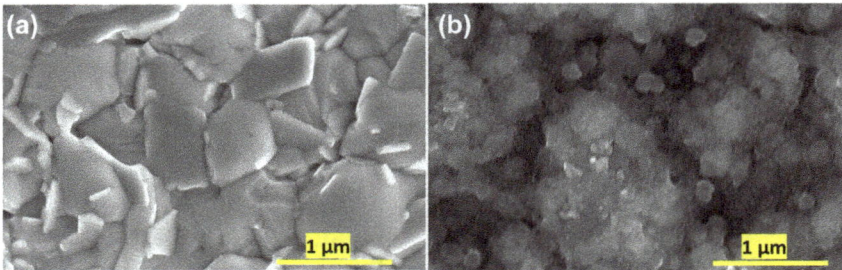

Figure 5.18: SEM of (a) surface morphology of a VTD CdTe film and (b) a CdTe film after barium copper sulfide deposition. Adapted with permission from Ref. [28].

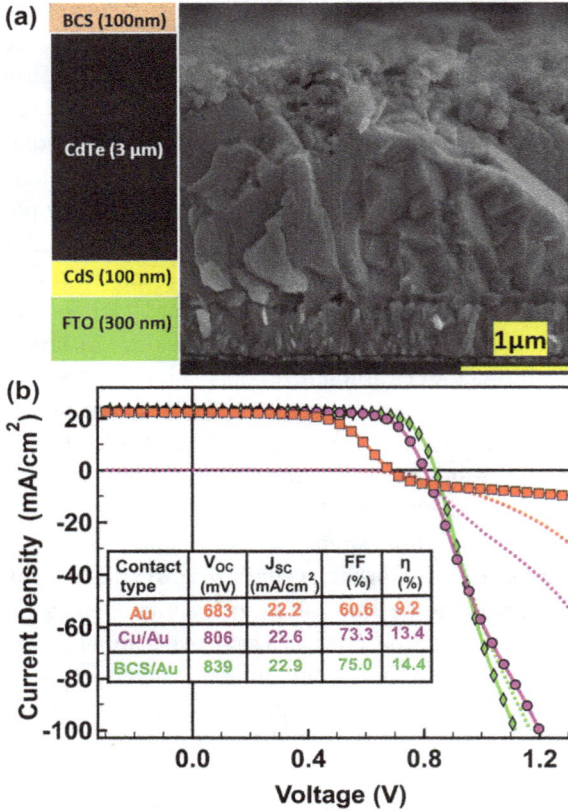

Figure 5.19: Characteristics of solar cells.
(a) Schematic diagram and cross-sectional SEM of a VTD deposited CdTe solar cell with barium copper sulfide as HTL layer, and (b) J–V characteristics of CdS/CdTe solar cell devices with and without barium copper sulfide thin films as interface layers. Adapted with permission from Ref. [28].

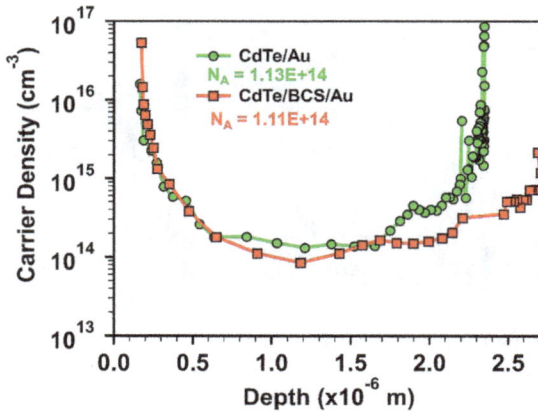

Figure 5.20: C–V measurement used to determine the apparent carrier density of the CdTe film stack with BCS/Au and Au only interface layer. Adapted with permission from Ref. [28].

sulfide interfacial layer, C–V analysis were executed at room temperature (Figure 5.20). The study also discussed semi-transparent CdTe solar cells with barium copper sulfide as the layer for hole transport and ITO (indium tin oxide) as an ending electrode. For the front side illumination, the semi-transparent CdTe solar cells showed conversion efficiency of around 13 and 1.2% efficiency for backside illumination. It indicated that great charge carrier recombination yielded near to the rear CdTe/BCS/ITO contact.

In the last 20 years, several studies were reported the principal mechanism of degradation in the case of CdTe solar cells. The influence of impurities and the material's defects were the main problems for reliability. It was reported in many studies that the migration of Cu from the metal has a profound action in the degradation of device [29–34].

Bertoncello et al. investigated the constancy of CdTe based solar cells which was disclosed to high-temperature storage [35]. Several solar cells with various Cu thicknesses in the contacts and CdTe absorber were considered during this study. The results pointed that the copper metal contact has an important purpose in the cell degradation. The schematic representation of solar cell samples used for the study is shown in Figure 5.21.

During thermal stress, the EQE plots were evaluated on solar cells with various thicknesses of the CdTe layer (Figure 5.22). The cell having the leaner CdTe layer had both the strongest degradation and the lowest EQE efficiency. This might be because of the generation of the copper ions at the back for thinner CdTe layered solar cells. The schematic diagram of potential mechanisms that can induce the cells degradation is shown in Figure 5.23.

Because of their simplicity and low cost, colloidal arrangements have been employed in a great degree as optoelectronic and thermoelectric generator devices [36],

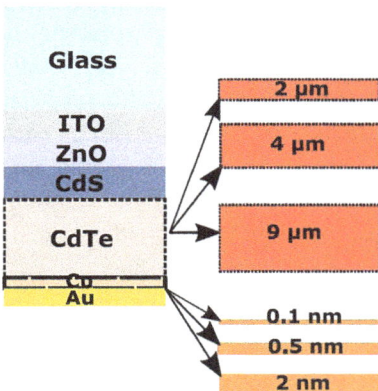

Figure 5.21: Schematic representation of samples that are used for the study. Six different kinds of cells: three with different thicknesses of CdTe and three with different thicknesses of Cu. Adapted with permission from Ref. [35].

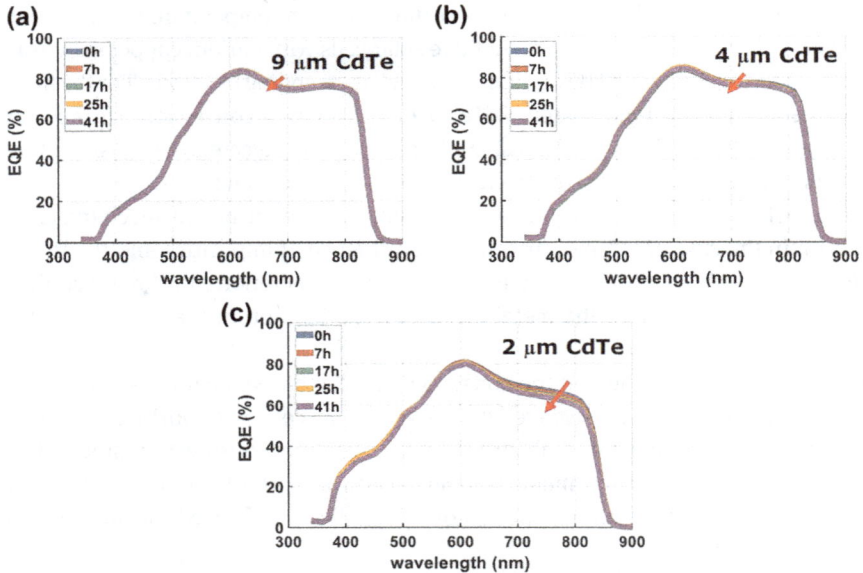

Figure 5.22: The EQE chart of the cell (a) with 9 μm of CdTe, (b) with 4 μm of CdTe, (c) with 2 μm of CdTe. These cells have a 0.5 nm thick Cu layer. Generally, the figures show that by increasing the CdTe thickness, the stability improves. Adapted with permission from Ref. [35].

biological markers in medicine [37, 38], and photovoltaic thin films solar cells [39]. For example, copper zinc tin sulfide cells [40] and bifacial cells [41] have been achieved employing colloidal arrangements. The glass/TCO/CdS/CdTe/back-contact is the traditional configuration of CdTe-based solar cells. By using chemical techniques, the glass/TCO/CdS system, the n-type materials of solar cell are commonly posited. By using spray pyrolysis method, the transparent conductive oxide is usually placed [42]. By using chemical bath deposition method, the cadmium sulfide film is traditionally placed (CBD) [42]. By using sublimation method, the CdTe/back contact (p-type solar cell) heterojunction is usually synthesized. Regarding the price of p-material and back contact deposition substituting it into chemicals methods could cut down the final system price which is entirely potential as prior observed.

Recently, Arce-Plaza et al. reported the usage of the colloidal-gel method for the preparation of the material and the Dr. Blade method in CdTe films placement [43]. To assemble the photovoltaic device, a conformation of Glass/SnO$_2$:F/CdS/CdTe/Ag was used. The solar cells manufacturing process is shown in Figure 5.24.

The optical characterization data of the CdTe thin layers are presented in Figure 5.25. J–V measurements were used to characterize the solar cells (Figure 5.26).

CdTe thin-film devices achieve great efficiency of conversion via the chlorine activation method. Unlike other thin-film devices, for example, CuInGaSe$_2$ [44] and

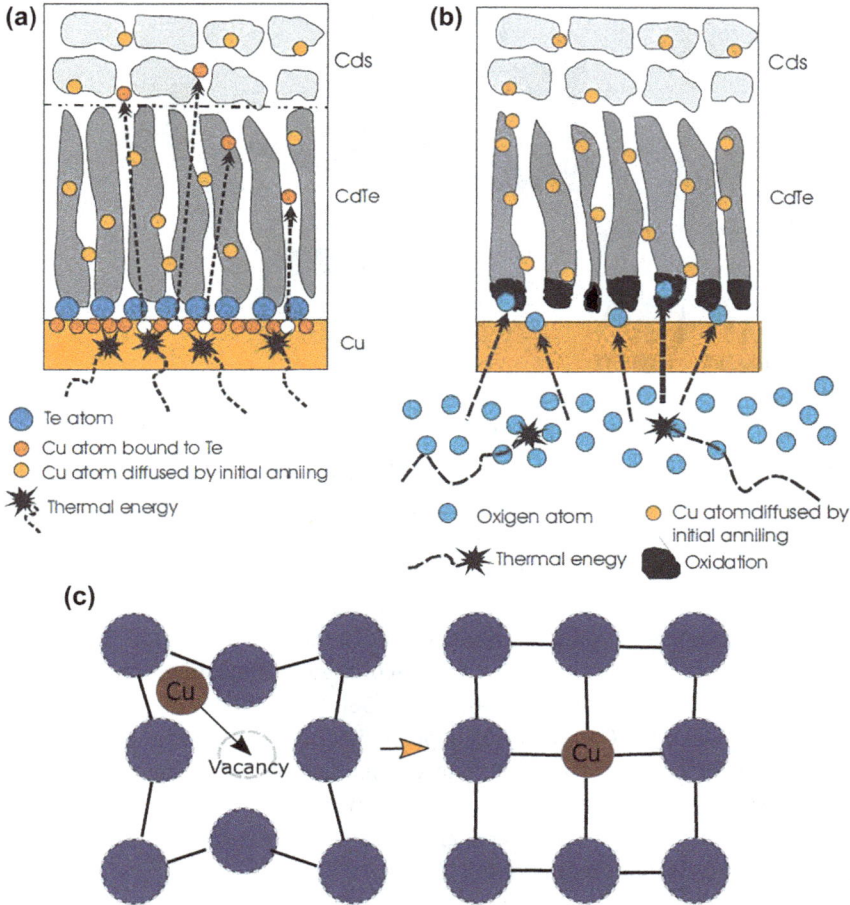

Figure 5.23: Schematic representation of possible phenomena that can cause degradation of cells: (a) decomposition of Cu₂Te in CuTe with the migration of copper in CdTe cells; (b) oxidation mechanism of tellurium atoms at the CdTe/Cu surface; (c) representation of iteration between copper and Cd atoms. Adapted with permission from Ref. [35].

Cu_2ZnSnS_4 [45]. To improve layer quality and increase the grain size, in a Cl-containing atmosphere high-temperature post-treatment is executed in a regular manner. Cl activation process is necessary for achieving great-efficiency solar cells, as the process takes away inherent structural defects [46] and improves the thin-film properties. To enhance the CdTe solar cell's efficiency, several chlorine activation methods have been reported. In most cases, chlorine activation therapy is accomplished employing Cl-containing gases such as Freon [47] and HCl [48] or using Cl salts such as $MgCl_2$, $CdCl_2$, NaCl, and NH_4Cl [49–54]. Among the activation methods, the most usual activation

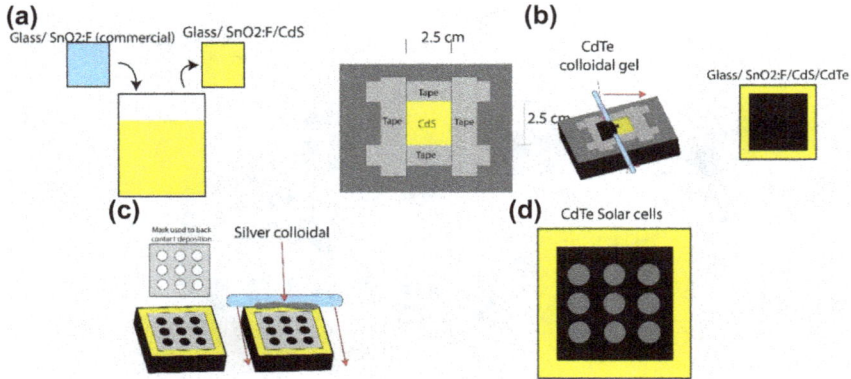

Figure 5.24: Solar cells manufacturing process. Adapted with permission from Ref. [43].

method is on the top of CdTe thin films build a $CdCl_2$ layer either by immersing the films in a solution (methanol) and annealing them in a specific temperature or by thermal evaporation. Usage of a gas containing chlorine, for example, Freon, has the advantage of annulling both carcinogens and in the activation method the usage of wet treatment. In that way, the scalability and reproducibility of the production of the solar cell increase [55]. Same as the dipping treatment of $CdCl_2$, the heat therapy of the cells at high-temperature in a vacuum chamber in a Cl-containing atmosphere makes grain boundary passivation events and recrystallization [56]. An efficiency of approximately 15% was reported when the CdTe cell was deposited at high temperature by a closed-space sublimation process and was experienced to method of gas treatment [57]. At the same time, the CdTe cells placed by the physical vapor deposition method at a lower temperature resulted in small dimension grains in CdTe cells. Hence, annealing is crucial to increase the grain size in CdTe.

Romeo et al. reported lower efficiency during the Freon treatment compared to the samples addressed with $CdCl_2$ utilizing a methanol dipping method [58]. This could be resolved to some extent by utilizing the Freon process, accompanied by the $CdCl_2$ dipping method. At the same time, if one treatment method is executed more overly than the other one, the total efficiency decreases [59]. When heat treatment is executed in compounding with the $CdCl_2$ dipping method and Freon method, the crystal growth direction was different [47].

Recently, Kim et al. reported improvement in the functioning of CdTe solar cells by employing a double treatment process that aggregates the Freon treatment and $CdCl_2$ dipping methods [60]. The double treatment sample showed higher efficiency compared to that of a single-treated cell.

The SIMS (secondary ion mass spectroscopy) analysis data entered for various depths of the device are presented in Figure 5.27. The SEM micrographs showed that after treatment by diverse methods the packing ratios and grain sizes alter in samples (Figure 5.28).

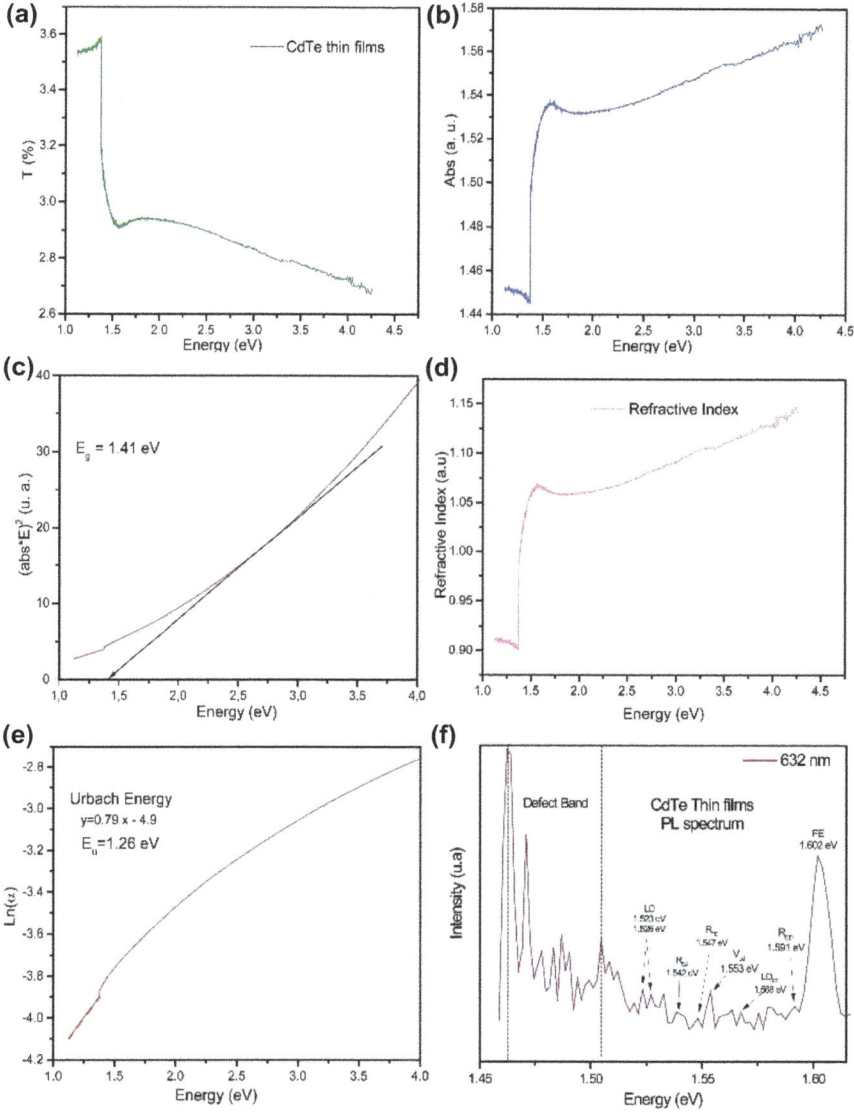

Figure 5.25: Optical CdTe thin films characterization (a) transmission spectrum (b) absorption spectrum, (c) Tauc plot and, (d) Refractive index (b) photoluminescent spectrum. Adapted with permission from Ref. [43].

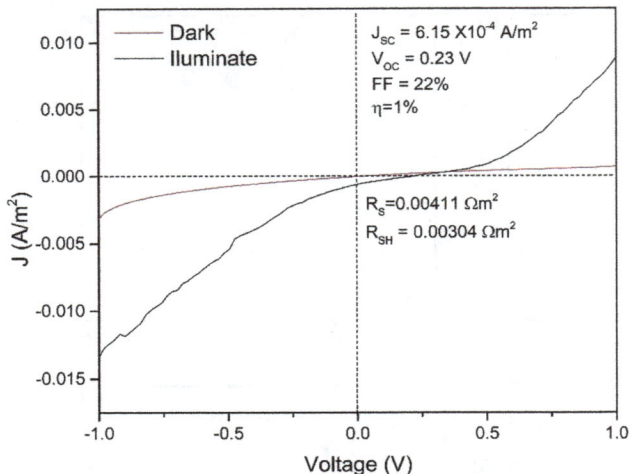

Figure 5.26: CdS/CdTe solar cell J–V characterization. Adapted with permission from Ref. [43].

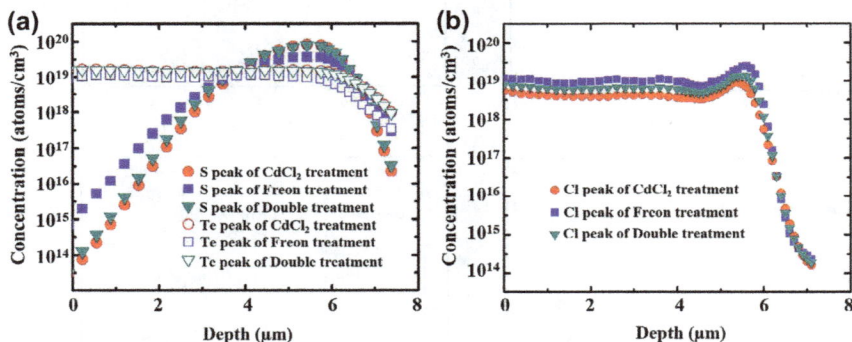

Figure 5.27: SIMS depth profiles of the S, Te peaks (a), and Cl peak (b) in the back-contact metal electrode (Au) to front contact electrode (TCO) for each sample treated with methanol dipping $CdCl_2$, Freon gas, and $CdCl_2$ + Freon. Adapted with permission from Ref. [60].

5.3 Tellurium doping in various solar cells

Triple junction solar cells are widely used for terrestrial and space solar electricity usages. These systems generally blend three solar cells in tandem interconnected with two tunnel junctions to allow for transport of carrier in the device. For triple-junction solar cells, Te doped InGaP is an appropriate substance for the *n* face of tunnel junction solar cells. Ebert et al. reported in triple junction the Te doping of InGaP for tunnel junction usages [61]. The study showed that before InGaP growth Te pre-doping of the

(a)

(b)

(c)

(d)

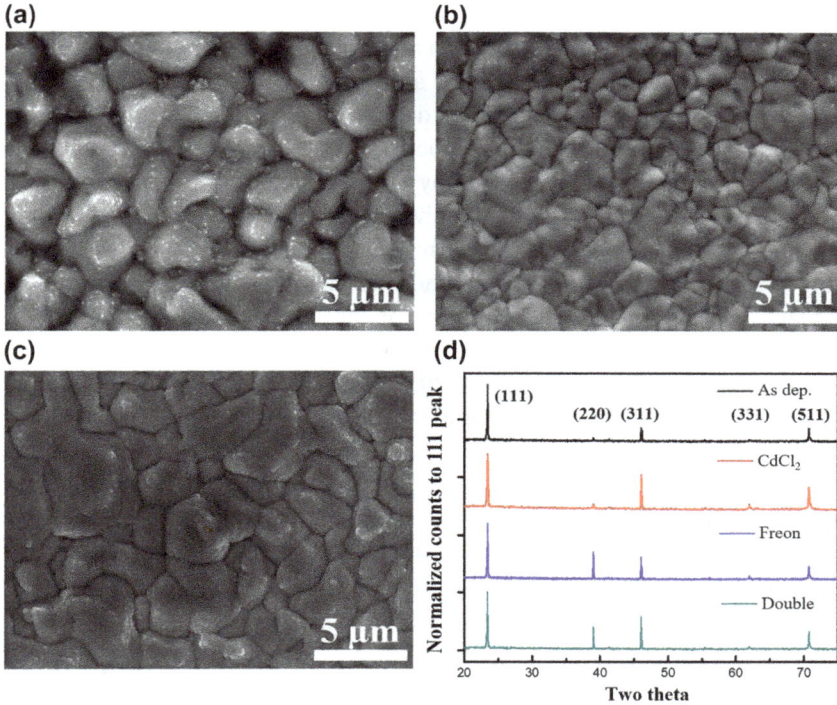

Figure 5.28: SEM images of (a) CdCl$_2$, (b), Freon, and (c) (CdCl$_2$ + Freon) treated samples, (d) shows XRD data of the As dep. sample, CdCl$_2$ dipping treatment sample, Freon treatment sample, and the double treatment sample. Adapted with permission from Ref. [60].

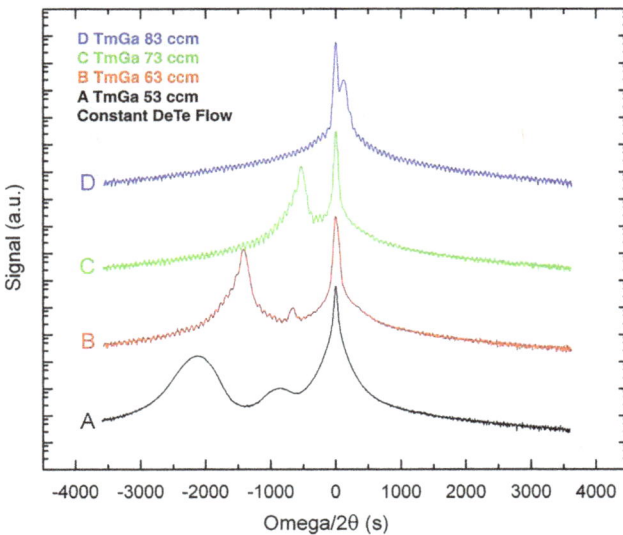

Figure 5.29: X-ray rocking curves for four different InGaP compositions with the same tellurium doping. Adapted with permission from Ref. [61].

film provided for an abrupt turn-on and that utilizing a growth pause at higher growth temperatures afterward InGaP provided for an abrupt turn-off in the profile of doping. A doping level of more than 1×10^{19} cm^{-3} was attained over 4in. diameter germanium wafers in a uniform manner. The study also demonstrated that InGaP layers can be strained by high Te doping and to produce smooth epitaxial layers modification of mole fraction of indium in InGaP is needed. X-ray rocking plots for four diverse InGaP constitutions with the similar Te doping are shown in Figure 5.29.

SIMS profile of Te in a three-layer InGaP system is shown in Figure 5.30. The AFM and scanning electron microscope images of the top layers are presented in Figure 5.31.

By utilizing a facile sol-gel process, groups of ZnO nanoparticles with Te and Ga dual-doping were prepared [62]. Their performance as the photoanode material in dye-sensitized solar cells (DSSCs) was analyzed. The XPS analysis of Ga–Te dual doped and pure ZnO systems were analyzed to look into the chemical bonding states and to key out the O, Zn, Te, and Ga elements. The typical XPS spectra of pure ZnO and $Ga_{0.50}Te_{0.50}ZnO$ films are shown in Figure 5.32. The surface morphologies of the films are presented in Figure 5.33. Figure 5.34 depicts the change in photovoltaic parameters as a function of time. The study showed that the Ga and/or Te incorporation into the ZnO system helps superior characteristics to finally heighten the dye-sensitized solar cell performance.

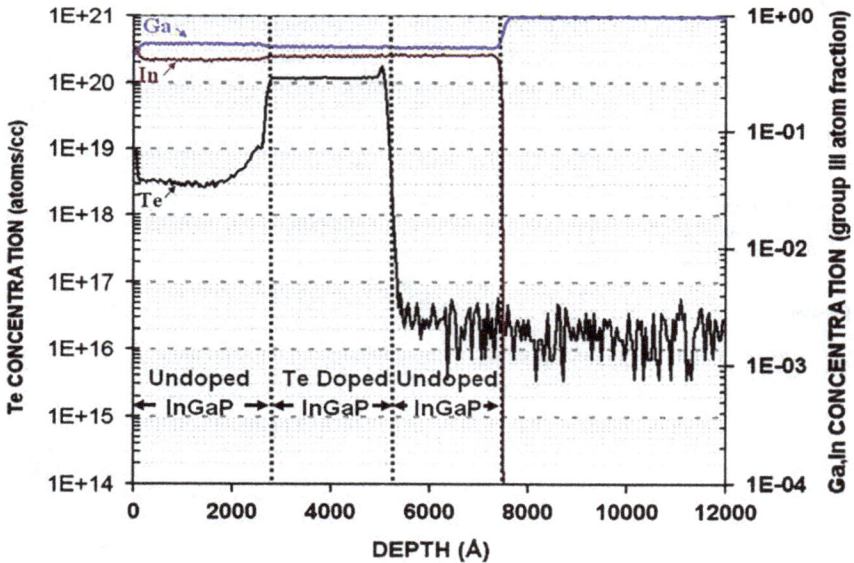

Figure 5.30: SIMS profile showing tellurium profile in InGaP using pre-doping pre-growth process and post-growth elevated temperature growth pause. Adapted with permission from Ref. [61].

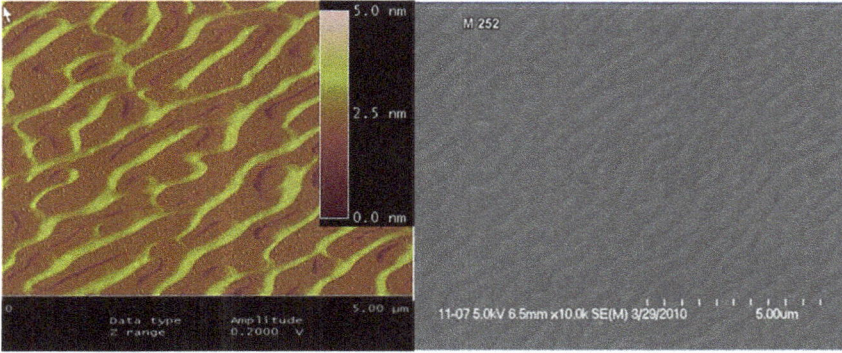

Figure 5.31: AFM 5 × 5 µm² scan showing the smooth surface of Te: InGaP (left) and SEM image of the same surface showing the smooth surface with growth steps (right). Adapted with permission from Ref. [61].

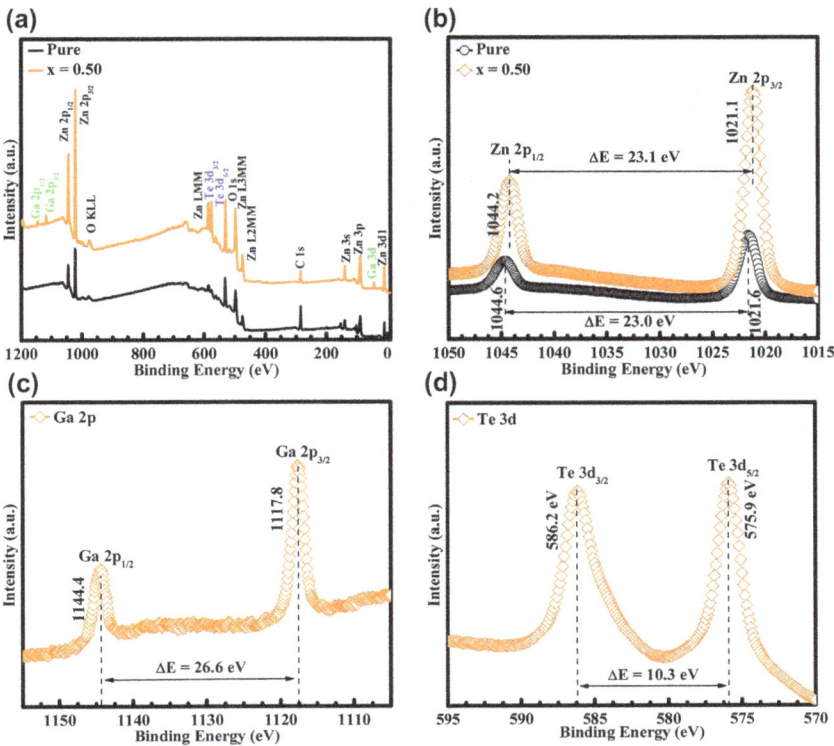

Figure 5.32: Characteristics of films.
(a) Complete XPS spectra of pure ZnO and Ga$_{0.50}$Te$_{0.50}$ZnO films. High–resolution spectra of samples for the elements of (b) Zn, (c) Ga, (d) Te. Adapted with permission from Ref. [62].

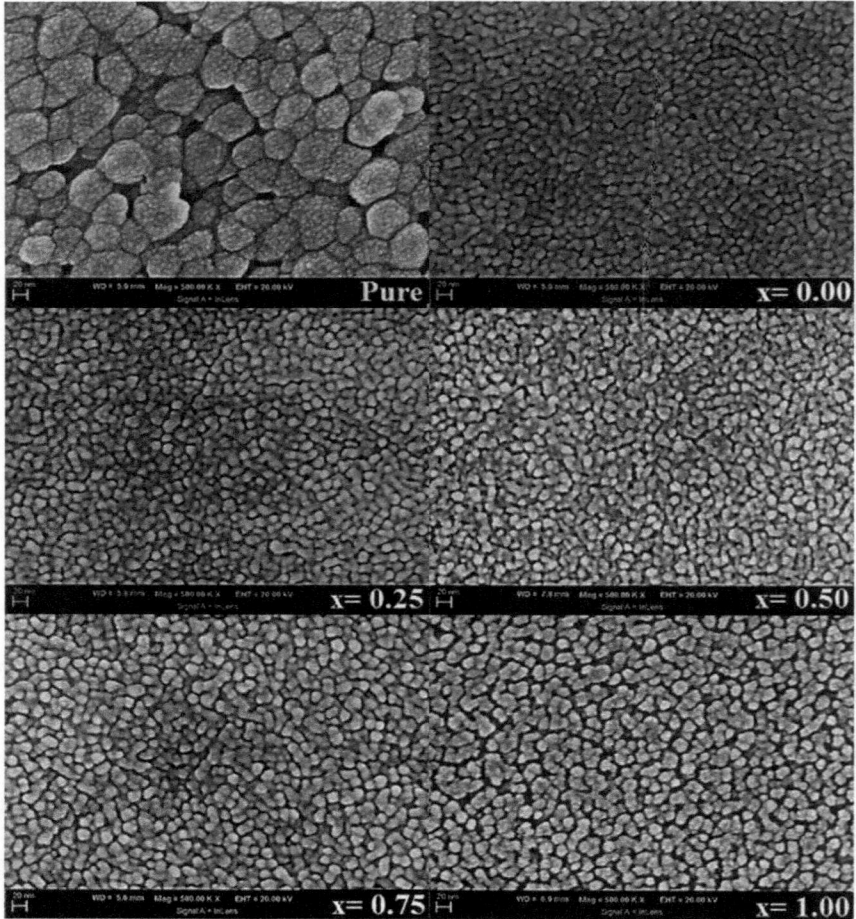

Figure 5.33: FE–SEM images of pure ZnO and Ga_xTe_{1-x} ZnO films. Adapted with permission from Ref. [62].

5.4 Conclusions

Tellurium is a promising and important functional material because of its unique chemical and physical characteristics. This article has discussed the synthesis and properties of Te-based solar cells. Te-based solar cells have shown promising results. The unique properties and morphologies of these solar cells may also contribute to the great improvement in this field.

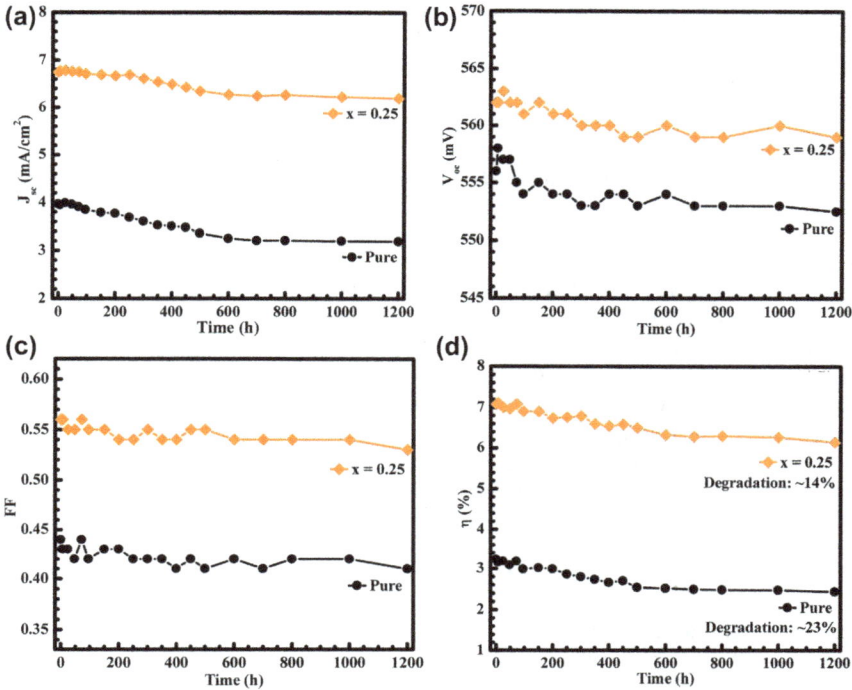

Figure 5.34: Cell parameter decays for 1200 h: (a) J_{sc}, (b) V_{oc}, (c) FF, (d) Efficiency. Adapted with permission from Ref. [62].

Acknowledgements: AD and BKB are grateful to Prince Mohammad Bin Fahd University for support.

References

1. Ohyama H, Aramoto T, Kumazawa S, Higuchi H, Arita T, Shibutani S, et al. 16.0% efficient thin-film CdS/CdTe solar cells. In: Conf rec twenty sixth IEEE photovolt spec conf-1997. IEEE, Anaheim, CA, USA; 1997:343–6 pp.
2. Britt J, Ferekides C. Thin-film CdS/CdTe solar cell with 15.8% efficiency. Appl Phys Lett 1993;62: 2851–2.
3. Geisthardt RM, Topič M, Sites JR. Status and potential of CdTe solar-cell efficiency. IEEE J Photovolt 2015;5:1217–21.
4. Ferekides CS, Marinskiy D, Viswanathan V, Tetali B, Palekis V, Selvaraj P, et al. High efficiency CSS CdTe solar cells. Thin Solid Films 2000;361–362:520–6.
5. Oladeji IO, Chow L, Ferekides CS, Viswanathan V, Zhao Z. Metal/CdTe/CdS/Cd$_{1-x}$Zn$_x$S/TCO/glass: a new CdTe thin film solar cell structure. Sol Energy Mater Sol Cells 2000;61:203–11.
6. Tiwari AN, Khrypunov G, Kurdzesau F, Bätzner DL, Romeo A, Zogg H. CdTe solar cell in a novel configuration. Prog Photovoltaics Res Appl 2004;12:33–8.

7. Lin T, Huang F, Liang J, Wang Y. A facile preparation route for boron-doped graphene, and its CdTe solar cell application. Energy Environ Sci 2011;4:862–5.
8. Wolden CA, Abbas A, Li J, Diercks DR, Meysing DM, Ohno TR, et al. The roles of ZnTe buffer layers on CdTe solar cell performance. Sol Energy Mater Sol Cells 2016;147:203–10.
9. Oladeji IO, Chow L. Synthesis and processing of CdS/ZnS multilayer films for solar cell application. Thin Solid Films 2005;474:77–83.
10. Abouelfotouh FA, Al Awadi R, Abd-Elnaby MM. Thin film $Cd_xZn_{1-x}S$/Si hybrid photovoltaic system. Thin Solid Films 1982;96:169–73.
11. Reddy KTR, Reddy PJ. Studies of $Zn_xCd_{1-x}S$ films and $Zn_xCd_{1-x}S$/CuGaSe$_2$ heterojunction solar cells. J Phys Appl Phys 1992;25:1345–8.
12. Ram PR, Thangaraj R, Agnihotri OP. Thin film CdZnS/CuInSe$_2$ solar cells by spray pyrolysis. Bull Mater Sci 1986;8:279–84.
13. Hossain MS, Rahman KS, Karim MR, Aijaz MO, Dar MA, Shar MA, et al. Impact of CdTe thin film thickness in $Zn_xCd_{1-x}S$/CdTe solar cell by RF sputtering. Sol Energy 2019;180:559–66.
14. Yang R, Wang D, Wan L, Wang D. High-efficiency CdTe thin-film solar cell with a mono-grained CdS window layer. RSC Adv 2014;4:22162–71.
15. Schaffner J, Motzko M, Tueschen A, Swirschuk A, Schimper H-J, Klein A, et al. 12% efficient CdTe/CdS thin film solar cells deposited by low-temperature close space sublimation. J Appl Phys 2011;110:064508.
16. Vásquez CH, Aguilera MLA, Trujillo MAG, Márquez JMF, Olarte DJ, Hernández SG, et al. Enhancement of CdS/CdTe solar cells by the interbuilding of a nanostructured Te-rich layer. Mater Res Express 2017;4:086403.
17. Venkatesh R, Banapurmath NR, Ramesh K, Venkatesh A, Khandake SA, Kurade PR, et al. Enhancement of open circuit voltage of CdTe solar cell. Mater Today Proc 2020;27:117–9.
18. Alfadhili FK, Phillips AB, Liyanage GK, Gibbs JM, Jamarkattel MK, Heben MJ. Controlling band alignment at the back interface of cadmium telluride solar cells using ZnTe and Te buffer layers. MRS Adv 2019;4:913–9.
19. Chou HC, Rohatgi A, Thomas EW, Kamra S, Bhat AK. Effects of Cu on CdTe/CdS heterojunction solar cells with Au/Cu contacts. J Electrochem Soc 1995;142:254.
20. Bastola E, Bhandari KP, Subedi I, Podraza NJ, Ellingson RJ. Structural, optical, and hole transport properties of earth-abundant chalcopyrite (CuFeS$_2$) nanocrystals. MRS Commun 2018;8:970–8.
21. Zhang M-J, Lin Q, Yang X, Mei Z, Liang J, Lin Y, et al. Novel p-type conductive semiconductor nanocrystalline film as the back electrode for high-performance thin film solar cells. Nano Lett 2016;16:1218–23.
22. Li J, Diercks DR, Ohno TR, Warren CW, Lonergan MC, Beach JD, et al. Controlled activation of ZnTe: Cu contacted CdTe solar cells using rapid thermal processing. Sol Energy Mater Sol Cells 2015;133: 208–15.
23. Moore A, Song T, Sites J. Improved CdTe solar-cell performance with an evaporated Te layer before the back contact. MRS Adv 2017;2:3195–201.
24. Bastola E, Alfadhili FK, Phillips AB, Heben MJ, Ellingson RJ. Wet chemical etching of cadmium telluride photovoltaics for enhanced open-circuit voltage, fill factor, and power conversion efficiency. J Mater Res 2019;34:3988–97.
25. Watthage SC, Phillips AB, Liyanage GK, Song Z, Gibbs JM, Alfadhili FK, et al. Selective Cd removal from CdTe for high-efficiency Te back-contact formation. IEEE J Photovoltaics 2018;8:1125–31.
26. Phillips AB, Subedi KK, Liyanage GK, Alfadhili FK, Ellingson RJ, Heben MJ. Understanding and advancing bifacial thin film solar cells. ACS Appl Energy Mater 2020;3:6072–8.
27. Liyanage GK, Phillips AB, Alfadhili FK, Ellingson RJ, Heben MJ. The role of back buffer layers and absorber properties for >25% efficient CdTe solar cells. ACS Appl Energy Mater 2019;2:5419–26.

28. Subedi KK, Bastola E, Subedi I, Bista SS, Rijal S, Jamarkattel MK, et al. Semi-transparent p-type barium copper sulfide as a back contact interface layer for cadmium telluride solar cells. Sol Energy Mater Sol Cells 2020;218:110764.

29. Lalitha S, Sathyamoorthy R, Senthilarasu S, Subbarayan A, Natarajan K. Characterization of CdTe thin film—dependence of structural and optical properties on temperature and thickness. Sol Energy Mater Sol Cells 2004;82:187–99.

30. Artegiani E, Major JD, Shiel H, Dhanak V, Ferrari C, Romeo A. How the amount of copper influences the formation and stability of defects in CdTe solar cells. Sol Energy Mater Sol Cells 2020;204: 110228.

31. Kosyachenko LA, Savchuk AI, Grushko EV. Dependence of efficiency of thin-film CdS/CdTe solar cell on parameters of absorber layer and barrier structure. Thin Solid Films 2009;517:2386–91.

32. Artegiani E, Menossi D, Shiel H, Dhanak V, Major JD, Gasparotto A, et al. Analysis of a novel $CuCl_2$ back contact process for improved stability in CdTe solar cells. Prog Photovoltaics Res Appl 2019; 27:706–15.

33. Guo D, Akis R, Brinkman D, Sankin I, Fang T, Vasileska D, et al. One-dimensional reaction-diffusion simulation of Cu migration in polycrystalline CdTe solar cells. In: 2014 IEEE 40th Photovolt spec conf PVSC. IEEE, Denver, CO, USA; 2014:2011–5 pp. https://doi.org/10.1109/PVSC.2014.6925321.

34. Ma J, Wei S-H, Gessert TA, Chin KK. Carrier density and compensation in semiconductors with multiple dopants and multiple transition energy levels: case of Cu impurities in CdTe. Phys Rev B 2011;83:245207.

35. Bertoncello M, Casulli F, Barbato M, Artegiani E, Romeo A, Trivellin N, et al. Influence of CdTe solar cell properties on stability at high temperatures. Microelectron Reliab 2020;114:113847.

36. Fan F-J, Yu B, Wang Y-X, Zhu Y-L, Liu X-J, Yu S-H, et al. Colloidal synthesis of $Cu_2CdSnSe_4$ nanocrystals and hot-pressing to enhance the thermoelectric figure-of-merit. J Am Chem Soc 2011; 133:15910–3.

37. Wilson ER, Parker LM, Orth A, Nunn N, Torelli M, Shenderova O, et al. The effect of particle size on nanodiamond fluorescence and colloidal properties in biological media. Nanotechnology 2019; 30:385704.

38. Tekin E, Smith PJ, Hoeppener S, van den Berg AMJ, Susha AS, Rogach AL, et al. Inkjet printing of luminescent CdTe nanocrystal–polymer composites. Adv Funct Mater 2007;17:23–8.

39. Liu J, Shi Z, Yu Y, Yang R, Zuo S. Water-soluble multicolored fluorescent CdTe quantum dots: synthesis and application for fingerprint developing. J Colloid Interface Sci 2010;342:278–82.

40. Su Z, Sun K, Han Z, Cui H, Liu F, Lai Y, et al. Fabrication of Cu_2ZnSnS_4 solar cells with 5.1% efficiency via thermal decomposition and reaction using a non-toxic sol–gel route. J Mater Chem A 2013;2: 500–9.

41. Sarswat PK, Free ML. Demonstration of a sol–gel synthesized bifacial CZTS photoelectrochemical cell. Phys Status Solidi 2011;208:2861–4.

42. Agbim EG, Ikhioya IL, Ekpunobi AJ. Syntheses and characterization of fluorine doped tin oxide using spray pyrolysis technique. IOSR J Appl Phys 2019;11:70–8.

43. Arce-Plaza A, Sánchez-Rodríguez FJ, Courel M, Pérez-Cuapio R, Alvarado JA, Roque J, et al. CdTe colloidal-gel: synthesis and thin films deposition applied to solar cells. Mater Sci Semicond Process 2021;131:105871.

44. Repins I, Contreras MA, Egaas B, DeHart C, Scharf J, Perkins CL, et al. 19·9%-efficient ZnO/CdS/ CuInGaSe₂ solar cell with 81·2% fill factor. Prog Photovoltaics Res Appl 2008;16:235–9.

45. Ahmed S, Reuter KB, Gunawan O, Guo L, Romankiw LT, Deligianni H. A high efficiency electrodeposited Cu_2ZnSnS_4 solar cell. Adv Energy Mater 2012;2:253–9.

46. Moutinho HR, Al-Jassim MM, Abulfotuh FA, Levi DH, Dippo PC, Dhere RG, et al. Studies of recrystallization of CdTe thin films after CdCl/sub 2/treatment [solar cells]. In: Conf rec twenty sixth IEEE photovolt spec conf-1997. IEEE, Anaheim, CA, USA; 1997:431–4 pp.

47. Salavei A, Rimmaudo I, Piccinelli F, Zabierowski P, Romeo A. Study of difluorochloromethane activation treatment on low substrate temperature deposited CdTe solar cells. Sol Energy Mater Sol Cells 2013;112:190–5.

48. Qu Y, Meyers PV, McCandless BE. HCl vapor post-deposition heat treatment of CdTe/CdS films. In: Conf rec twenty fifth IEEE photovolt spec conf – 1996. IEEE, Washington, DC, USA; 1996:1013–6 pp. https://doi.org/10.1109/PVSC.1996.564303.

49. Bayhan H. Investigation of the effect of $CdCl_2$ processing on vacuum deposited CdS/CdTe thin film solar cells by DLTS. J Phys Chem Solid 2004;65:1817–22.

50. Hiie J. CdTe:$CdCl_2$:O_2 annealing process. Thin Solid Films 2003;431–432:90–3.

51. Niles DW, Waters D, Rose D. Chemical reactivity of $CdCl_2$ wet-deposited on CdTe films studied by X-ray photoelectron spectroscopy. Appl Surf Sci 1998;136:221–9.

52. Williams BL, Major JD, Bowen L, Keuning W, Creatore M, Durose K. A comparative study of the effects of nontoxic chloride treatments on CdTe solar cell microstructure and stoichiometry. Adv Energy Mater 2015;5:1500554.

53. Potlog T, Ghimpu L, Gashin P, Pudov A, Nagle T, Sites J. Influence of annealing in different chlorides on the photovoltaic parameters of CdS/CdTe solar cells. Sol Energy Mater Sol Cells 2003;80: 327–34.

54. Potter MDG, Halliday DP, Cousins M, Durose K. A study of the effects of varying cadmium chloride treatment on the luminescent properties of CdTe/CdS thin film solar cells. Thin Solid Films 2000; 361–362:248–52.

55. Romeo N, Bosio A, Romeo A. An innovative process suitable to produce high-efficiency CdTe/CdS thin-film modules. Sol Energy Mater Sol Cells 2010;94:2–7.

56. Romeo A, Buecheler S, Giarola M, Mariotto G, Tiwari AN, Romeo N, et al. Study of CSS- and HVE-CdTe by different recrystallization processes. Thin Solid Films 2009;517:2132–5.

57. Rios-Flores A, Arés O, Camacho JM, Rejon V, Peña JL. Procedure to obtain higher than 14% efficient thin film CdS/CdTe solar cells activated with HCF_2Cl gas. Sol Energy 2012;86:780–5.

58. Romeo A, Artegiani E, Menossi D. Low substrate temperature CdTe solar cells: a review. Sol Energy 2018;175:9–15.

59. Rimmaudo I, Salavei A, Artegiani E, Menossi D, Giarola M, Mariotto G, et al. Improved stability of CdTe solar cells by absorber surface etching. Sol Energy Mater Sol Cells 2017;162:127–33.

60. Kim S, Song J-Y, Kim D, Hong J, Cho IJ, Kim YH, et al. Effect of novel double treatment on the properties of CdTe solar cells. Energy Rep 2021;7:1396–403.

61. Ebert C, Pulwin Z, Byrnes D, Paranjpe A, Zhang W. Tellurium doping of InGaP for tunnel junction applications in triple junction solar cells. J Cryst Growth 2011;315:61–3.

62. Akin S, Erol E, Sonmezoglu S. Enhancing the electron transfer and band potential tuning with long-term stability of ZnO based dye-sensitized solar cells by gallium and tellurium as dual-doping. Electrochim Acta 2017;225:243–54.

Muhammad Waqar Ashraf*, Syed Iqleem Haider,
Amber Rehana Solangi and Almas Fatima Memon

6 Toxicity of tellurium and its compounds

Abstract: Tellurium (Te) is widely used in industry because of its unique physicochemical properties. In the general population, foodstuff like meat, dairy products, and cereals is the major source of tellurium exposure. In the occupational environment, inhalational exposure predominates. Due to its exceptional properties as a metalloid, Te is broadly used in the industry. For example, Te is used as an alloy for solar panels, phase change optical magnetic disks, and Peltier devices. Recently, alloys of Te with cadmium, zinc, and other metals are used for nanomaterials, such as quantum dots. Thus, it is suggested that there is an existence of risk of exposure to Te in everyday life. Commercial Te is mostly obtained from slimes of electrolytic copper refineries. Te concentration in the slimes can extend up to 10% or more. Slight levels of its organic compounds may also be absorbed via skin. Not much information is available to prove Te as carcinogenic but its toxicity is well established. The present paper will review the toxicity of Te and its compounds.

Keywords: industry; tellurium; toxicity.

Tellurium (Te) is a scarce element found in nature and its properties are similar to the elements that are identified as toxic to humans. Over the years, utilization of Te due to its exceptional properties as a metalloid has emerged in various fields including electronics, catalysis, glass industry, metallurgy, semiconductor manufacturing, solar panels, and rubber industry. Tellurium has been used as an antibacterial element before the discovery of antibiotics. Recently it has gained a lot of attention as it is being used in nanotechnological advancements. Tellurium quantum dots are used as biosensors for biological detection systems. Whereas, other related nanostructures of Te are used in imaging, labeling, targeted drug delivery systems and are also used for antifungal and antibacterial properties. Besides, Te nanostructures also demonstrate free radical scavenging, novel antioxidant, and lipid-lowering properties. The growing use of Te and its compounds in various applications indicates that in near future it will be introduced into the environment and will cause various health problems as it accumulates in the body. Recent reports have demonstrated that enhanced

*Corresponding author: Muhammad Waqar Ashraf, Mathematics & Natural Sciences, Prince Mohammad Bin Fahd University, Azizeyah, Al-Khobar, 31952, Saudi Arabia,
E-mail: mashraf@pmu.edu.sa
Syed Iqleem Haider, Chemistry, University College Hyderabad, Hyderabad, Pakistan
Amber Rehana Solangi, Center of Excellence in Analytical Chemistry, University of Sindh, Jamshoro, Pakistan
Almas Fatima Memon, Chemistry, Government College University, Hyderabad, Pakistan

As per De Gruyter's policy this article has previously been published in the journal Physical Sciences Reviews. Please cite as: M. W. Ashraf, S. I. Haider, A. R. Solangi and A. F. Memon "Toxicity of tellurium and its compounds" *Physical Sciences Reviews* [Online] 2022. DOI: 10.1515/psr-2021-0112 | https://doi.org/10.1515/9783110735840-006

contamination of environment with Te has a fundamental connection to oncological, neurodegenerative, and autoimmune diseases. Thus, the knowledge of the toxicity of tellurium-containing compounds is important. Current review presents an overview of the exposure, absorption, and the toxicity of tellurium and its compounds.

6.1 Introduction

Tellurium (Te) belongs to the chalcogen family (group 6 A of the periodic table). It is a true metalloid and shares few chemical properties with other family members such as Selenium, Sulfur, and Oxygen [1]. It was initially discovered in 1782 from the ores of gold in the district of Transylvania, by a Romanian mining official named Franz Joseph Mueller von Reichenstein [2, 3]. Later, in 1798, it was named Tellurium (taken from a Latin word "tellus" meaning Earth) by Martin Heinrich Klaproth [4, 5]. Tellurium is one of the rarest elements on the Earth and is considered as a non-essential trace element. Nevertheless, a normal human body contains >0.5 g of Te surpassing the level of all other trace elements in humans, excluding rubidium, zinc, and iron [6, 7]. In the Earth's crust its abundance is around 0.027 ppm which is similar to elements such as gold and silver [8]. Te is rarely found in its elemental form as it is usually found as tellurides of gold such as Krennerite ($AuTe_2$), calaverite ($AuTe_2$), and sylvanite ($AgAuTe_4$). It is also found in the form of tellurides of more common metals. Tellurides are oxidized near the erath surface resulting in the formation of natural tellurate and tellurite minerals [9]. Te is also found in coal [10, 11]. These days, most of the Te is gained as a by-product in the process of mining and refining of copper, and its yearly production is around 220 tonnes per year mostly from Japan, the United States, China Canada, Sweden, Peru, and Russia. Two foremost Te producers of the world are Sweden and China, interpreting around 15% of worldwide production [12]. Based on its metal-like properties, Te naturally exists in various oxidation states including elemental tellurium Te (0), telluride Te (II), tellurate Te (VI), and tellurite Te (IV) [13].

　Due to its exclusive properties as a metalloid, tellurium is used in larger quantities in industries. To get some desired electrical characteristics, Te being a semiconductor is generally added to silver, gold, tin, or copper. Small amounts of Te are mainly added to stainless steel and copper alloys to make them easier to mill and machine. It is also added to lead to boost its toughness and resistance to sulphuric acid. Te is one of the chief constituents in blasting caps and is also utilized in metal oxidizing solutions to tarnish or blacken metals, as a pigment to color ceramics and glass, and in the vulcanization of rubber. On the agricultural side, it is being widely used as a component of insecticides [14]. It is also used as an alloy with antimony (Sb) and germanium (Ge), in DVD recordable disks (DVD-RW), and digital versatile disk-random access memory (DVD-RAM) [15]. Throughout the last 20 years, Te has been linked with the production of new materials including fluorescent Te-based quantum dots used as photovoltaic products, novel probes, telluride nanoparticles, nanotubes and clusters used in electronics and medicine [16], and some other compounds used in nanoscience such as solar panels [17, 18]. In the early 1930s, it was used medicinally in the treatment of leprosy and syphilis [19] and recently as an adjuvant in the treatments of cancer [20]. People employed in the manufacture of the above-mentioned products may be exposed to Te. Blackadder et al. [21] have reported the adverse effects experienced by workers that were exposed to the fumes of melting alloy of tellurium and copper. Humans' exposure across water and soil will likely rise as Te-based products make their way into landfills. Accumulation of Te in plants also occurs, mostly in the members of the *Alium* family such as garlic and onions, and thus consumption of accumulated Te demonstrates a viable risk [22].

The increased demand for Te because of its wide usage in everyday life has elevated alarms about environmental and human health problems as Te is considered as toxic to humans by which various organs can be affected [11]. Some reports have indicated that the bioavailability, toxicity, and the transport in environment mainly depend on its oxidation state. For instance, as compared to tellurate Te (VI), tellurite Te (IV) is 10 times more toxic [23]. Over the years, many Te compounds have been the subject of interest. A study done by Meotti et al. [24] reports that diphenyl ditelluride (DPDT) shows both hepatic and renal toxicity, and is the most toxic compound of the organochalcogenides. In several studies, the toxicity of DPDT has been examined both *in vivo* and *in vitro*. Carlton and Kelly [25] observed the toxic effects of tellurium tetrachloride ($TeCl_4$) on male Pekin ducks at concentrations of 500 and 100 ppm in the diet. As a result, 70% of the ducks were found dead four weeks after feeding. It was found that the inorganic metabolite of tellurium (dimethyl tellurium) induce peripheral neuropathy in rats. This compound is an inhibitor of squalene monooxygenase, a significant enzyme in the synthesis of cholesterol. It has been observed that amendment in this enzyme blocks the formation of the myelin sheath and results in the buildup of squalene in Schwann cells causing paralysis and segmental demyelination [26]. Thus, there is a desire need to handle Te and its compounds carefully as medical indices of tellurium toxicity are detected at incredibly low concentrations [27]. Besides, the release of inorganic Te compounds in the environment might also lead to severe problems due to its chronic and acute toxicity [28].

As a consequence of the less exposure of animals and man with this element and its compounds, Te toxicity has attained significantly less consideration than that of Selenium (Se). Generally, it seems that Te is less toxic as compared to Se but there are few exemptions to this rule. The interest in the toxicity of Te and its compounds have been largely increased due to the recent productions of semiconducting Te compounds for their application in thermoelectric.

6.1.1 Exposure of Tellurium

Elemental Te and soluble tellurites (mainly potassium or sodium) are absorbed through the gastrointestinal tract of mammals to 10–15%. As Te and its dioxide are instantly absorbed through the lungs, thus the absorption of Te generally occurs through the repiratory tract in the course of electrolytic refining of heavy metals [29]. Whereas the liquid volatile Te esters are immediately absorbed via intact skin [21]. In contradiction of the above-discussed forms, tellurium hexafluorides and hydrogen tellurides are demonstrated to be extremely toxic triggering cardiac and respiratory complications. Te is reduced to tellurides and becomes methylated after its absorption by the organisms. It is excreted mainly in the feces and to a lesser extent in the urine. A typical garlic-like odor is caused by the exhalation of dimethyl telluride, which is a volatile compound [29].

6.1.1.1 Humans

Exposure to hydrogen telluride and tellurium vapors may lead to respiratory tract irritation, causing pneumonia and bronchitis [30]. Whereas, exposure to tellurium hexafluoride may cause blue-black skin staining and dermatitis. Incorporation of soluble tellurium may produce scarce breath smell. It has been reported that the formation of dimethyl telluride might be the reason for the garlic-like smell, likely by hindering the squalene epoxidase enzyme [27].

6.1.1.2 Animals

Lobar pneumonia and desquamatory bronchitis were observed in rats due to the inhalation of tellurium dioxide and elemental tellurium [31]. Whereas, inhalation of increased levels of tellurium hexafluoride caused pulmonary edema in guinea pigs, rabbits, mice, and rats [32]. Black deposits in tissues of lungs and loss of body weight were detected in male Harlan Wistar albino rats, six months after a single endotracheal instillation of tellurium and its dioxide [33]. Demyelination and quick reparative remyelination in the white matter, optic nerve, and cerebral cortex were detected due to tellurium poisoning in adult rats. The results reveal ultrastructural fluctuations to few axons and impairment to glial cells and myelin sheath [34].

6.1.1.3 Plants

Depending upon some factors such as the surrounding environment and the occurrence of the element in the soil, plants demonstrate a high variation in the amounts of the metalloid that they can hold. It has been demonstrated that in the plant's surface, there is tremendously less abundance of the metal, approximately around 0.27 ppm [17]. Extensive research was carried out in a study reported by Cowgill et al. [35]. Thousands of samples were collected from various places in the USA, and it was observed that plants that are identified to accumulate selenium were also able to accumulate tellurium ranging to around 1 ppm concentration. The metabolism of Te in plants was demonstrated by Anan et al. [36]. Garlic that was cultivated in aquatic media

and subjected to sodium tellurate, was selected as a plant model because of its identified accumulation of Se. High-performance liquid chromatography (HPLC) coupled
with inductively coupled plasma mass spectrometry (ICP-MS) was used to identify the
metabolites containing Te in the aqueous extracts of the garlic leaves. Two metabolites,
cysteine S-methyltellurosulfide and Te-methyltellurocysteine oxide, were discovered.
Thus, it was assumed that in spite of the unidentified function of Te in the biology and
biochemistry of these species, it might be detoxified by using the usual ways of Se,
aiming to eliminate it from the polluted areas, permitting an appropriate plant
development.

6.1.2 Metabolism of Tellurium and its compounds

On entering into cells, the compounds of Te can stimulate (i) variations in glutathione metabolism, (ii) replacement of metal in enzymes, (iii) variations in the integrity
of cellular membrane structures, and (iv) oxidative stress [37, 38]. Their chemical
attraction to proteins and non-protein thiols and their capability to produce cellular
oxidative stress by the Fenton reaction is the common feature of these metalloids. Their
interface with the cell thiolome induces oxidative stress [39, 40], representing the entity
of the cellular thiol pool. Tellurium oxyanions (TeO_3^{2-}) are involved in the thiol: redox
system of the cell and interfere with thiol: redox enzymes (thioredoxin reductase and
glutathione reductase) and their metabolites (glutaredoxin, thioredoxin, and glutathione) [41]. Glutathione which takes part in TeO_3^{2-} to Te (Te 0) reduction accompanied
by reactive oxygen species (ROS) formation is the main target of TeO_3^{2-} cellular processing [42, 43].

The metabolic pathway of Te is not fully understood yet but even then, the body of
a human being is known to metabolize and excrete Te. After the absorption of TeO_3^{2-}
and TeO_4^{2-}, both are reduced in the liver. TeO_3^{2-} is methylated and results in dimethyl
tellurium (($CH_3)_2Te$) and lastly trimethyl tellurium (($CH_3)_3Te)^+$. Dimethyl tellurium may
bind to hemoglobin, and then accumulate in the blood cells [44]. Ba et al. [17] have
reported that these methylated species are found in the lungs, spleen, and kidney and
are considered as the most abundant form of Te in the human body. Te finally leaves the
human body via breath and urine as volatile ($CH_3)_2Te$, which causes the garlic-like
odor.

6.1.2.1 Absorption

The absorption of tellurium and its compounds is indicated by rashes or burns on the
skin, trailed by a garlic-like smell in the breath. Blackadder et al. [21] described two
cases of unintentional exposure of tellurium hexafluoride gas when about 50 g was
escaped from a cylinder into the laboratory. Both of them presented a significant garlic-
like breath odor. Whereas one of them complained of bluish-black coloration in the

skin of the neck, face, and fingers which took a long time to vanish. The authors assumed the skin absorption of volatile tellurium esters.

In humans, the intestinal absorption of tellurium is not much identified. An intestinal absorption of 25% ± 10% was detected for soluble salts of tellurium, during a study conducted on healthy male volunteers [45]. Steinberg et al. [46] have reported an increase in urinary tellurium of the workers who were exposed to tellurium fumes and it was interpreted that absorption took place through the respiratory tract. In a study conducted by Mead and Gies [47], it was found that tellurium dioxide was absorbed through the gastrointestinal tract of dogs. De Meio [48] has reported a three-week study in which rats were given huge dosages of elemental tellurium and it was observed that around 25% was absorbed through the gastrointestinal tract. Further experiments on animals presented similar results and it was detected that 25% of orally taken tellurium and its dioxide was absorbed in the gut [27, 49]. The gastrointestinal tellurium absorption is accomplished within 2 h after its ingestion. In rats, the chief absorption occurs in the jejunum and duodenum, and the entire absorption was assessed to be in the range of 10–25% [50]. In a study reported by Wright and Bell [51], sodium tellurite was administered to swine (barrows) and sheep (wethers) and it was observed that 24% of the Tellurium was absorbed from the colon in these species.

6.1.3 Distribution

In rat experimentation, it has been observed that around 90% of tellurium in the blood goes into erythrocytes, possibly attached to hemoglobin [52]. Rest is attached to plasma proteins. Te can pass both the blood-brain barriers and the placenta [53, 54]. The maximum tissue amounts have been observed in the kidney; amounts found in the lung, spleen, and heart are around 10–30% of the kidney amounts; and amounts in the liver are around 50% of those in the spleen, heart, and lungs. Estimated amounts in cardiac muscles are found to be around 20 times greater than in the skeletal muscles. Intracerebrally injected tellurium is accumulated into the gray matter of the nervous system. Besides, Te has also accumulated overtimes in bones, which bear greater than 90% of the entire body load [7].

6.1.3.1 Excretion

The pattern of excretion depends on the mode of administration and chemical forms of Te and its compounds. Orally consumed tellurium salts are mostly eliminated unabsorbed in the feces whereas parenterally administered tellurium is mainly eliminated in the urine [55]. Tellurium is transported to the intestine by biliary excretion [49]. Slight amounts can also be expelled in sweat and milk. In a study conducted by DeMeio and Henriques [56] on dogs, rabbits, and rats, it was observed that the trace concentration expelled by the respiratory tract in 24 h was less than 0.1% of the injected quantity. The maximum amount was detected in the kidney, with the liver holding comparatively a small amount of the tellurium. The entire quantity expelled within 5–6 days was only 20% of the injected quantity. This makes the accumulation of tellurium in the body seem possible. Slight amounts, perhaps around 0.1% of the absorbed elemental tellurium and tellurite are respired, apparently as dimethyl telluride, producing a typical garlic-like smell of sweat and breath [48].

6.1.4 Toxicity of elemental tellurium

Tellurium is rarely present in the environment but is considered to be very toxic as it exhibits properties parallel to those of heavy elements recognized as toxic to humans [57]. Contrasting its chalcogen members selenium and sulfur, tellurium does not hold a biological function. Unlike selenium, it is a non-essential element. It is not an essential part of any enzyme or natural protein in eukaryotes or prokaryotes. Due to its presence in the environment, it enters into the biological system [58]. Elemental tellurium is the least toxic form of tellurium. When the salts of tellurium are infused into the animal body, both dimethyl telluride and tellurium are produced and tellurium is expelled all the way through the urine, lungs, and sweat. If the salts of tellurium are taken orally, these are reduced to elemental tellurium in the intestinal tract by bacterial action [29].

Amdur [59] has reported the poisoning of three laboratory workers by incidental contact to tellurium-containing fumes formed via the incorporation of elemental tellurium into molten copper. Metallic taste, garlic odor, headache, and minor epigastric distress were the main complaints of the workers. Not a single person experienced vomiting or nausea. On examination, all were observed to have minor epigastric tenderness. Whereas, one of the victims with the greatest contact had numerous loose bowel movements.

Steinberg et al. [60] examined the impacts of tellurium exposure on iron foundry workers. 98 workers were exposed to 0.01–0.1 mg Te/m^3 for 22 months. 84 out of the 98 workers complained about the garlic odor of breath whereas garlic odor in the sweat of 30 exposed workers was observed. The exposed workers also showed tellurium excretion in the urine with concentrations ranging from 0.01 to 0.06 mg/L. Hemoglobin, blood cell counts, and routine investigations were all normal. Besides garlic odor of breath and sweat, workers also complained about occasional nausea, dryness of mouth, loss of appetite, somnolence, and the metallic taste of the mouth. Somnolence was not observed at a concentration of less than 0.01 mg of Te/L of urine. Also, the metallic taste of the mouth was not observed with consistency until the concentration of Te in urine was at least 0.01 mg/L. The maximum concentration of Te in the air at which no garlic breath odor appears is given as 0.01–0.02 mg Te/L^3 of air, and the maximum permissible concentration is given as 0.01–0.1 mg/cm^3 of air (maximum limit value).

DeMeio [48] reported the effect of orally given elemental tellurium on the growth of rats at a dose of 1500 ppm in the diet. No effect on the rate of growth was observed at

750 ppm and 375 ppm. Anatomically and pathologically, no irregularities were noticed even at 1500 ppm. Garlic-like odor in the breath was observed in all the rats, but there was no association with the tellurium concentration in the food. This shows that the elemental tellurium is converted into a soluble compound in the body of animals. Out of the entire amount of tellurium consumed, 63–84% was expelled in the feces. No change was detectable in the respiratory quotient of the kidney and liver tissue of rats getting tellurium.

Rats given a food comprising 1.25% elemental tellurium preliminary on postnatal day 20 produced garlic smell in 48 h and generally developed hindlimb paresis in 72 h. The rats demonstrated gradual rises in blood-nerve barrier penetrability 24–72 h after exposure. However, the blood-brain barrier remained unaffected. Other effects contained enhanced numbers of cytoplasmic excrescences, intracytoplasmic membrane-delimited clear vacuoles, intracytoplasmic lipid droplets in myelinating Schwann cells after 24 h; endoneurial edema after 72 h; and axon demyelination after 48 h. The formation of cholesterol was abruptly restricted after exposure to Te for 12 h [61]. A diet containing elemental Te given to weanling rats might develop peripheral neuropathy characterized by minimal axonal degeneration and segmental demyelination [62].

Laden and Porter [63] found brief, peripheral demyelination generated by the distraction of cholesterol formation in Schwann cells secondary to the hindrance of squalene monooxygenase in weanling rats given a food comprising 1% elemental Te. Other experiments on weanling rats have also presented the same results [64].

A group of growing rats given a food comprising 1.1% Te develop primary demyelination of the peripheral nerves, followed by rapid remyelination. This demyelination causes repression of the messenger RNA (mRNA) expression for myelin-specific proteins. Tellurium exposure was trailed by a rise in total RNA in the sciatic nerve, which could not be described by cellular proliferation. The enhanced concentration of ribosomal RNA (rRNA) may characterize a Schwann cell response to poisonous insult and may relate to the enhanced levels of protein formation need throughout remyelination. Whereas the steady-state concentrations of mRNA, analyzed by Northern blot analysis, for myelin basic protein and P0 were noticeably reduced. Throughout the consequent duration of remyelination, transcript levels increased and approached around normal levels 30 days after the beginning of exposure of Te [65].

In weanling rats, paralysis of the hindlimbs can be caused by ingestion of Te because of the segmented demyelination of the sciatic nerves bilaterally. Swollen Schwann cells, dislocated myelin sheaths, broadened endoneurial spaces, and a few cases of axonal degeneration have been reported [66].

6.1.5 Toxicity of tellurium compounds

The mechanism of Te toxicity and its effects on human health has been poorly explored till to date. The compounds of Te may enter into the body by ingestion or inhalation but none of the pathways is reported. Nausea, the characteristic garlic-like smell of the breath, caustic gastrointestinal tract injury, black staining of the oral mucosa and skin, metallic taste, and vomiting are the proven indications of the absorption of metal-oxidizing solutions containing considerable amounts of Te [67]. As compared to inorganic Te compounds, organotellurium compounds are generally less toxic. They overcome various metabolic conversions in the human body and their pharmacokinetic and pharmacological profiles are different from each other. Direct ingestion of Te compounds has been demonstrated to cause chronic or acute intoxication. Whereas breathing in of Tellurium dust may invade the human body via the lungs. On entering into the body, the compounds of Te may produce noxiousness in various ways, chiefly by intense interface with enzymes and proteins that hold cysteine [68]. Shie and Deeds [69] presented the first industrial report on the indications of absorption of Te in men working in a lead refinery. Thirteen workers were exposed to Te, probably in the form of tellurium dioxide and hydrogen telluride. Seven out of thirteen workers presented indications of tellurium intoxication with signs including metallic taste and dryness of the mouth, and garlic-like smell of sweat, breath, and urine. Five of them displayed decreased production of sweat, and three of them had nausea, somnolence, depression, anorexia, and dry itchy skin.

Two instances of tellurium intoxication caused by professional exposure to tellurium vapors have been stated. Symptoms included black-green staining of the mucosa of the nasopharynx and tongue, amnesia, garlic breath odor, pallor of the skin, shivering, coughing, and general weakness. Both pulse rate and temperature were raised and modest leukocytosis, neutrophilia, and leukopenia were detected [30].

Blackadder and Manderson [21] reported a case where two postgraduate chemists accidentally inhaled tellurium hexafluoride gas escaped from a cylinder in the laboratory. Both of them displayed typical signs of intoxication, particularly the stink of garlic was observed from excreta and in the breath. One of the victims presented scarce bluish-black staining on the skin of the neck, face, and fingers. None of them showed any symptoms of permanent damage and were spontaneously recovered without any treatment.

A study was reported by Berriault and Lightfoot [70] on workers of a silver refinery in Canada. Logistic regression models using age at sampling, urinary tellurium concentration, and employment duration as predictor variables displayed a link between the reporting of garlic smell and concentration of Te in the urine. The possibility of garlic reporting was raised as workers reached urinary tellurium concentration surpassing 1 µmol/mol creatinine.

Tellurium poisoning is rare and is almost limited to professionaly exposed workers. Just a few instances of non-occupational intoxications have been stated up to now. Keall et al. [71] have reported three cases of accidental intoxication of tellurium. Instead of sodium iodide, a solution of sodium tellurite was mistakenly given to the patients during retrograde pyelography. For two of them, the estimated dose was almost 2 g (30 mg/kg). Both of them displayed signs of vomiting, renal pain, stupor, cyanosis, irregular breathing, and died almost 6 h after injection. A sharp garlic-like smell was produced from all tissues during the autopsy and deposition of black Te was detected in the mucosa of the ureter and the bladder. Clogging was found in the liver, kidney, spleen, and lungs. The third patient who did not receive a larger dose recovered without showing other symptoms except characteristic garlic-like breath odor.

The major signs during tellurium intoxication include a metallic taste in the mouth, dryness of the mouth, suppression of sweating, loss of appetite, and a strong garlic-like odor of the sweat, urine, and breath. These signs were noticed in a woman aged 37 years who were non occupationally exposed to Te intoxication [72].

Vij and Hardej [73] have reported the toxicity of tellurium in transformed and non-transformed human colon cells. The results of the study showed that both tellurium tetrachloride and diphenyl ditelluride are cytotoxic to colon cell lines. Diphenyl ditelluride caused apoptosis through the inherent pathway in all tested cells, whereas tellurium tetrachloride resulted in necrosis.

Yarema et al. [67] reported the two cases of children who accidentally ingested metal oxidizing solutions holding considerable amounts of Te. Clinical features comprised garlic odor breath, vomiting, and black staining of the oral mucosa. One of them developed a corrosive injury to the esophagus. Both patients were recovered without serious consequences, which is characteristic of Te toxicity.

Critical and subcritical studies of systematic intoxication of Te compounds in rats display various symptoms including gastrointestinal disturbances, hind leg paralysis, listlessness, weight loss, anorexia, somnolence, reduced locomotor activity, occasionally epilation, and changes in the fur [31, 48, 74].

Studies on rabbits, rats, and Pekin ducks exposed to diverse Te compounds such as sodium tellurite, tellurium tetrachloride, and tellurium dioxide, presented changes in the liver extending from simple cellular swelling to hydropic and fatty degeneration and cell necrosis. Additional variations were also observed including impairment to glycogen function, protein metabolism, and detoxifying functions, as represented by a dose-related decrease of galactose tolerance, reduced albumin-globulin ration in serum, inhibition of cholinesterase, hippuric acid excretion, and urinary bilirubin excretion [25, 75].

The molecular basis of Te toxicity has been thoroughly investigated in prokaryotic cells. Such as, various genetic determinants of Te resistance have been recognized in diverse bacterial strains [76, 77]. Furthermore, a study has been reported in which numerous factors involved in the bacterial resistance against Te and the metabolic pathways of Te toxicity have been identified [78].

The pathway of Te uptake in *Escherichia coli* cells was successfully demonstrated by Elias et al. [79] and it was indicated that a vital route of the entrance of Te into *E. coli* is facilitated by the PitA phosphate transporter which is a class of cell membrane transporters. Tellurium behaves as a general oxidizing agent in prokaryotic cells, thus its toxicity seems to be facilitated across the generation of reactive oxygen species and specifically the superoxide anion [76, 80, 81].

A description of the toxicity of Te compounds may be related to the chemistry of tellurium and sulfur. Nevertheless, tellurium's high affinity to Selenium is associated with its biochemical actions within the body. Thus, one more reason for the toxicity of Te can be linked with the binding between Te compounds and Se that can be found in some selenium enzymes and proteins, leading to substantial toxicity for human cells. As a result, it has been assumed that Te compounds are more likely to bind with Se than sulfur since it shows some specificity for cellular selenium proteins [28]. Furthermore, unwanted connections between Te compounds and Se- and S-containing biomolecules might have severe consequences. The compounds of Te can deteriorate the antioxidant defense of the cell at the same time that actively produces reactive oxygen species (ROS). Such ROS are linked to various molecules such as ozone, hydrogen peroxide, or super-oxides [82]. Tellurium also persuades the overproduction of these species that cause mitochondrial dysfunction, and as a result, the death of cells takes place because of a process of apoptosis (self-programmed death) [83, 84].

The toxicity of Se and Te is linked with the production of analogical secondary metabolites such as dimethyl selenide ($(CH_3)_2Se$) and dimethyl telluride ($(CH_3)_2Te$) both of which have a significant garlic smell. Their toxicity is exerted through a strong interface with cysteine-containing enzymes and proteins [17].

Short-lived radioactive isotope, ^{132}Te, is one of the most toxic forms of Te. It is a part of nuclear disaster contamination and was the third most released radionuclide succeeding the Fukushima Daiichi incident, with the overall released activity of 180 petabecquerels (PBq) only less than that from ^{133}Xe and ^{131}I [85]. It is believed that exposure to ^{132}Te contributes to the high occurrence of thyroid cancers in people that are directly exposed to radiation after the Fukushima Daiichi nuclear incident [86].

Though less toxic than their bioavailable and soluble oxyanion counterparts, nanoparticles of Te and Se are themselves toxic to several microorganisms. Numerous Te-resistant microorganisms actively produce Te nanoparticles from Te oxyanions as a detoxifying mechanism. The toxicity mechanism for non-resistant microorganisms (i) may contribute to functional damage of cell membranes by altering their composition [87], and (ii) is believed to be due to the reaction of nanoparticles with intracellular thiols, generating reactive oxygen species (ROS) resulting in oxidative stress, similarly like the oxyanions of Te and Se [88].

Another type of toxic Te-containing nanoparticle emerges in the form of Cd-Te quantum dots, which are now considered cytotoxic [89, 90]. The main factor in their toxicity is the generation of ROS when Cd-Te nanoparticles directly interact with the mitochondria, plasma membrane, and nucleus of cells [90]. Other factors of toxicity include the release of soluble Cd^{2+} ions and the whole intracellular distribution of the quantum dots [89]. Thus, for safe usage, Cd-Te quantum dots must be firmly encapsulated in the protective coatings (e.g. glutathione) [91] and should have sometimes been replaced by quantum dots produced from other materials [92, 93].

6.1.6 Treatment of tellurium poisoning

The medical diagnosis of workers exposed to tellurium should include vigilant CNS and skin examination, which may be followed by urine analysis, pulmonary function tests, and chest radiography. In case of exposure to tellurium hexafluoride, pulse oximetry or arterial blood gases should be monitored and hematological parameters should also be checked. For the treatment of Te intoxication in case of oral exposure, activated charcoal as a slurry should be administered. The normal dosage is 25–100 g in adults, 25–50 g in children, and 1 g/kg in infants. Treatment is not required in case of inhalational exposure, but the patient should immediately be taken to fresh air. In case of exposure to eyes, they should be washed with plenty of water for at least 15 min. The affected should be moved to a health care facility if pain, irritation, photophobia, or lacrimation occurs. Whereas in case of skin exposure, the infected clothing should be taken off, and the subjected area should be carefully cleaned with soap and water [30].

6.2 Conclusions

Tellurium is one of the rarest elements on the Earth and is regarded as toxic to humans. Its toxicity depends upon its chemical form and oxidation state. The increased demand for Te due to its wide use in everyday life has developed distress about environmental and human health problems. To date, the toxicity of tellurium and its compounds has only been investigated superficially and very little is known about its intoxication. As tellurium is extensively used in daily life activities, humans are becoming more exposed to this element. Therefore, thorough studies are required to investigate its toxicity and the associated pathologic and metabolic processes in the near future.

References

1. Issa YM, Abdel-Fattah HM, Abdel-Moniem NB. A review on the chemical analysis of tellurium and its removal from real and biological samples. Int J Res Pharm Chem 2016;6:192–206.
2. Bagnall KW. The chemistry of selenium. Amsterdam: Tellurium Pol Elsevier; 1966:200 p.
3. Cunha RLOR, Gouvea IE, Juliano L. A glimpse on biological activities of tellurium compounds. An Acad Bras Cienc 2009;81:393–407.
4. Patel CKN. Efficient phase-matched harmonic generation in tellurium with a CO_2 laser at 10.6 μ. Phys Rev Lett 1965;15:1027.
5. Engman L. Synthetic applications of organotellurium chemistry. Acc Chem Res 1985;18:274–9.
6. Cohen BL. Anomalous behavior of tellurium abundances. Geochem Cosmochim Acta 1984;48: 203–5.
7. Schroeder HA, Buckman J, Balassa JJ. Abnormal trace elements in man: tellurium. J Chron Dis 1967; 20:147–61.

8. Turner RJ, Borghese R, Zannoni D. Microbial processing of tellurium as a tool in biotechnology. Biotechnol Adv 2012;30:954–63.
9. Rosing MT. On the evolution of minerals. Nature 2008;456:456–8.
10. Moscoso-Pérez C, Moreda-Piñeiro J, López-Mahía P, Muniategui-Lorenzo S, Fernández-Fernández E, Prada-Rodríguez D. As, Bi, Se (IV), and Te (IV) determination in acid extracts of raw materials and by-products from coal-fired power plants by hydride generation-atomic fluorescence spectrometry. At Spectrosc Connect 2004;25:211–6.
11. Matusiewicz H, Krawczyk M. Determination of tellurium by hydride generation with in situ trapping flame atomic absorption spectrometry. Spectrochim Acta Part B At Spectrosc 2007;62:309–16.
12. Belzile N, Chen Y-W. Tellurium in the environment: a critical review focused on natural waters, soils, sediments and airborne particles. Appl Geochem 2015;63:83–92.
13. Ogra Y. Biology and toxicology of tellurium explored by speciation analysis. Metallomics 2017;9:435–41.
14. Emsley J. Nature's building blocks: an AZ guide to the elements. London, United Kingdom: Oxford University Press; 2011.
15. Ogra Y, Kobayashi R, Ishiwata K, Suzuki KT. Identification of urinary tellurium metabolite in rats administered sodium tellurite. J Anal At Spectrom 2007;22:153–7.
16. Wachter J. Metal telluride clusters—from small molecules to polyhedral structures. Eur J Inorg Chem 2004;2004:1367–78.
17. Ba LA, Döring M, Jamier V, Jacob C. Tellurium: an element with great biological potency and potential. Org Biomol Chem 2010;8:4203–16.
18. Deng Z, Zhang Y, Yue J, Tang F, Wei Q. Green and orange CdTe quantum dots as effective pH-sensitive fluorescent probes for dual simultaneous and independent detection of viruses. J Phys Chem B 2007;111:12024–31.
19. Frazer AD. Tellurium in the treatment of syphilis. Lancet 1930;216:133–4.
20. Hayun M, Saida H, Albeck M, Peled A, Haran-Ghera N, Sredni B. Induction therapy in a multiple myeloma mouse model using a combination of AS101 and melphalan, and the activity of AS101 in a tumor microenvironment model. Exp Hematol 2009;37:593–603.
21. Blackadder ES, Manderson WG. Occupational absorption of tellurium: a report of two cases. Occup Environ Med 1975;32:59–61.
22. Larner AJ. How does garlic exert its hypocholesterolaemic action? The tellurium hypothesis. Med Hypotheses 1995;44:295–7.
23. Huang C, Hu B. Speciation of inorganic tellurium from seawater by ICP-MS following magnetic SPE separation and preconcentration. J Separ Sci 2008;31:760–7.
24. Meotti FC, Borges VC, Zeni G, Rocha JB, Nogueira CW. Potential renal and hepatic toxicity of diphenyl diselenide, diphenyl ditelluride and Ebselen for rats and mice. Toxicol Lett 2003;143:9–16.
25. Carlton WW, Kelly WA. Tellurium toxicosis in Pekin ducks. Toxicol Appl Pharmacol 1967;11:203–14.
26. Stangherlin EC, Favero AM, Zeni G, Rocha JB, Nogueira CW. Exposure of mothers to diphenyl ditelluride during the suckling period changes behavioral tendencies in their offspring. Brain Res Bull 2006;69:311–7.
27. Taylor A. Biochemistry of tellurium. Biol Trace Elem Res 1996;55:231–9.
28. Garberg P, Engman L, Tolmachev V, Lundqvist H, Gerdes RG, Cotgreave IA. Binding of tellurium to hepatocellular selenoproteins during incubation with inorganic tellurite: consequences for the activity of selenium-dependent glutathione peroxidase. Int J Biochem Cell Biol 1999;31:291–301.
29. Cerwenka EA, Jr., Cooper WC. Toxicology of selenium and tellurium and their compounds. Arch Environ Health 1961;3:189–200.
30. Gerhardsson L. Tellurium. In: Handbook on the toxicology of metals. Amsterdam: Elsevier; 2015:1217–28 pp.

31. Sandratskaia SE. Experimental studies on the characteristics of tellurium as an industrial poison. Gig Tr Prof Zabol 1962;6:44–50.
32. Kimmerle G. Comparative studies on the inhalation toxicity of sulfur-, selenium-and tellurium hexafluoride. Arch Toxikol 1960;18:140.
33. Geary DL, Myers RC, Nachreiner DJ, Carpenter CP. Tellurium and tellurium dioxide: single endotracheal injection to rats. Am Ind Hyg Assoc J 1978;39:100–9.
34. Smiałek M, Gajkowska B, Otrebska D. Electron microscopy studies on the neurotoxic effect of sodium tellurite in the central nervous system of the adult rat. J Hirnforsch 1994;35:223.
35. Cowgill UM. The tellurium content of vegetation. Biol Trace Elem Res 1988;17:43–67.
36. Anan Y, Yoshida M, Hasegawa S, Katai R, Tokumoto M, Ouerdane L, et al. Speciation and identification of tellurium-containing metabolites in garlic, Allium sativum. Metallomics 2013;5: 1215–24.
37. Kim D-H, Kanaly RA, Hur H-G. Biological accumulation of tellurium nanorod structures via reduction of tellurite by Shewanella oneidensis MR-1. Bioresour Technol 2012;125:127–31.
38. Valdivia-González M, Pérez-Donoso JM, Vásquez CC. Effect of tellurite-mediated oxidative stress on the Escherichia coli glycolytic pathway. Biometals 2012;25:451–8.
39. Rubino FM. Toxicity of glutathione-binding metals: a review of targets and mechanisms. Toxics 2015;3:20–62.
40. Kell DB. Towards a unifying, systems biology understanding of large-scale cellular death and destruction caused by poorly liganded iron: Parkinson's, Huntington's, Alzheimer's, prions, bactericides, chemical toxicology and others as examples. Arch Toxicol 2010;84:825–89.
41. Turner RJ, Weiner JH, Taylor DE. Neither reduced uptake nor increased efflux is encoded by tellurite resistance determinants expressed in Escherichia coli. Can J Microbiol 1995;41:92–8.
42. Turner RJ, Aharonowitz Y, Weiner JH, Taylor DE. Glutathione is a target in tellurite toxicity and is protected by tellurite resistance determinants in Escherichia coli. Can J Microbiol 2001;47:33–40.
43. Kessi J, Hanselmann KW. Similarities between the abiotic reduction of selenite with glutathione and the dissimilatory reaction mediated by Rhodospirillum rubrum and Escherichia coli. J Biol Chem 2004;279:50662–9.
44. Kobayashi A, Ogra Y. Metabolism of tellurium, antimony and germanium simultaneously administered to rats. J Toxicol Sci 2009;34:295–303.
45. Kron T, Hansen C, Werner E. Renal excretion of tellurium after peroral administration of tellurium in different forms to healthy human volunteers. J Trace Elem Electrolytes Health & Dis 1991;5:239–44.
46. Steinberg HH, Massari SC, Miner AC, Rink R. Industrial exposure to tellurium: atmospheric studies and clinical evaluation. J Ind Hyg Toxicol 1942;24:183–92.
47. Mead LD, Gies WJ. Physiological and toxicological effects of tellurium compounds, with a special study of their influence on nutrition. Am J Physiol Content 1901;5:104–49.
48. De Meio RH. Tellurium. I. The toxicity of ingested elementary tellurium for rats and rat tissues. J Ind Hyg Toxicol 1946;28:229–32.
49. Hollins JG. The metabolism of tellurium in rats. Health Phys 1969;17:497–505.
50. Slouka V, Hradil J. Kinetics of orally administered radiotellurium in the rat. Hradec Kralove, Czech: Military-Medical Research Inst.; 1970.
51. Wright PL, Bell MC. Comparative metabolism of selenium and tellurium in sheep and swine. Am J Physiol Content 1966;211:6–10.
52. Agnew WF, Cheng JT. Protein binding of tellurium-127m by maternal and fetal tissues of the rat. Toxicol Appl Pharmacol 1971;20:346–56.
53. Agnew WF. Transplacental uptake of 127mtellurium studied by whole-body autoradiography. Teratology 1972;6:331–7.
54. Agnew WF, Fauvre FM, Pudenz PH. Tellurium hydrocephalus: distribution of tellurium-127m between maternal, fetal, and neonatal tissues of the rat. Exp Neurol 1968;21:120–31.

55. Durbin PW. Metabolic characteristics within a chemical family. Health Phys 1960;2:225–38.
56. DeMeio RH, Henriques FC Jr. Tellurium. 4. Excretion and distribution in tissues studied with radioactive isotope. J Biol Chem 1947;169:609–23.
57. Issa YM, Abdel-Fattah HM, Shehab OR, Abdel-Moniem NB. Determination and speciation of tellurium hazardous species in real and environmental samples. Int J Electrochem Sci 2016;11: 7475–98.
58. Sadeh T. Biological and biochemical aspects of tellurium derivatives. Org Selenium Tellurium Compd 1987;2:367–76.
59. Amdur ML. Tellurium. Accidental exposure and treatment with BAL in oil. Occup Med (Lond) 1947;3: 386–91.
60. Sax NI. Handbook of dangerous materials. New York: Reinhold; 1951.
61. Bouldin TW, Earnhardt TS, Goines ND, Goodrum J. Temporal relationship of blood-nerve barrier breakdown to the metabolic and morphologic alterations of tellurium neuropathy. Neurotoxicology 1989;10:79.
62. Goodrum JF, Earnhardt TS, Goines ND, Bouldin TW. Lipid droplets in Schwann cells during tellurium neuropathy are derived from newly synthesized lipid. J Neurochem 1990;55:1928–32.
63. Laden BP, Porter TD. Inhibition of human squalene monooxygenase by tellurium compounds: evidence of interaction with vicinal sulfhydryls. J Lipid Res 2001;42:235–40.
64. Chugh A, Ray A, Gupta JB. Squalene epoxidase as hypocholesterolemic drug target revisited. Prog Lipid Res 2003;42:37–50.
65. Toews AD, Lee SY, Popko B, Morell P. Tellurium-induced neuropathy: a model for reversible reductions in myelin protein gene expression. J Neurosci Res 1990;26:501–7.
66. Pun TWC, Odrobina E, Xu Q, Lam TY, Munro CA, Midha R, et al. Histological and magnetic resonance analysis of sciatic nerves in the tellurium model of neuropathy. J Peripher Nerv Syst 2005;10: 38–46.
67. Yarema MC, Curry SC. Acute tellurium toxicity from ingestion of metal-oxidizing solutions. Pediatrics 2005;116:e319–21.
68. Medina-Cruz D, Tien-Street W, Vernet-Crua A, Zhang B, Huang X, Murali A, et al. Tellurium, the forgotten element: a review of the properties, processes, and biomedical applications of the bulk and nanoscale metalloid. Racing Surf 2020;1:723–83.
69. Shie MD, Deeds FE. The importance of tellurium as a health hazard in industry. A preliminary report. Publ Health Rep 1920;35:939–54.
70. Berriault CJ, Lightfoot NE. Occupational tellurium exposure and garlic odour. Occup Med (Lond) 2011;61:132–5.
71. Keall JHH, Martin NH, Tunbridge RE. Accidental poisoning due to sodium tellurite. Br J Ind Med 1946;3:175.
72. Müller R, Zschiesche W, Steffen HM, Schaller KH. Tellurium-intoxication. Klin Wochenschr 1989; 67:1152–5.
73. Vij P, Hardej D. Evaluation of tellurium toxicity in transformed and non-transformed human colon cells. Environ Toxicol Pharmacol 2012;34:768–82.
74. Amdur MT. Tellurium oxide. An animal study in acute toxicity. Arch Indust Heal 1958;17:665–7.
75. De Mkio RH, Jetter WW. Tellurium. III. The toxicity of ingested tellurium dioxide for rats. J Ind Hyg Toxicol 1948;30:53–8.
76. Taylor DE. Bacterial tellurite resistance. Trends Microbiol 1999;7:111–5.
77. Zannoni D, Borsetti F, Harrison JJ, Turner RJ. The bacterial response to the chalcogen metalloids Se and Te. Adv Microb Physiol 2007;53:1–312.
78. Chasteen TG, Fuentes DE, Tantaleán JC, Vásquez CC. Tellurite: history, oxidative stress, and molecular mechanisms of resistance. FEMS Microbiol Rev 2009;33:820–32.

79. Elías AO, Abarca MJ, Montes RA, Chasteen TG, Pérez-Donoso JM, Vásquez CC. Tellurite enters Escherichia coli mainly through the PitA phosphate transporter. Microbiologyopen 2012;1: 259–67.
80. Sandoval JM, Verrax J, Vásquez CC, Calderon PB. A comparative study of tellurite toxicity in normal and cancer cells. Mol Cell Toxicol 2012;8:327–34.
81. Summers AO, Jacoby GA. Plasmid-determined resistance to tellurium compounds. J Bacteriol 1977; 129:276–81.
82. Nathan C, Cunningham-Bussel A. Beyond oxidative stress: an immunologist's guide to reactive oxygen species. Nat Rev Immunol 2013;13:349–61.
83. Brenneisen P, Reichert AS. Nanotherapy and reactive oxygen species (ROS) in cancer: a novel perspective. Antioxidants 2018;7:31.
84. Kim KS, Lee D, Song CG, Kang PM. Reactive oxygen species-activated nanomaterials as theranostic agents. Nanomedicine 2015;10:2709–23.
85. Gil-Díaz T. Tellurium radionuclides produced by major accidental events in nuclear power plants. Environ Chem 2019;16:296–302.
86. Drozdovitch V, Kryuchkov V, Chumak V, Kutsen S, Golovanov I, Bouville A. Thyroid doses due to Iodine-131 inhalation among Chernobyl cleanup workers. Radiat Environ Biophys 2019;58: 183–94.
87. Pi J, Yang F, Jin H, Huang X, Liu R, Yang P, et al. Selenium nanoparticles induced membrane bio-mechanical property changes in MCF-7 cells by disturbing membrane molecules and F-actin. Bioorg Med Chem Lett 2013;23:6296–303.
88. Zonaro E, Lampis S, Turner RJ, Qazi SJ, Vallini G. Biogenic selenium and tellurium nanoparticles synthesized by environmental microbial isolates efficaciously inhibit bacterial planktonic cultures and biofilms. Front Microbiol 2015;6:584.
89. Chen N, He Y, Su Y, Li X, Huang Q, Wang H, et al. The cytotoxicity of cadmium-based quantum dots. Biomaterials 2012;33:1238–44.
90. Lovrić J, Cho SJ, Winnik FM, Maysinger D. Unmodified cadmium telluride quantum dots induce reactive oxygen species formation leading to multiple organelle damage and cell death. Chem Biol 2005;12:1227–34.
91. Zheng Y, Gao S, Ying JY. Synthesis and cell-imaging applications of glutathione-capped CdTe quantum dots. Adv Mater 2007;19:376–80.
92. Li L, Daou TJ, Texier I, Kim Chi TT, Liem NQ, Reiss P, et al. Highly luminescent CuInS2/ZnS core/shell nanocrystals: cadmium-free quantum dots for in vivo imaging. Chem Mater 2009;21:2422–9.
93. Pons T, Pic E, Lequeux N, Cassette E, Bezdetnaya L, Guillemin F, et al. Cadmium-free CuInS2/ZnS quantum dots for sentinel lymph node imaging with reduced toxicity. ACS Nano 2010;4:2531–8.

Garima Pandey* and Sangeeta Bajpai*

7 Accessing the environmental impact of tellurium metal

Abstract: Tellurium is gaining technical significance because of being a vital constituent for the growth of green-energy products and technologies. Owing to its unique property of interchangeable oxidation states it has a tricky though interesting chemistry with basically unidentified environmental effects. The understanding of environmental actions of tellurium has significant gaps for instance, its existence and effects in various environmental sections related to mining, handling and removal and disposal methods. To bridge this gap it is required to assess its distinctive concentrations in the environment together with proper knowledge of its environmental chemistry. This in turn significantly requires developing systematic diagnostic schemes which are sensitive enough to present statistics in the concentrations which are environmentally relevant. The broad assessment of available statistics illustrates that tellurium is being found in a very scarce concentrations in various environmental sections. Very less information is available for the presence and effects of tellurium in air and natural water resources. Various soil and lake sediment analysis statistics indicate towards the presence of tellurium in soil owing to release of dust, ash and slag during mining and manufacturing practices. Computing the release and behavior of tellurium in environment needs a thorough assessment of its anthropogenic life cycle which in turn will facilitate information about its existing and prospective release in the environment, and will aid to handle the metal more sensibly.

Keywords: anthropogenic life cycle; diagnostic schemes; green-energy products; scarce concentrations.

7.1 Introduction

The demand for tellurium has inflated to a great extent due to its usage in star-photovoltaics globally. Computing the effect of element onto the surroundings requires a life-cycle wide examination of the element-cycle (parallel to the natural element-cycles). Tellurium is one of the rare elements (eight times scarce than gold) on Earth. Majority of the rocks contain grains of the metal in the form of a silver-white colored

*Corresponding authors: Garima Pandey, Department of Chemistry, SRM Institiute of Science and Technology, Delhi NCR Campus, Modinagar 201204, Ghaziabad, Uttar Pradesh, India; and Sangeeta Bajpai, Department of Chemistry, Amity School of Applied Sciences, Amity University, Uttar Pradesh, India, 226008, E-mail: garimapandey.pandey8@gmail.com (G. Pandey), sbajpai1@amity.edu (S. Bajpai). https://orcid.org/0000-0002-4615-9119 (G. Pandey)

As per De Gruyter's policy this article has previously been published in the journal Physical Sciences Reviews. Please cite as: G. Pandey and S. Bajpai "Accessing the environmental impact of tellurium metal" *Physical Sciences Reviews* [Online] 2022. DOI: 10.1515/psr-2021-0113 | https://doi.org/10.1515/9783110735840-007

brittle material amounting around three parts per billion amount of the total tellurium [1, 2]. Grains of native tellurium seem in rocks as a brittle, silvery-white substance, but it usually occurs in telluride minerals of platinum, silver and gold. Tellurium is a metalloid exhibiting the characteristics of metals and non-metals both. The distinct isolation of element was made possible only after 15 years of its discovery (in the year 1780, from the gold ores) and was named tellurium, after the Latin phrase "tellus," meaning as "the fruit of the Earth" [3–6]. Traditionally, tellurium is been exploited as an additive in the alloys of iron, lead and copper. Owing to its low-abundance, not much is known about its toxicity or its impact on human beings and various ecosystems (Tables 7.1 and 7.2).

7.2 Where does tellurium come from? sources of tellurium

Tellurium (Te) is a metalloid from sixth group of periodic table, which is been used for alloying, for manufacturing solar-panels, glass optical-fibers, magnetic-discs and nanomaterials, as a pigment and as catalyst for ceramic industry [7, 8]. Though tellurium rarely exists as elemental state, it predominantly exists in Earth crust as alloys of silver, copper and gold. The concentration of this rare-earth element ranges from 3–205 ppb in Earth crust and 0.06–1 ppm in hydro-thermal samples. TeO_3^{2-} and TeO_4^{2-} are the most commonly occurring, soluble and most toxic tellurite oxy-anions having tellurium in +4 and +6 oxidation states [9–11].

Contrary to the belief of considering tellurium as a foreign-element in the bodies of living-beings, tellurium has been found in substantial amounts in the human bone tissues, urine and blood. Additionally, tellurium was also established to form

Table 7.1: Commonly occurring chemical forms of tellurium.

Commonly occurring forms of tellurium	$Te(C_2H_5)_2$, H_2Te, $Te(CH_3)_2$, H_6TeO_6, H_2TeO_3, $H_2O_3Te.Na$, $TeBr_2$, $TeCl_2$, TeO_2, $TeBr_4$, $TeCl_4$, TeF_4, TeI_4, TeO_3

Table 7.2: Common sources of tellurium.

Anode slimes at copper refineries
Seafloor volcanogenic massive sulfide (VMS) deposits
Discreet telluride minerals
Recycling solar cells
Structures of sulfide minerals in most gold deposits
Undersea mineral recovery (ferromanganese nodules on the ocean floor)

structural part of some amino-acids, such as tellurocysteine and telluromethionine in some bacterial proteins, yeast and fungi, suggesting some possible biological role for the metalloid [12, 13].

7.3 How do we use tellurium?

Tellurium is primarily been used for the production of photovoltaic-films to be used in solar-cells. The alloying of tellurium with cadmium produces a compound with enhanced the electrical-conductivity (Table 7.3) [14–16].

Table 7.3: Uses of tellurium

As an alloying component with cadmium	Enhances electrical conductivity
As an additive to steel, copper, and lead alloys	Improves tensile-strength, ductility and helps in prevention of sulfuric acid corrosion in thermoelectric equipments
As a vulcanizing agent in the chemical industry	Durable products, improves heat-resistance of rubber
As a coloring agent in copying machines	Improves coloring of ceramics and glass
As an additive to gasoline	Prevents engine-knocks in automobiles
As an additive to integrated-circuits, laser-diodes, and medical-instruments	Improved durability, electrical conductivity and enhanced performance

7.4 Assessment studies on effects of tellurium

Metals are vital for maintaining the materials-base of current technology and are expected to be of growing-importance in the transition towards the sustainable-technologies. With the growing technological advances, the demand for tellurium, the rare-element, is also increasing [17, 18]. Tellurium metal is useful for making solar-panels for green-energy technologies, for the production of rubber, plays an important role in electronic equipment, but the toxic nature of some of its forms is making researchers anxious about the likely environmental contaminations. Owing to their prevalent uses, it is very much important and needed to understand the environmental-implications of tellurium and its derivatives, all-through their life-cycle in order to build conversant choices of raw-materials, methods, designs and stay away from shifting of environmental-burdens [19].

Life cycle assessment (LCA) can be utilized as a significant tool to compute the system-wide environmental-load of services, and technologies involving the usage of tellurium based products. A significant research on the life-cycle impacts of metals is obtainable from a range of articles and scientific reports being published based on life cycle inventory (LCI) data-bases [20–22]. However, many LCI data are reported is either the pre-allocated one or is reported at system process level itself. The available LCI data are also do not always represent the global metals production-route of tellurium and its various forms and may sometimes be correspond to the inventory data of obsolete technologies.

Though tellurium occurs in slight concentrations in the oceans and in Earth crust, it is often released in the environment as a by-product of metal-production and mining of coal and burning of oil. The environmental deposition data of tellurium was majorly sourced from the studies of the lake-sediments around the coal burning and metal production sites. The existing information about the existence and behavior of tellurium and its compounds in the environmental domain is quite inadequate, and it straight away necessitates the upgrading of analytical methods. Still following studies by researchers have been performed to assess the environmental impacts of tellurium [23–25].

1. Nuss (2019) presented an appropriate assessment of life-cycle of the anthropogenic-cycle of tellurium linking its environmental inputs during the whole process of production and uses. The author highlighted the need of implementing appropriate end-of-life management-approaches for the photovoltaic units after the end of their operational-life.

2. Filella and May [26] identified gaps and proposed the 'best' constant-values by performing the critical evaluation of thermodynamics and equilibrium-data for tellurium.

3. [27] reviewed the interaction of tellurium with prokaryotic organisms and highlighted the lagging behind of the biological studies of tellurium. Tellurium-interaction with the prokaryotes is the most studied area, covering an array of aspects, together with the recent attention towards bio-nanotechnology. Tellurium is found in substantial amounts as a fission-product in nuclear-facilities.

4. Gil-Díaz [28] pointed towards the existence, impacts and importance of tellurium radio-nuclides after evaluating the environmental releases from the nuclear power-plants at Chernobyl and Fukushima-Daiichi.

5. [29] performed the physiochemical categorization of tellurium found in tellurium-enriched mine tailings and the results obtained pointed towards minimal bio-accessibility of the element.

6. A non-negligible bio-accumulation of tellurium was observed in the wild oysters by Gil-Díaz et al. [30].

Not much information is available about the impending environmental and human health-impacts of the tellurium release and exposure. For an all-inclusive assessment a sustainable metal-management system has to be implemented to cover various aspects of its life cycle with a goal of assuring sustainability of social, economic and environmental spheres. Broad study of quite a few impact-categories of immediate consideration (toxicity, a-biotic resource exhaustion, excess land consumption), should be performed in future studies when site-specific inventory-data for sophisticated LCA modeling studies on tellurium and its derivatives will become available and accessible [31–34].

7.5 Effects of tellurium

Te has its reach up to the liver, kidneys, testes, and to the bones in human body. Moreover, the chlorides of tellurium may cause cytotoxicity by getting accumulated in human brain-cells, called astrocytes [35]. It is been observed that the plants which have the characteristic behavior of accumulating the selenium metal, can also accumulate the tellurium metal. Consequently, there is a threat of Te poisoning, after consuming plant products grown in contaminated area [36]. Te has been found to be present in vegetables like onion, garlic, citrus-fruits, cereals, potatoes, leguminous plants and also is been found in few baby-food samples [37–39].

7.5.1 Potential environment impacts

The assessment of impending environmental effects of technologically significant element like tellurium requires substantial concentration of the element (and its compounds) in environmental samples to analyze the chemical-processes governing its environmental behavior [40, 41]. The comprehensive assessment of the available data illustrates that the values for tellurium-concentrations in various environmental samples (from different environmental components) are very scarce, predominantly in natural-water samples, where a consistent evaluation of tellurium concentration is can't even be produced. Figures available for air are less abundant than for the natural-water. Notable and presentable concentration-figures do exist for soil-samples indicating towards the geological sources of tellurium. Certain studies of soil-examinations from urban areas and lake-sediments close to tellurium sources point towards its distribution and spread in the environment because of the human-activities; though the distant transport in atmosphere is yet to be proved [42–44]. Tellurium element along with its compounds (cadmium telluride being the most likely) should be characterized as strongly-toxic substance to the aquatic-ecosystem. Though, in some literature it is suggested that this opinion is exaggeratedly cautious. Recent toxicity-testing in aquatic ecosystems did not show any fatal or sub-fatal effect of cadmium telluride on zebra fish. The environmental impacts tellurium compounds are not much investigated and explored and such studies have recently gained momentum [26, 28, 30, 45].

7.5.2 Potential health impacts

Providentially, the compounds of tellurium are not commonly been encountered by the humans and animals. The likelihood of tellurium acting as a health hazard in industries was first noticed in the fall of 1918 [46–49]. Tellurium is found to be present in the dust and fumes coming out of the furnaces of industries probably as it is the oxide and hydrogen telluride. They are mildly toxic in nature but intake in even very small amount can causes awful-smelling breath and dreadful body-odor. Therefore it is very much necessary to take certain precautions while handling and working with them. Though tellurium is not carcinogenic in nature still the treatment of tellurium-poisoning is a difficult process because the chelating-agents used for treatment result into increase in the toxicity of tellurium. Tellurium compounds might sometimes behave as an embryo-toxic substance. Inhalation of dusts containing tellurium, released during metal-refining process is one of the likely sources of workplace-exposure to tellurium compounds [45, 50, 51].

Routes of exposure: the substance can enter inside the body as aerosol through following routes
1. Respiratory tract
2. Alimentary tract
3. Nasopharynx
4. Absorption through skin (tellurous acid)

Fate of tellurium inside the body – tellurium undergoes through following chemical transformations inside the body [52, 53]
1. Firstly the tellurium salts are reduced to metallic-tellurium (this step plays the major role in toxic effects of tellurium).
2. Next to reduction, methyl telluride is formed (this synthesis of methyl telluride is an uncommon example of synthesis of a methyl compound inside the animal body).

Excretion [54–56]
1. Most of the consumed substance is exterminated from the body in the metallic form through feces
2. It is also eliminated as methyl telluride through urine, dermal-secretions, exhaled-breath and feces.
3. Small amounts are also expelled in their soluble through bile and urine.

The health effects of tellurium are summed up in Table 7.4

Tellurites particularly are found to be harmful for prokaryotic species by-far more lethal than silver, lead and mercury. Though the strong antibacterial-action of tellurium is known for many years (since 1930s), the chemistry underlying tellurite toxicity is still unspoken. In-spite of being highly toxic to bacteria, tellurite resistant bacteria have also been isolated in the year 1980 [57, 58]. The tellurium found in bacterial cells induces black intra-cellular deposits resulting into the black-coloration of bacterial colonies [59]. Indeed, micro-organisms play an essential role in instituting the environmental-fate of an element. They might alter the bio-availability and the

Table 7.4: Health impacts of tellurium

Inhalation of dispersed airborne particles	Garlic-odor in breath (tellurium-breath), drowsiness, nausea, headache, dryness of mouth, metallic taste in mouth, suppressed secretion of saliva
Short-term exposure	Irritation of the eyes, dryness and itching of skin, irritation in respiratory tract, effects on liver and central nervous system
Ingestion	Pain in stomach, mild gastroenteritis, weight loss, diarrhea, suppression of sweat, cytotoxicity
Chemical effects	Toxic-fumes (hydrogen telluride gas) on heating, risk of fire when brought in contact of halogens, has combustible nature

mobility of the element in the environment and for this reason affect the intrinsic-toxicity of that element [60, 61]. Tellurium, in its elemental state by itself is non-toxic, but its pro-oxidant nature of its tellurite oxyanion is highly toxic to bacterial-cells [57]. Biological-effects of Te and its derivatives on the central nervous-system are not limited to the peripheral nerves only (Table 7.5).

As of now no cases of severe tellurium-poisoning have been reported but patients suffering with tellurium poisoning may show signs of aggressive vomiting, intestinal-hemorrhages, loss of sense and reflexes, acute depression, and paralysis of the central-nervous-system, with un-consciousness, and at last death in spasm. Associated with these symptoms patients may also show signs of obliteration of the red-blood-cells, declining hemoglobin-content, severe nephritis and gastro-enteritis [36, 57, 62, 63].

Table 7.5: Toxicity data of tellurium and compounds following prolonged exposure to substantial amount.

Animal-species	Te derivative	Result of prolonged exposure to substantial amount
Rat	Te, TeO_2	Weight-loss, irritation in respiratory-tract, fur-loss, haemolysis, deterioration of Schwann-cells, growth-retardation, paresis, redness, short-term paralysis of hind-legs, anuria or oligouria, necrosis of kidney and liver
Rat, mouse and guinea-pig	$Te(C_2H_5)_2$	Dermatitis, weight-loss, haemolytic-effects, effects on heart, serum-proteins, liver-enzymes and central nervous system and other organs
Pigs and weanling	$TeCl_4$	Necrosis of skeletal and cardiac muscles, decreased blood GSH-activity
Rats and rabbits	Na_2TeO_3	Necrosis of intestine and liver effects on CNS; distorted brain-morphology
Rats	K_2TeO_3	Enhanced-activity, impaired-growth, no consequences observed on the learning-ability

7.6 Concluding remarks

Though the toxicity level and effects of some of the compounds of tellurium are known, still the very little concentrations in almost all the environment hinder in the process of its assessment. The existence of tellurium in sediments, air-particulate matters, soil and natural-water validates its bio-mobility. Overcoming the existing methodical restrictions is critical for progressing the environmental assessment. Further research is required to have a comprehensive assessment of the effects of tellurium and its compounds, their redox-transformations, volatility and assimilations in environment and its components.

References

1. Ba LA, Döring M, Jamier V, Jacob C. Tellurium: an element with great biological potency and potential. Org Biomol Chem 2010;8:4203–16.
2. Belzile N, Chen Y-W. Tellurium in the environment: a critical review focused on natural waters, soils, sediments and airborne particles. Appl Geochem 2015;63:83–92.
3. Hein JR, Koschinsky A, Halliday AN. Global occurrence of tellurium-rich ferromanganese crusts and a model for the enrichment of tellurium. Geochim Cosmochim Acta 2003;67:1117–27.
4. Hu Z, Gao S. Upper crustal abundances of trace elements: a revision and update. Chem Geol 2008; 253:205–21.
5. Huang C, Hu B. Speciation of inorganic tellurium from seawater by ICP-MS following magnetic SPE separation and preconcentration. J Separ Sci 2008;31:760–7.
6. Jabłónska-Czapla M, Grygoýc K. Speciation and fractionation of less-studied technology-critical elements (Nb, Ta, Ga, In, Ge, Tl, Te): a review. Pol J Environ Stud 2021;30:1477–86.
7. Baesman SM, Bullen TD, Dewald J, Zhang D, Curran S, Islam FS, et al. Formation of tellurium nanocrystals during anaerobic growth of bacteria that use Te oxyanions as respiratory electron acceptors. Appl Environ Microbiol 2007;73:2135–43.
8. Baesman SM, Stolz JF, Kulp TR, Oremland RS. Enrichment and isolation of *Bacillus beveridgei* sp. nov., a facultative anaerobic haloalkaliphile from Mono Lake, California, that respires oxyanions of tellurium, selenium, and arsenic. Extremophiles 2009;13:695–705.
9. Baturin GN, Dubinchuk VT, Azarnova LA, Mel'nikov ME. Species of molybdenum, thallium, and tellurium in ferromanganese crusts of oceanic seamounts. Oceanology 2007;47:415–22.
10. Baturin GN. Tellurium and thallium in ferromanganese crusts and phosphates on oceanic seamounts. Dokl Earth Sci 2007;413:331–5.
11. Chen YW, Alzahrani A, Deng TL, Belzile N. Valence properties of tellurium in different chemical systems and its determination in refractory environmental samples using hydride generation–atomic fluorescence spectroscopy. Anal Chim Acta 2016;905:42–50.
12. Chiou KY, Manuel OK. Tellurium and selenium in aerosols. Environ Sci Tech 1986;20:987–91.
13. Cunha RL, Gouvea IE, Juliano LA. Glimpse on biological activities of tellurium compounds. Ann Acad Bras Cienc 2009;81:393–407.
14. Wiklund JA, Kirk JL, Muir DCG, Carrier J, Gleason A, Yang F, et al. Widespread atmospheric tellurium contamination in Industrial and remote regions of Canada. Environ Sci Techn 2018;52:6137–45.

15. Kashiwabara T, Oishi Y, Sakaguchi A, Sugiyama T, Usui A, Takahashi Y. Chemical processes for the extreme enrichment of tellurium into marine ferromanganese oxides. Geochim Cosmochim Acta 2014;131:150–63.
16. Kavlak G, Graedel TE. Global anthropogenic tellurium cycles for 1940–2010. Resour Conserv Recycl 2013;76:21–6.
17. Bustamante ML. Criticality of Byproduct Materials: Assessing Supply Risk, Environmental Impact, and Strategic Policy Response for Tellurium Thesis. Rochester Institute of Technology; 2016.
18. Chiou KY, Manuel OK. Determination of tellurium and selenium in atmospheric aerosol samples by graphite furnace atomic absorption spectrometry. Anal Chem 1984;56:2721–3.
19. Dickson RS, Glowa GA. Tellurium behavior in the Fukushima Dai-ichi nuclear power plant accident. J Environ Radioact 2019;204:49.
20. Duan L-Q, Song J-M, Yuan H-M, Li X-G, Li N, Ma J. Selenium and tellurium fractionation, enrichment, sources and chronological reconstruction in the East China Sea. Estuar Coast Shelf Sci 2014;143: 48–57.
21. Filella M, Reimann C, Rodushkin I, Marc B, Katerina R. Tellurium in the environment: current knowledge and identification of gaps. Environ Chem 2019;16:215–28.
22. Fthenakis VM. Life cycle impact analysis of cadmium in CdTe PV production. Renew Sustain Energy Rev 2004;8:303–34.
23. Fujiwara K, Takahashi T, Kinouchi T, Fukutani S, Takahashi S, Watanabe T, et al. Transfer factors of tellurium and cesium from soil to radish (Raphanus sativus var. sativus) and komatsuna (Brassica rapa var. perviridis). Jpn J Health Phys 2017;52:192–9.
24. García-Figueroa A, Lavilla I, Bendicho C. Speciation of CdTe quantum dots and Te(IV) following oxidative degradation induced by iodide and headspace single-drop microextraction combined with graphite furnace atomic absorption spectrometry. Spectrochim Acta, Part B 2019;158:105631.
25. Harada T, Takahashi Y. Origin of the difference in the distribution behavior of tellurium and selenium in a soil–water system. Geochim Cosmochim Acta 2008;72:1281–94.
26. Filella M, May PM. The aqueous chemistry of tellurium: critically-selected equilibrium constants for the low-molecular-weight inorganic species. Environ Chem 2019;16:289–95.
27. Presentato A, Turner RJ, Vásquez CC, Yurkov V, Zannoni D. Tellurite-dependent blackening of bacteria emerges from the dark ages. Environ Chem 2019;16:266–88.
28. Gil-Díaz T. Tellurium radionuclides produced by major accidental events in nuclear power plants. Environ Chem 2019;16:296–302.
29. Hayes SM, Ramos NA. Surficial geochemistry and bioaccessibility of tellurium in semiarid mine tailings. Environ Chem 2019;16:251–65.
30. Gil-Díaz T, Schäfer J, Dutruch L, Bossy C, Pougnet F, Abdou M, et al. Tellurium behaviour in a major European fluvial-estuarine system (Gironde, France): fluxes, solid/liquid partitioning, and bioaccumulation in wild oysters. Environ Chem 2019;16:229–42.
31. Kolesnikov SI. Impact of contamination with tellurium on biological properties of ordinary chernozem, soil and sediment contamination. Int J 2019;28:792–800.
32. Kumar A, Holuszko M, Espinosa DCR. E-Waste: an overview on generation, collection, legislation and recycling practices. Resour Conserv Recycl 2017;122:32–42.
33. Mead LD, Gies WJ. Physiological and toxicological effects of tellurium compounds. Am J Physiol 1901;5:104–49.
34. Larner AJ. Biological effects of tellurium: a review. Trace Elem Electrolytes 1995;12:26–31.
35. Babula P, Adam V, Opatrilova R, Zehnalek J, Havel L, Kizek R. Uncommon heavy metals, metalloids and their plant toxicity: a review. Environ Chem Lett 2008;6:189–213.
36. Cowgill UM. The tellurium content of vegetation. Biol Trace Elem Res. 1988;17:43–67.
37. Liu Y, He M, Chen B, Hu B. Simultaneous speciation of inorganic arsenic, selenium and tellurium in environmental water samples by dispersive liquid–liquid microextraction combined with

electrothermal vaporization inductively coupled plasma mass spectrometry. Talanta 2015;142: 213–20.

38. Maltman C, Yurkov V. The effect of tellurite on highly resistant freshwater aerobic anoxygenic phototrophs and their strategies for reduction. Microorganisms 2015;3:826–38.

39. Maltman C, Donald L, Yurkov V. Two distinct periplasmic enzymes are responsible for tellurite/ tellurate and selenite reduction by strain ER-Te-48 isolated from a deep sea hydrothermal vent tube worms at the Juan de Fuca Ridge black smokers. Arch Microbiol 2017;199:1113–20.

40. McLennan SM. Relationships between the trace element composition of sedimentary rocks and upper continental crust. G-cubed 2001;2:1021.

41. Meyer J, Schmidt A, Michalke K, Hensel R. Volatilisation of metals and metalloids by the microbial population of an alluvial soil. Syst Appl Microbiol 2007;30:229–38.

42. Missen OP, Ram R, Mills SJ, Etschmann B, Reith F, Shuster J, et al. Love is in the Earth: a review of tellurium (bio)geochemistry in surface environments. Earth Sci Rev 2020;204:103150.

43. Molina RC, Burra R, Pérez-Donoso JM, Elías AO, Muñoz C, Montes RA, et al. Simple, fast, and sensitive method for quantification of tellurite in culture media. Appl Environ Microbiol 2010;76: 4901–4.

44. Philip N, Blengini GA. Towards better monitoring of technology critical elements in Europe: coupling of natural and anthropogenic cycles. Sci Total Environ 2018;613–614:569–78.

45. Philip N. Losses and environmental aspects of a byproduct metal: tellurium. Environ Chem 2019; 16:243–50.

46. Ollivier PRL, Bahrou AS, Marcus S, Cox T, Church TM, Hanson TE. Volatilization and precipitation of tellurium by aerobic, tellurite-resistant marine microbes. Appl Environ Microbiol 2008;74: 7163–73.

47. Ottosson L-G, Logg K, Ibstedt S, Sunnerhagen P, Käll M, Blomberg A, et al. Sulfate assimilation mediates tellurite reduction and toxicity in Saccharomyces cerevisiae. Eukaryot Cell 2010;9: 1635–47.

48. Ou X, Wang C, He M, Chen B, Hu B. Online simultaneous speciation of ultra-trace inorganic antimony and tellurium in environmental water by polymer monolithic capillary microextraction combined with inductively coupled plasma mass spectrometry. Spectrochim Acta B 2020;168: 105854.

49. Pasi A-E, Glänneskog H, Mark R, Foreman S.-J, Ekberg C. Tellurium behavior in the containment sump: dissolution, redox, and radiolysis effects. Nucl Technol 2021;207:217–27.

50. Perkins WT. Extreme selenium and tellurium contamination in soils – an eighty year-old industrial legacy surrounding a Ni refinery in the Swansea Valley. Sci Total Environ 2011;412–413:162–9.

51. Pinel-Raffaitin R, Pécheyran C, Amouroux D. New volatile selenium and tellurium species in fermentation gases produced by composting duck manure. Atmos Environ 2008;42:7786–94.

52. Qin HB, Takeichi Y, Nitani H, Terada Y, Takahashi Y. Tellurium distribution and speciation in contaminated soils from abandoned mine tailings: comparison with selenium. Environ Sci Technol 2017;51:6027–35.

53. Rajwade J, Paknikar K. Bioreduction of tellurite to elemental tellurium by Pseudomonas mendocina MCM B-180 and its practical application. Hydrometallurgy 2003;71:243–8.

54. Rathgeber C, Yurkova N, Stackebrandt E, Beatty JT, Yurkov V. Isolation of tellurite- and selenite-reducing bacteria from hydrothermal vents of the Juan de Fuca Ridge in the Pacific Ocean. Appl Environ Microbiol 2002;68:4613–22.

55. Ródenas-Torralba E, Cava-Montesinos P, Morales-Rubio A, Cervera ML, De La Guardia M. Multicommutation as anenvironmentally friendly analytical tool in the hydride generation atomic fluorescence determination of tellurium in milk. Anal Bioanal Chem 2004;379:83–9.

56. Su CK, Cheng TY, Sun YC. Selective chemical vaporization of exogenous tellurium for characterizing the time-dependent biodistribution and dissolution of quantum dots in living rats. J Anal Atom Spectrom 2015;30:426–34.
57. Taylor A. Biochemistry of tellurium. Biol Trace Elem Res 1996;55:231–9.
58. Vávrová S, Struhářnanská E, Turna J, Stuchlík S. Tellurium: a rare element with influence on prokaryotic and eukaryotic biological systems. Int J Mol Sci 2021;22:5924.
59. Wiklund JA, Kirk JL, Muir DCG, Carrier J, Gleason A, Yang F, et al. Widespread atmospheric tellurium contamination in Industrial and remote regions of Canada. Environmental Science & Technology 2018;52:6137–45.
60. Wojcieszek J, Szpunar J, Lobinski R. Speciation of technologically critical elements in the environment using chromatography with element and molecule specific detection. Trends Anal Chem 2018;104:42–53.
61. Yang G, Zheng J, Tagami K, Uchida S. Rapid and sensitive determination of tellurium in soil and plant samples by sector-field inductively coupled plasma mass spectrometry. Talanta 2013;116:181–7.
62. Zhang Q, Liu Y, He M, Bai M, Xu W, Zhao C. Ore prospecting model and targets for the Dashuigou tellurium deposit, Sichuan Province, China. Acta Geochim 2018;37:578–91.
63. Chiou KY, Manuel OK. Chalcogen element in snow: relation to emission source. Environ Sci Technol 1988;22:453–56.

Bubun Banerjee*, Aditi Sharma, Gurpreet Kaur, Anu Priya,
Manmeet Kaur and Arvind Singh

8 Latest developments on the synthesis of bioactive organotellurium scaffolds

Abstract: This review deals with the latest developments on the synthesis of biologically promising organotellurim scaffolds reported during last two decades.

Keywords: bioactivity; diaryl tellurides; organotellurium scaffolds; tellurium containing heterocycles.

8.1 Introduction

Organotellurium compounds found to possess a wide range of pharmacological activities. Figure 8.1 represents a glimpse of organotellurium scaffolds having a wide range of biological activities. Ammonium trichloro (dioxoethylene-*O,O'*)tellurate (AS101) (**I**) may be the most significant synthetic organotellurium compounds having a wide range of biological activities [1–11]. Yosef et al. [12] demonstrated that octa-O-*bis*-(R, R)-tartarate ditellurane (SAS) (**II**) can selectively inactivate cysteine proteases. Compound **III** and **IV** are used as antioxidants [13]. The compound **V** possesses significant cysteine protease inhibitor activity [14]. Compound **VI** exhibited as kidney-type glutaminase (KGA) inhibitors [15]. Very recently, in 2021, compound **VII** has found to possess antitumor activity [16]. Compound **VIII** showed glutathione peroxidase-like catalysis of the reduction of hydroperoxides [17]. Potent anticancer activity was exhibited by compound **XI** [18]. By following Passerini and Ugi reaction protocols, Shaaban et al. [19] synthesized a number of tellurium-containing scaffolds (**X–XV**) having significant cytotoxic activity. The new organotellurium compound **XVI** showed significant antioxidant activity and was found more effective than solketal ((2,2-dimethyl-1,3-dioxolan-4-yl)methanol) to inhibit induced lipid peroxidation [20]. Degrandi et al. [21] evaluated cytotoxicity, mutagenicity and genotoxicity of diphenyl ditelluride (**XVII**) in several biological models (Figure 8.2). On the other hand, organotellurium compounds were found effective for various nonconventional transformations which include dehalogenation of vicinal dibromides [22], metal insertion [23], silyltelluride-mediated radical coupling reactions [24] and oxidative addition [25]

*Corresponding author: Bubun Banerjee, Department of Chemistry, Akal University, TalwandiSabo, Bathinda, Punjab 151302, India, E-mail: banerjeebubun@gmail.com. https://orcid.org/0000-0001-7119-9377
Aditi Sharma, Gurpreet Kaur, Anu Priya, Manmeet Kaur and Arvind Singh, Department of Chemistry, Akal University, TalwandiSabo, Bathinda, Punjab 151302, India

As per De Gruyter's policy this article has previously been published in the journal Physical Sciences Reviews. Please cite as: B. Banerjee, A. Sharma, G. Kaur, A. Priya, M. Kaur and A. Singh "Latest developments on the synthesis of bioactive organotellurium scaffolds" *Physical Sciences Reviews* [Online] 2022. DOI: 10.1515/psr-2021-0115 | https://doi.org/10.1515/9783110735840-008

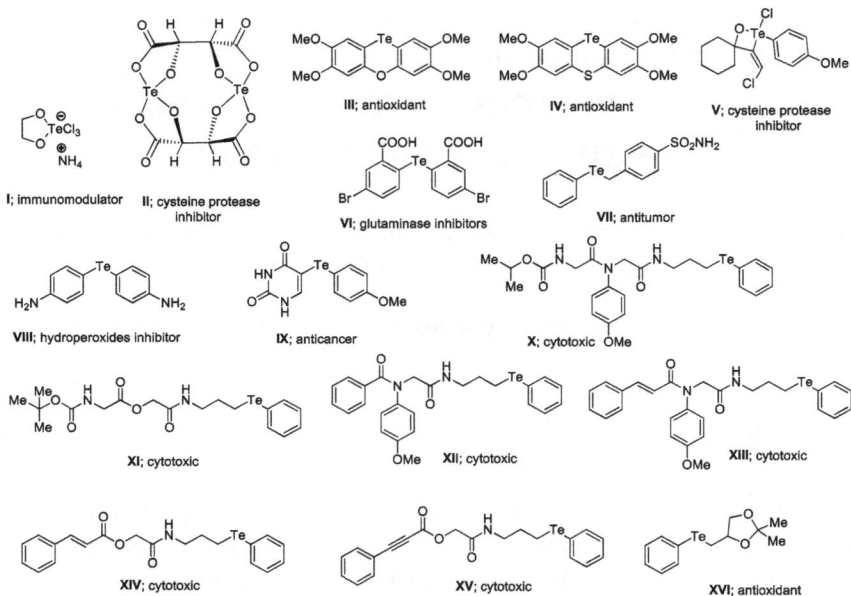

Figure 8.1: Glimpse of organotellurium scaffolds having a wide range of biological activities.

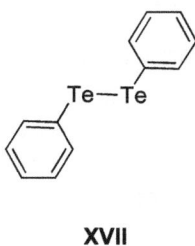

XVII

Figure 8.2: Diaryl ditelluride as cytotoxicity and genotoxicity agent.

reactions. Because of these beneficial activities, organotellurium compounds have been received immense importance since long back. In 1987, Hu et al. [26] synthesized novel organotellurium compounds through the aminotellurinylation of olefins. In 1992, Yoshida et al. [27] prepared a series of tellurium substituted novel O-heterocycles via the intramolecular ring closure reactions of unsaturated hydroxyl. Next year, a series of ortho-amino substituted novel organotellurium compounds was synthesized by Al-Rubaie et al. [28]. In 1996, Dabdoub et al. [29] synthesized ketene butyltelluro (phenylseleno)acetals via the Al/Te exchange reactions. Very next year, the same group [30] also synthesized structurally diverse organotellurium compounds through the formation of zirconated vinyl telluride intermediates. In 1999, Ferraz et al. [31] synthesized another series of biologically promising tellurium substituted novel O-heterocycles. In the same year, Ogawa et al. [32] carried out photo-irradiated regioselective thiotelluration of vinylcyclopropanes. Various research groups synthesized structurally diverse highly functionalized vinyl tellurides in good yields [33–36]. In continuation of our strong interest with organochalcogenides [37, 38] and bioactive scaffolds [39–45], in this review, we have summarized the latest developments on the synthesis of structurally diverse biologically promising organotellurium compounds reported during last two decades.

8.2 Synthesis of organotellurium scaffolds

8.2.1 Synthesis of diaryl tellurides

Andersson and co-researchers [46] studied lipid peroxidation of various diaryl tellurides and found promising efficacy (Figure 8.3). Roy et al. [47] prepared a heterogeneous catalyst by anchoring cupric acetate onto polystyrene functionalized with pyridine thiosemicarbazone ligand (Cu@PSTC). After well characterized the material by using FT-IR, SEM, EDAX, TGA and EPR techniques, they employed this material as an efficient catalysts for the synthesis of a wide variety of unsymmetrical diaryl tellurides (**3**) via the phenyl tellurylation of aryl boronic acids (**1**) using diphenyl ditelluride (**2**) in polyethylene glycol (PEG-600) at 80 °C (Figure 8.4). The catalyst was recovered and reused for six further runs with almost same catalytic efficiency.

The same synthesis was also carried out by Prof. Ranu and his research group [48] using magnetically separable nano-$CuFe_2O_4$ as catalyst in PEG-400 and DMSO mixture as solvent at 100 °C (Figure 8.5). Along with diaryl tellurides, several aryl-heteroaryl, aryl-styrenyl, aryl-alkenyl, aryl-allyl, aryl-alkyl and aryl-alkynyl tellurides were also synthesized in excellent yields under this developed protocol. The catalyst was recycled for eight runs without any significant loss in its catalytic activities. Wang et al. [49] reported another facile protocol for the efficient synthesis of unsymmetrical diaryl

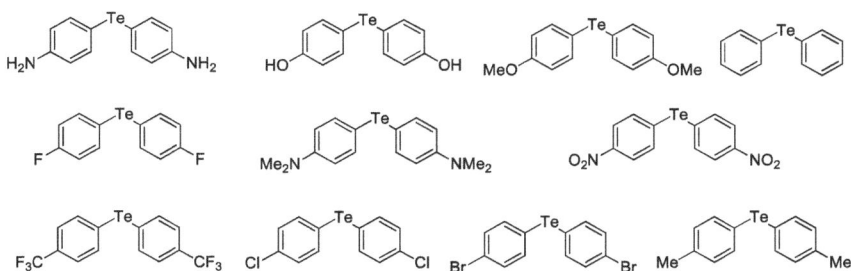

Figure 8.3: Glimpse of diaryl tellurides showed prominent lipid peroxidation efficacy.

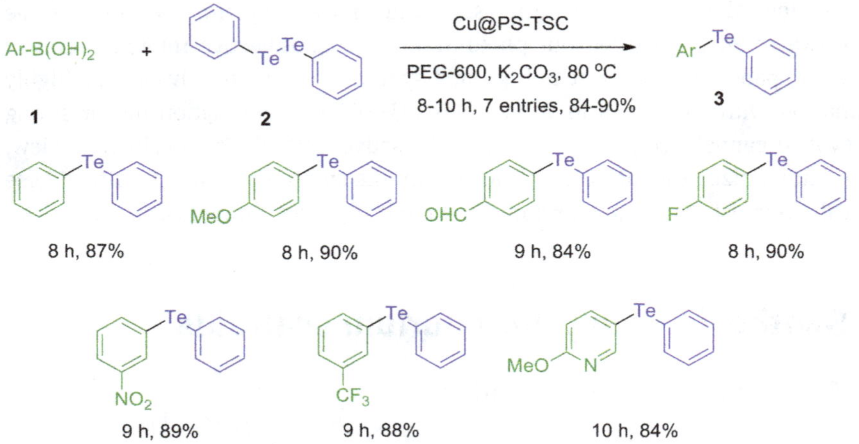

Figure 8.4: Synthesis of unsymmetrical diaryl tellurides from aryl boronic acids in PEG-600.

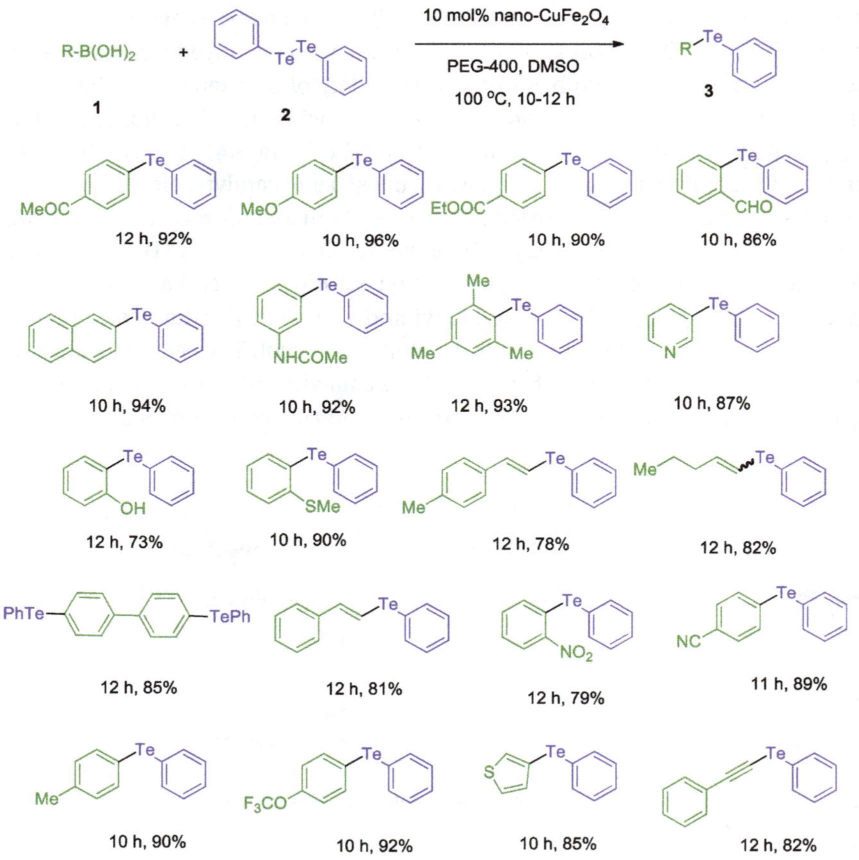

Figure 8.5: Synthesis of unsymmetrical diaryl or aryl-alkyl tellurides from various boronic acids.

Figure 8.6: Iron-catalyzed synthesis of unsymmetrical diaryl tellurides from various boronic acids.

tellurides (**3**) in excellent yields using a catalytic amount of iron powder as catalyst in DMSO at 130 °C (Figure 8.6).

Prof. Ranu and his research group [50] demonstrated microwave-assisted rapid and facile protocol for the efficient synthesis of a series of unsymmetrical diaryl tellurides (**3**) from the reactions of aryl diazonium fluoroborate (**4**) and diphenyl ditelluride (**2**) using zinc dust as catalyst in dimethyl carbonate (Figure 8.7). Under microwave-irradiated conditions, all the reactions were accomplished within just 35 min and afforded excellent yields. The same reactions required much higher times under conventional heating conditions at 80 °C. The same group [51] also developed a photo-catalyzed novel strategy for the direct conversion of aryl or heteroarylamines (**5**) to the corresponding diaryl tellurides (**3**). In the absence of any metal catalysts, the reaction underwent in neutral medium via in situ diazotization of amines (**5**) by using tert-butyl nitrite followed by phenyl tellurylation with diphenyl ditelluride (**2**) (Figure 8.8). Yamago et al. [52] prepared a series of structurally diverse aryl-alkyl tellurides (**3**) from the reactions of silyl tellurides (**6**) and organic halides (**7**) in acetonitrile (Figure 8.9). Silyl halides were generated as byproducts. All the reactions were preceded smoothly and afforded the corresponding products in good to excellent yields. The reaction couldn't take place in non-polar solvents such as toluene, THF etc. Under this optimized conditions, alkyl bromides were found most reactive than alkyl chlorides or iodides.

8.2.2 Synthesis of alkenyl vinyl tellurides

Okoronkwo et al. [53] synthesized a series of alkenyl vinyl tellurides (**10**) starting from various vinyl tellurides (**8**) and several alkynyl iodides (**9**) via cuprous iodide catalyzed carbon–tellurium cross-coupling reaction in dimethyl formamide at room temperature (Figure 8.10). Under the same reaction conditions, unsymmetrical dialkenyl tellurides (**12**) were also synthesized in moderate yields from the reactions of alkynyl tellurides (**11**) and alkynyl iodides (**9**) (Figure 8.11). Synthesized compounds were screened for

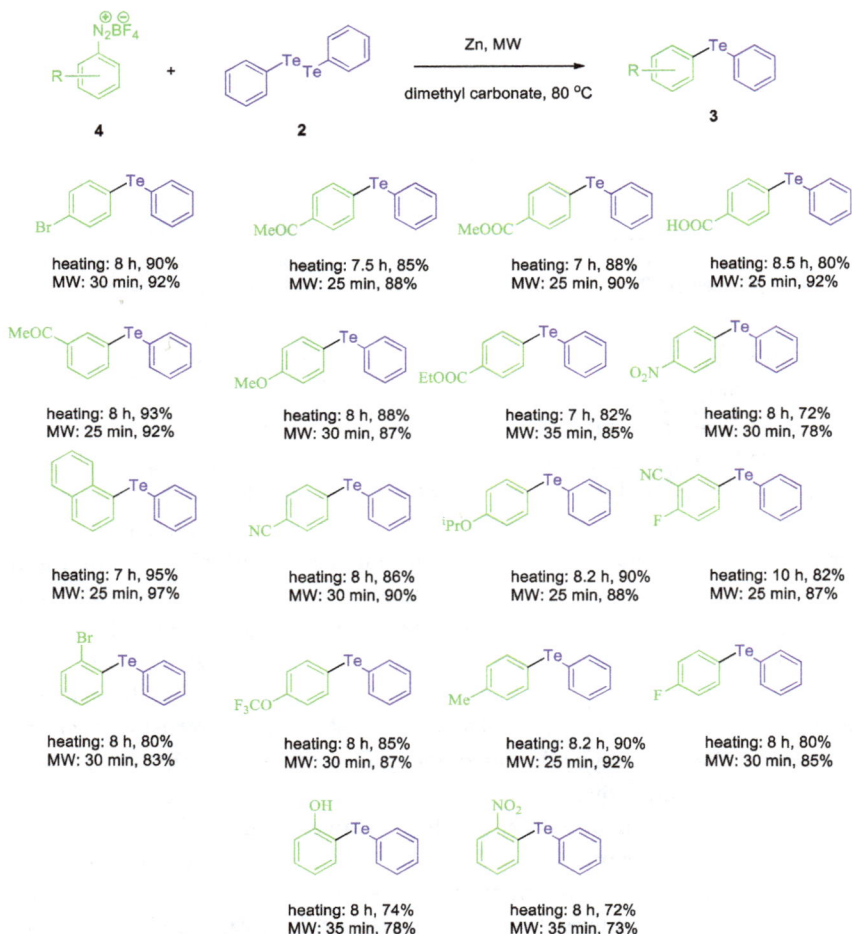

Figure 8.7: Microwave-assisted synthesis of unsymmetrical diaryl tellurides from aryl diazonium fluoroborate.

antidepressant activity and some of them showed prominent efficacy. Plausible mechanism is shown in Figure 8.12.

8.2.3 Synthesis of tellurodibenzoic acids

Hou et al. [54] prepared tellurodibenzoic acid derivatives (**15**) starting from 2-aminobenzoic acid derivatives (**13**). Diazotization of **1** afforded the corresponding

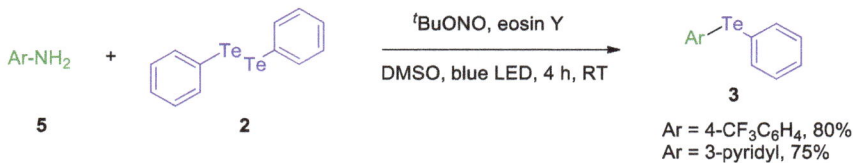

Figure 8.8: Photo-catalyzed synthesis of unsymmetrical diaryl tellurides from aryl amines.

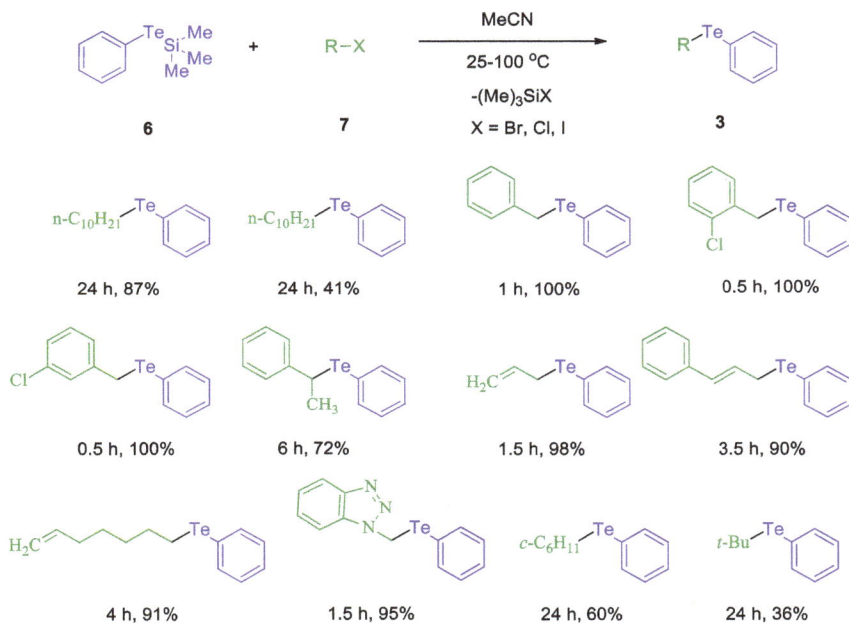

Figure 8.9: Catalyst-free synthesis of various tellurides from aryl silyl tellurides in acetonitrile.

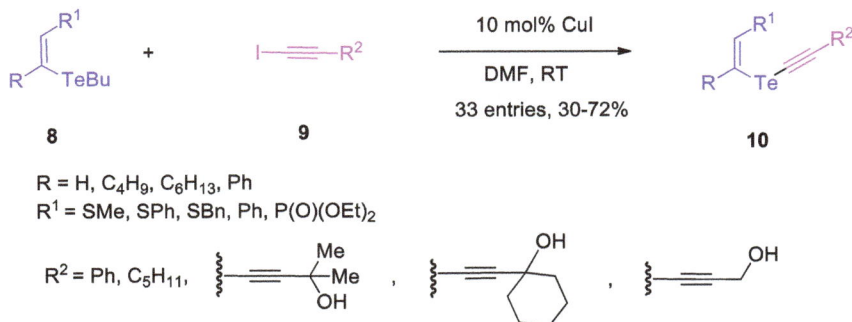

Figure 8.10: Synthesis of alkenyl vinyl tellurides from the reactions of vinyl tellurides and alkynyl iodides.

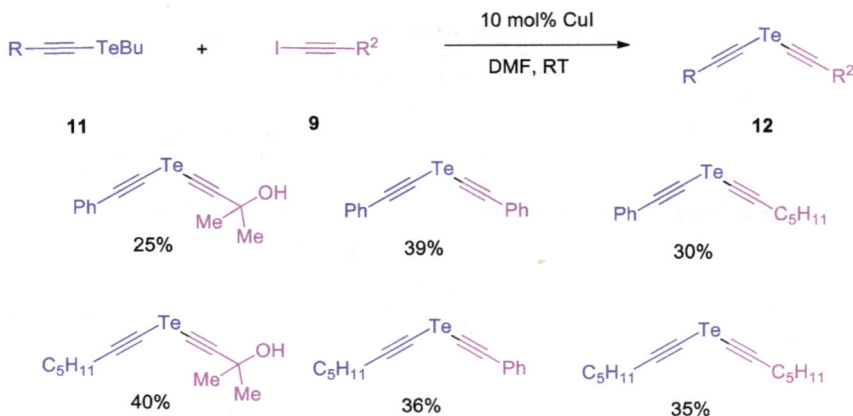

Figure 8.11: Synthesis of dialkenyl tellurides from the reactions of alkenyl tellurides and alkynyl iodides.

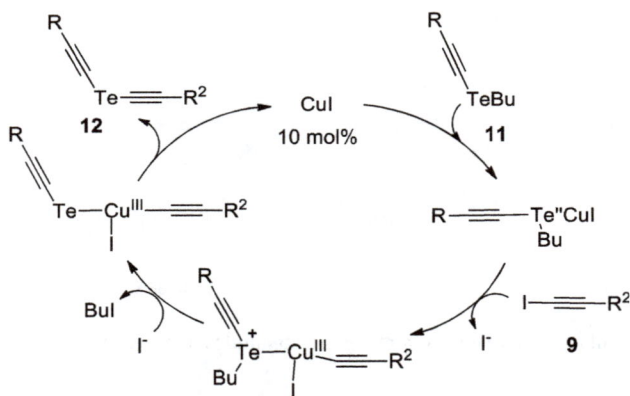

Figure 8.12: Plausible mechanism for the formation of dialkenyl tellurides.

intermediated **14**, which on reaction with sodium telluride produced the corresponding tellurodibenzoic acid derivatives (**15**) in moderate yields (Figure 8.13). Reaction of **15a** with excessive SOCl$_2$ generated the acyl chloride intermediate **16**, which on treating with alcohol (**17**) or amines (**18**) afforded the corresponding esters (**19**) or amides (**20**) in moderate yields (Figure 8.14). All the compounds were evaluated for the kidney type glutaminase (KGA) inhibitor activity.

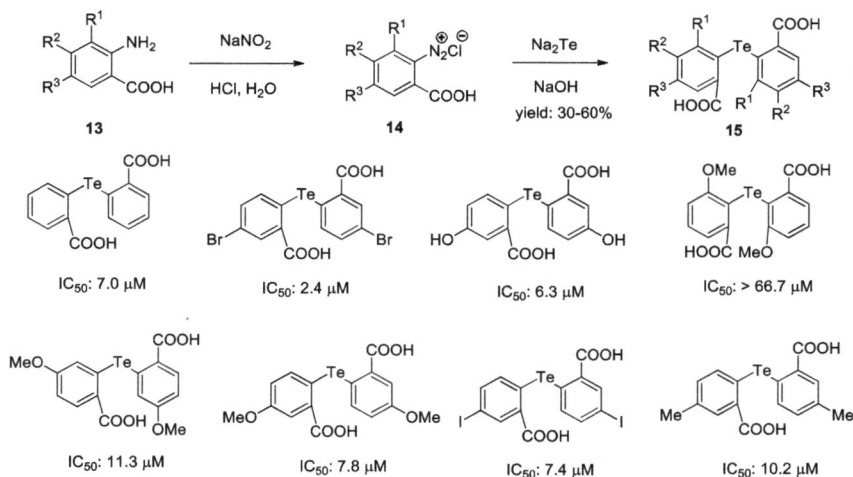

Figure 8.13: Synthesis of tellurodibenzoic acids starting from 2-aminobenzoic acid derivatives.

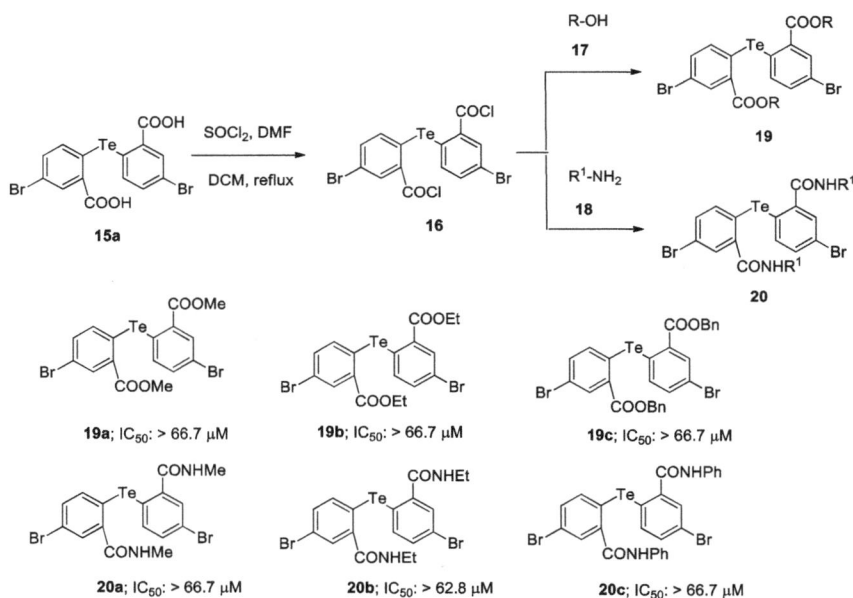

Figure 8.14: Synthesis of tellurodibenzoates and tellurodibenzamides.

8.2.4 Synthesis of 5-(aryltellanyl)pyrimidine-2,4(1H,3H)-diones

From the reaction between 2,4-bis(benzyloxy)-5-bromopyrimidine (**21**) and diaryl ditellurides (**2**) under basic medium, Engman et al. [55] synthesized 5-(aryltellanyl)pyrimidine-

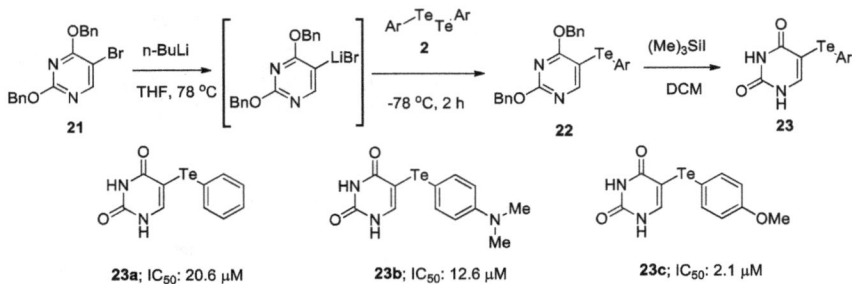

Figure 8.15: Synthesis of 5-(aryltellanyl)pyrimidine-2,4(1H,3H)-diones starting from 2,4-bis(benzyloxy)-5-bromopyrimidine.

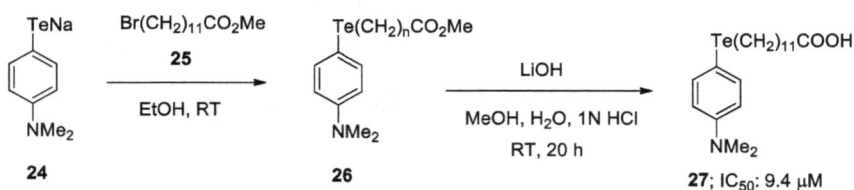

Figure 8.16: Synthesis of 12-((4-(dimethylamino)phenyl)tellanyl)dodecanoic acid as antitumor agent.

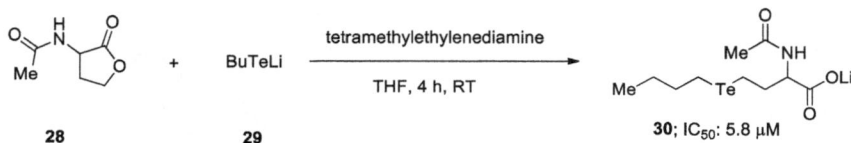

Figure 8.17: Synthesis of lithium 2-acetamido-4-(butyltellanyl)butanoate as antitumor agent.

2,4(1H,3H)-diones (**23**) through the formation of the compound **22** by following the pathway depicted in Figure 8.15. Among these synthesized compounds, compound **23c** showed significant antitumor activity by inhibiting the formation of thioredoxin reductase in the presence of thioredoxin and insulin. They have also prepared various structurally diverse novel organotellurium compounds which also showed the same antitumor activity with lower IC_{50} values (Figures 8.16–8.18). 12-((4-(Dimethylamino)phenyl)tellanyl)dodecanoic acid (**27**) was synthesized from the reaction of sodium 4-(dimethylamino)benzenetellurolate (**24**) and methyl 12-bromododecanoate (**25**) through the formation of compound 26 (Figure 8.16). Reactions of N-(2-oxotetrahydrofuran-3-yl)acetamide (**28**) and lithium butane-1-tellurolate (**29**) afforded the corresponding lithium 2-acetamido-4-(butyltellanyl)buta-noate (**30**) under basic conditions at room temperature (Figure 8.17). Starting from 2-(3-((4-(1,3-dioxoisoindolin-2-yl)butyl)amino)propyl)isoindoline-1,3-dione (**31**), N-(4-aminobutyl)-N-(3-aminopropyl)-4-(phenyltellanyl)butanamide (**33**) was synthesized by following the pathway depicted in Figure 8.18 through the formation of intermediate **32**.

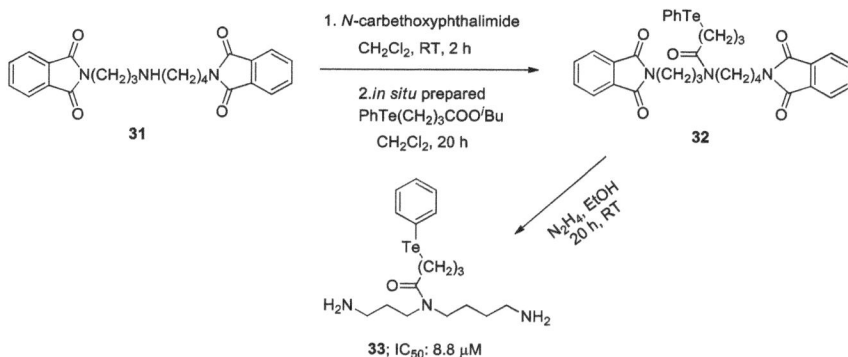

Figure 8.18: Synthesis of *N*-(4-aminobutyl)-*N*-(3-aminopropyl)-4-(phenyltellanyl)butanamide as antitumor agent.

8.2.5 Synthesis of *Te,N*-heterocycles

8.2.5.1 Synthesis of 1,3-benzotellurazoles

Smith et al. [56] synthesized 1,3-benzotellurazoles (**36**) from the reactions of bis (2-aminophenyl)ditelluride (**34**) and various isothiocyanates (**35**) in the presence of metallic mercury and rongalite in dimethyl formamide under refluxed conditions (Figure 8.19). Under this developed conditions various 2-benzoylamino-1,3-benzotellurazoles (**36a–36f**) as well as 2-arylamino-1,3-benzotellurazoles (36 g) were synthesized with moderate yields. It was proposed that the synthesis of 1,3-benzotellurazoles (**36**) was achieved through the formations of intermediate **37** and **38**. Plausible mechanism is depicted in Figure 8.20.

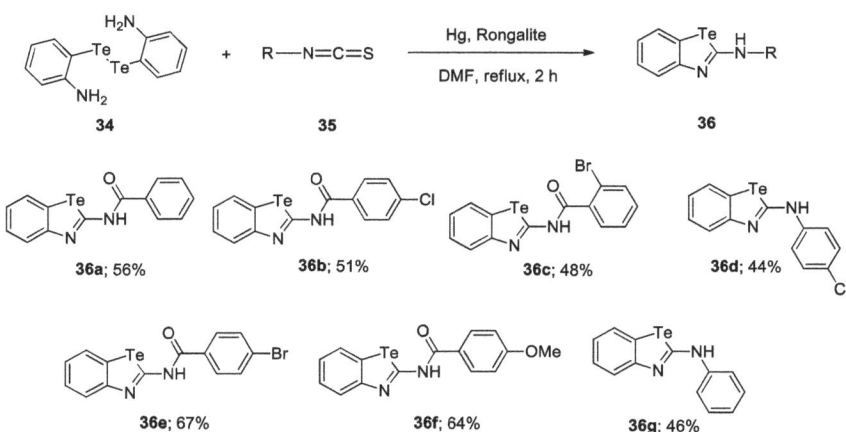

Figure 8.19: Synthesis of 1,3-benzotellurazoles starting from *bis*(2-aminophenyl) ditelluride.

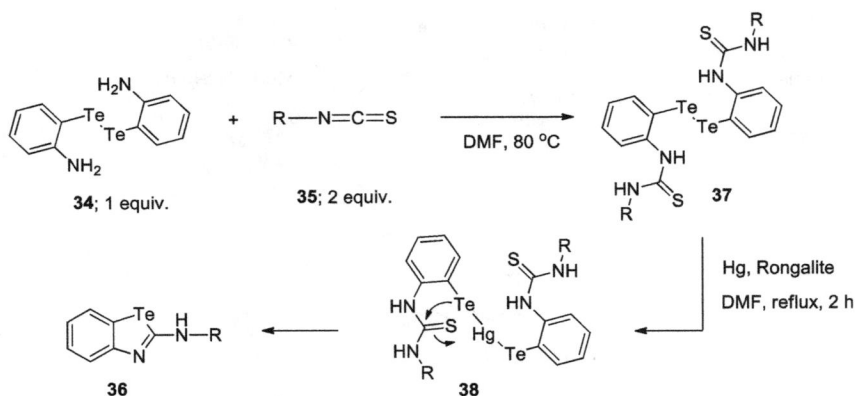

Figure 8.20: Plausible mechanism for the synthesis of 1,3-benzotellurazoles.

8.2.6 Synthesis Te,O-heterocycles

8.2.6.1 Synthesis of ammonium trichloro(dioxoethylene-O,O)tellurate

Trichloro (dioxoethylene-O,O)tellurate possess a wide range of biological efficacies including cysteine protease inhibitor activity [57]. Under microwave-assisted conditions, Vázquez-Tato et al. [58] synthesized ammonium trichloro (dioxoethylene-O,O) tellurate (**I**) from the reactions of Te (**IV**) chloride (**39**), ethylene glycol (**40**) and ammonium chloride (Figure 8.21). The synthesized compound showed prominent immunomodulator and antitumor activities.

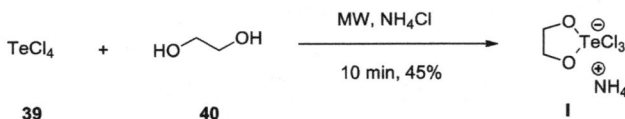

Figure 8.21: Microwave-assisted synthesis of ammonium trichloro (dioxoethylene-O,O)tellurate.

8.2.7 Synthesis of organotellurium substituted-heterocycles

8.2.7.1 Synthesis of 2,2-dimethyl-4-((aryltellanyl)methyl)-1,3-dioxolanes

Nobre et al. [59] synthesized 2,2-dimethyl-4-((aryltellanyl)methyl)-1,3-dioxolane (**42**) from the reactions of (2,2-dimethyl-1,3-dioxolan-4-yl)methyl 4-methylbenzenesulfonate (**41**) and diaryl ditellane (**2**) in the presence of sodium borohydride in polyethylene glycol at 50 °C (Figure 8.22). Antioxidant efficacy of the synthesized compounds was evaluated against 2,2-diphenyl-1-picrylhydrazyl (DPPH) assay.

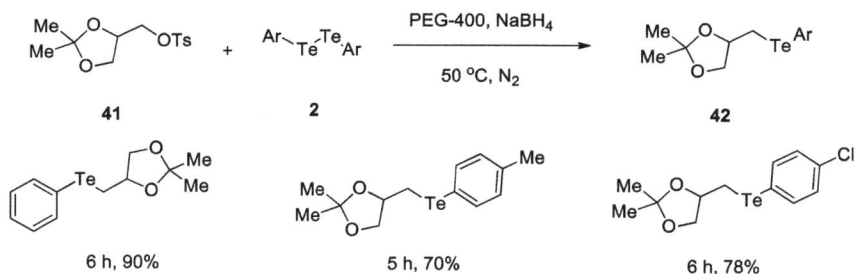

Figure 8.22: Synthesis of 2,2-dimethyl-4-((aryltellanyl)methyl)-1,3-dioxolanes in polyethylene glycol at 50 °C.

8.2.7.2 Synthesis of β-organotellurobutenolides

Xu et al. [60] synthesized β-organotellurobutenolides (**45**) from the reactions of substituted α-allenoic acids (**43**) and PhTeCl (**44**) as the tellurenylating agent (Figure 8.23). PhTeCl was generated in situ by the reactions of diphenyl ditelluride (**2**) and sulfuryl chloride in dry MeCN at room temperature under inert atmosphere.

8.2.7.3 Synthesis of organotellurium functionalized chrysin derivatives

Fonseca et al. [61] prepared organotellurium functionalized chrysin derivative (**49**) (Figure 8.24). At first, by following a reported method [62], the key intermediate 7-(2-bromoethoxy)-5-hydroxy-2-phenyl-4H-chromen-4-one (**48**) was synthesized by the reactions of chrysin (**46**) and 1,2-dibromoethane (**47**). Later on, the targeted compound was synthesized in good yields from the reaction of the compound 48 and diphenyl ditelluride (**2**) under refluxed conditions. Antioxidant activity of the synthesized compound was evaluated and it showed moderate efficacy.

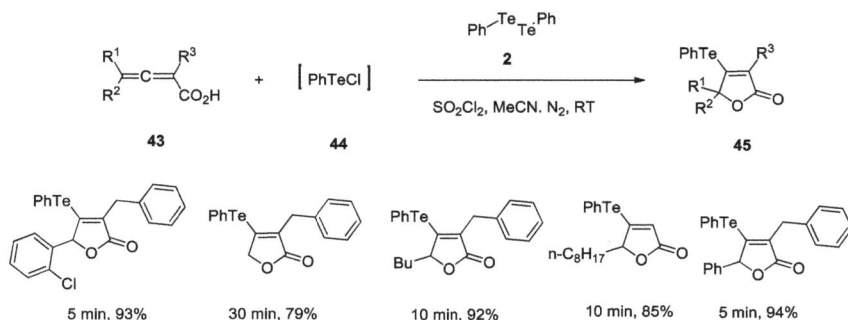

Figure 8.23: Synthesis of β-organotellurobutenolides by using *in situ* generated PhTeCl as the tellurenylating agent.

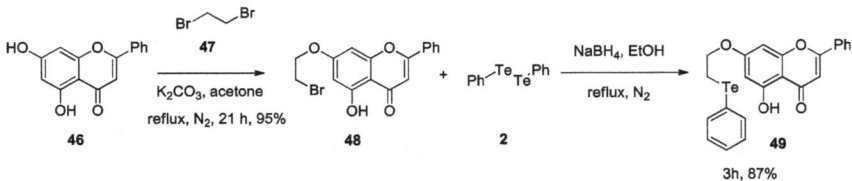

Figure 8.24: Synthesis of 5-hydroxy-2-phenyl-7-(2-(aryltellanyl)ethoxy)-4*H*-chromen-4-ones using diaryl ditellurides.

8.2.7.4 Synthesis of 3-(phenyltellanyl)octahydrobenzofuran

Starting from 7-oxabicyclo [4.1.0]heptanes (**50**), Ericsson and Engma [63] synthesized 3-(phenyltellanyl)octahydrobenzofuran (**54**) in good yield (Figure 8.25). At the first step, the ring-opening of epoxide (**50**) was carried out by using diphenyl ditelluride (**2**) to generate the corresponding compound **51**. In the next step, O-allylation of the compound **51** by using allyl bromide (**52**) provided the compound **53** which undergoes group-transfer cyclization under microwave-assisted conditions and afforded the corresponding 3-(phenyltellanyl)octahydro-benzofuran (**54**) within just 3 min.

8.2.7.5 Synthesis of bis(chloro-bis(2-thienyl)tellurium)oxide

Starting from bis(thienyl)tellurium dichloride (**55**), Singh et al. [64] demonstrated the synthesis of bis(chloro-bis(2-thienyl)tellurium)oxide (**57**) via the dimerization of the compound **56** in the presence of silver cyanide (Figure 8.26).

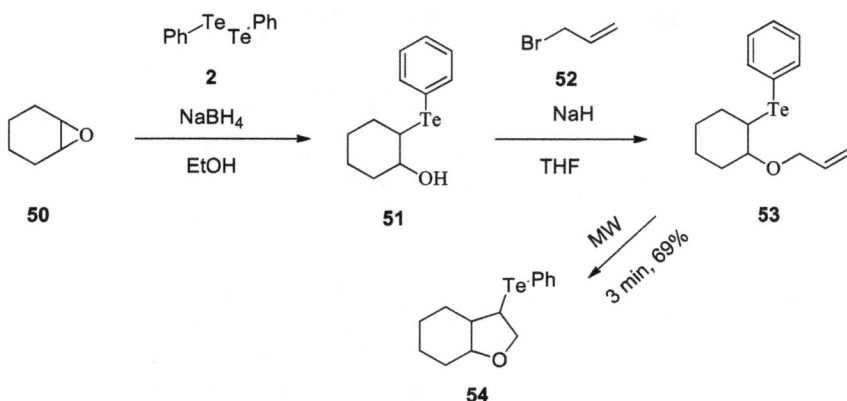

Figure 8.25: Synthesis of 3-(phenyltellanyl)octahydrobenzofuran starting from 7-oxabicyclo[4.1.0] heptanes.

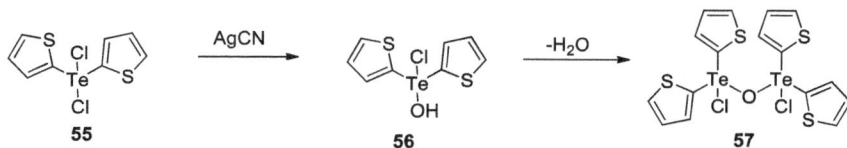

Figure 8.26: Synthesis of *bis*(chloro-*bis*(2-thienyl)tellurium)oxide starting from *bis*(thienyl)tellurium dichloride.

8.3 Conclusions

Various organotelluride derivatives have been found to possess significant biological efficacies. In this review, we have summarized latest developments on the synthesis of structurally diverse biologically promising organotellurium compounds which include unsymmetrical diaryl or aryl-alkyl tellurides, alkenyl vinyl tellurides, tellurodibenzoic acids, tellurodibenzoates, tellurodibenzamides, 1,3-benzotellurazoles, β-organo-tellurobutenolides and various tellurium substituted heterocycles.

Acknowledgments: Authors are thankful to Prof. Gurmail Singh, Vice-Chancellor, Akal University for his wholehearted encouragement and support. BB is grateful to Akal University and Kalgidhar Trust, Baru Sahib, India for providing laboratory facilities.

References

1. Albeck M, Tamari T, Sredni B. Synthesis and properties of ammonium trichloro(dioxyethylene-O,O')tellurate (AS-101). A new immunomodulating compound. Synthesis 1989;1989:635–36.
2. Sredni B, Caspi RR, Klein A, Kalechman Y, Danziger Y, Ben Yaakov M, et al. A new immunomodulating compound (AS-101) with potential therapeutic application. Nature 1987;330: 173–6.
3. Sredni B, Xu RH, Albeck M, Gafter U, Gal R, Shani A, et al. The protective role of the immunomodulator AS101 against chemotherapy-induced alopecia studies on human and animal models. Int J Cancer 1996;65:97–103.
4. Sredni B, Gal R, Cohen IJ, Dazard JE, Givol D, Gafter U, et al. Hair growth induction by the tellurium immunomodulator AS101: association with delayed terminal differentiation of follicular keratinocytes and ras-dependent up-regulation of KGF expression. Faseb J 2004;18:400–2.
5. Rosenblatt-Bin H, Kalechman Y, Vonsover A, Xu RH, Da JP, Shalit F, et al. The immunomodulator AS101 restores TH1Type of response suppressed by *Babesia rodhainiin* BALB/c mice. Cell Immunol 1998;184:12–25.
6. Kalechman Y, Gafter U, Da JP, Albeck M, Alarcon-Segovia D, Sredni B. Delay in the onset of systemic lupus erythematosus following treatment with the immunomodulator AS101: association with IL-10 inhibition and increase in TNF-alpha levels. J Immunol 1997;159:2658–67.

7. Kalechman Y, Gafter U, Gal R, Rushkin G, Yan D, Albeck M, et al. Anti-IL-10 therapeutic strategy using the immunomodulator AS101 in protecting mice from sepsis-induced death: dependence on timing of immunomodulating intervention. J Immunol 2002;169:384–92.

8. Kalechman Y, Gafter U, Weinstein T, Chagnac A, Freidkin I, Tobar A, et al. Inhibition of interleukin-10 by the immunomodulator AS101 reduces mesangial cell proliferation in experimental mesangioproliferative glomerulonephritis association with dephosphorylation of stat3. J Biol Chem 2004;279:24724–32.

9. Sredni B, Geffen R, Duan W, Albeck M, Shalit S, Lander HM, et al. Multifunctional tellurium molecule protects and restores dopaminergic neurons in Parkinson's disease models. Faseb J 2007;21:1870–83.

10. Sredni B, Albeck M, Tichler T, Shani A, Shapira J, Bruderman I, et al. Bone marrow-sparing and prevention of alopecia by AS101 in non-small-cell lung cancer patients treated with carboplatin and etoposide. J Clin Oncol 1995;13:2342–53.

11. Sredni B, Shani A, Catane R, Kaufman B, Strassmann G, Albeck M, et al. Predominance of TH1 Response in tumor-bearing mice and cancer patients treated with AS 101. J Natl Cancer Inst 1996; 88:1276–84.

12. Yosef S, Brodsky M, Sredni B, Albeck A, Albeck M. Octa-O-bis-(R,R)-Tartarate Ditellurane (SAS)-A novel bioactive organotellurium(iv) compound: synthesis, characterization, and protease inhibitory activity. Chem Med Chem 2007;2:1601–6.

13. Cunha RLOR, Gouvea IE, Juliano L. A glimpse on biological activities of tellurium compounds. An Acad Bras Cienc 2009;81:393–407.

14. Cunha RLOR, Urano ME, Chagas JR, Almeida PC, Bincoletto C, Tersariol ILS, et al. Tellurium-based cysteine protease inhibitors: evaluation of novel organotellurium(IV) compounds as inhibitors of human cathepsin B. Bioorg Med Chem Lett 2005;15:755–60.

15. Hou W, Zhou Y, Rui J, Bai R, Bhasin AKK, Ruan BH. Design and synthesis of novel tellurodibenzoic acid compounds as kidney-type glutaminase (KGA) inhibitors. Bioorg Med Chem Lett 2019;29: 1673–6.

16. Angeli A, Pinteala M, Maier SS, Toti A, Di Cesare Mannelli L, Ghelardini C, et al. Tellurides bearing benzensulfonamide as carbonic anhydrase inhibitors with potent antitumor activity. Bioorg Med Chem Lett 2021;45:128147.

17. Andersson C-M, Hallberg A, Brattsand R, Cotgreave IA, Engman L, Persson J. Glutathione peroxidase-like activity of diaryl tellurides. Bioorg Med Chem Lett 1993;3:2553–8.

18. Engman L, Al-Maharik N, McNaughton M, Birmingham A, Powis G. Thioredoxin reductase and cancer cell growth inhibition by organotellurium compounds that could be selectively incorporated into tumor cells. Bioorg Med Chem 2003;11:5091–100.

19. Shaaban S, Sasse F, Burkholz T, Jacob C. Sulfur, selenium and tellurium pseudopeptides: synthesis and biological evaluation. Bioorg Med Chem 2014;22:3610–9.

20. Nobre PC, Borges EL, Silva CM, Casaril AM, Martinez DM, Lenardão EJ, et al. Organochalcogen compounds from glycerol: synthesis of new antioxidants. Bioorg Med Chem 2014;22:6242–9.

21. Degrandi TH, de Oliveira IM, d' Almeida GS, Garcia CRL, Villela IV, Guecheva TN, et al. Evaluation of the cytotoxicity, genotoxicity and mutagenicity of diphenyl ditelluride in several biological models. Mutagenesis 2010;25:257–69.

22. Leonard KA, Zhou F, Detty MR. Chalcogen(IV)–chalcogen(II) redox cycles. 1. Halogenation of organic substrates with dihaloselenium(IV) and -tellurium(IV) derivatives. Dehalogenation of vicinal dibromides with diaryl tellurides. Organometallics 1996;15:4285–92.

23. Peppe C, de Andrade FM, Uhl W. On the reactions of the tetraindium cluster $In_4[C(SiMe_3)_3]_4$: insertion of the monomeric fragments $InC(SiMe_3)_3$ into chalcogen–chalcogen bonds. J Organomet Chem 2009;694:1918–21.

24. Yamago S, Miyazoe H, Nakayama T, Miyoshi M, Yoshida J. A diversity-oriented synthesis of α-amino acid derivatives by a silyltelluride-mediated radical coupling reaction of imines and isonitriles. Angew Chem Int Ed 2003;42:102–17.
25. Han L-B, Choi N, Tanaka M. The first example of facile oxidative addition of carbon–tellurium bonds to zero-valent Pt, Pd, and Ni complexes. J Am Chem Soc 1997;119:1795–6.
26. Xing HN, Yoshia A, Tetsuo O, Fumio O. Aminotellurinylation of olefins with benzenetellurinyl acetate and ethyl carbamate. Chem Lett 1987;16:1327–30.
27. Yoshida M, Suzuki T, Kamigata N. Novel preparation of highly electrophilic species for benzenetellurenylation or benzenesulfenylation by nitrobenzenesulfonyl peroxide in combination with ditelluride or disulfide. Application to intramolecular ring closures. J Org Chem 1992;57: 383–6.
28. Al-Rubaie AZ, Al-Salim NI, Al-Jadaan SAN. Synthesis and characterization of new organotellurium compounds containing an ortho-amino group. J Organomet Chem 1993;443:67–70.
29. Dabdoub MJ, Cassol TM, Barbosa SL. Synthesis of Ketene (phenylseleno)acetals by the Al/Te exchange reaction. Tetrahedron Lett 1996;37:831–4.
30. Dabdoub MJ, Begnini ML, Guerrero PG. Hydrozirconation of acetylenic chalcogenides. Synthesis and reactions of zirconated vinyl chalcogenide intermediates. Tetrahedron 1998;54:2371–400.
31. Ferraz HMC, Sano MK, Scalfo AC. Tellurium and iodine promoted cyclofunctionalization of alkenyl substituted β-keto esters. Synlett 1999;1999:567–8.
32. Ogawa A, Ogawa I, Obayashi R, Umezu K, Doi M, Hirao T. Highly selective thioselenation of vinylcyclopropanes with a $(PhS)_2$-$(PhSe)_2$ binary system and its application to thiotelluration. J Org Chem 1999;64:86–92.
33. Dabdoub MJ, Dabdoub VB, Pereira MA. Hydrochalcogenation of phenylthioacetylenes. Synthesis of mixed (Z)-trisubstituted 1,2-bis(organylchalcogeno)-1-alkenes. Tetrahedron Lett 2001;42: 1595–7.
34. Perin G, Jacob RG, Dutra LG, de Azambuja F, dos Santos GFF, Lenardão EJ. Addition of chalcogenolate anions to terminal alkynes using microwave and solvent-free conditions: easy access to bis-organochalcogen alkenes. Tetrahedron Lett 2006;47:935–8.
35. Princival C, Dos Santosa AA, Comasseto JV. Solventless and mild procedure to prepare organotellurium(IV) compounds under microwave irradiation. J Braz Chem Soc 2015;26:832–6.
36. Huang X, Liang CG, Xu Q, He QW. Alkyne-based, highly stereo- and regioselective synthesis of stereodefined functionalized vinyl tellurides. J Org Chem 2001;66:74–80.
37. Banerjee B, Koketsu M. Recent developments in the synthesis of biologically relevant selenium-containing scaffolds. Coord Chem Rev 2017;339:104–27.
38. Banerjee B, Ranu BC. 9. Selenoamides, selenazadienes, and selenocarbonyls in organic synthesis. In: Ranu BC, Banerjee B, editors Organoselenium Chemistry. Berlin, Boston: De Gruyter; 2020. pp. 347–80.
39. Kaur G, Devi P, Thakur S, Kumar A, Chandel R, Banerjee B. Magnetically separable transition metal ferrites: versatile heterogeneous nano-catalysts for the synthesis of diverse bioactive heterocyclic. ChemistrySelect 2019;4:2181–99.
40. Kaur G, Devi M, Kumari A, Devi R, Banerjee B. One-pot pseudo five component synthesis of biologically relevant 1,2,6-Triaryl-4-arylamino-piperidine-3-ene-3- carboxylates: a decade update. ChemistrySelect 2018;3:9892–910.
41. Banerjee B. [Bmim]BF_4: a versatile ionic liquid for the synthesis of diverse bioactive heterocyclic. ChemistrySelect 2017;2:8362–76.
42. Banerjee B. Bismuth(III) triflate: an efficient catalyst for the synthesis of diverse biologically relevant heterocyclic. ChemistrySelect 2017;2:6744–57.

43. Banerjee B. Recent developments on organo-bicyclo-bases catalyzed multicomponent synthesis of biologically relevant heterocyclic. Curr Org Chem 2018;22:208–33.

44. Banik BK, Banerjee B, Kaur G, Saroch S, Kumar R. Tetrabutylammonium Bromide (TBAB) catalyzed synthesis of bioactive heterocyclic. Molecules 2020;25:5918.

45. Banerjee B. Recent developments on ultrasound-assisted one-pot multicomponent synthesis of biologically relevant heterocyclic. Ultrason Sonochem 2017;35:15–35.

46. Andersson CM, Brattsand R, Hallberg A, Engman L, Persson J, Moldéus P, et al. Diaryl tellurides as inhibitors of lipid peroxidation in biological and chemical systems. Free Radic Res 1994;20: 401–10.

47. Roy S, Chatterjee T, Islam SM. Solvent selective phenyl selenylation and phenyl tellurylation of aryl boronic acids catalyzed by Cu(II) grafted functionalized polystyrene. Tetrahedron Lett 2015;56: 779–83.

48. Kundu D, Mukherjee N, Ranu BC. A general and green procedure for the synthesis of organochalcogenides by $CuFe_2O_4$ nanoparticle catalyzed coupling of organoboronic acids and dichalcogenides in PEG-400. RSC Adv 2013;3:117–25.

49. Wang M, Ren K, Wang L. Iron-catalyzed ligand-free carbon-selenium (or tellurium) coupling of arylboronic acids with diselenides and ditellurides. Adv Synth Catal 2009;351:1586–94.

50. Kundu D, Ahammed S, Ranu BC. Microwave-assisted reaction of aryl diazonium fluoroborate and diaryl dichalcogenides in dimethyl carbonate: a general procedure for the synthesis of unsymmetrical diaryl chalcogenides. Green Chem 2012;14:2024–30.

51. Kundu D, Ahammed S, Ranu BC. Visible light photocatalyzed direct conversion of aryl-/ heteroarylamines to selenides at room temperature. Org Lett 2014;16:1814–7.

52. Yamago S, Iida K, Yoshida J-i. A new, practical synthesis of organotellurium compounds from organic halides and silyl tellurides. Remarkable effects of polar solvents and leaving groups. Tetrahedron Lett 2001;42:5061–4.

53. Okoronkwo AE, Godoi B, Schumacher RF, Neto JSS, Luchese C, Prigol M, et al. Csp^3-tellurium copper cross-coupling: synthesis of alkynyl tellurides a novel class of antidepressive-like compounds. Tetrahedron Lett 2009;50:909–15.

54. Hou W, Zhou Y, Rui J, Bai R, Bhasin AKK, Ruan BH. Design and synthesis of novel tellurodibenzoic acid compounds as kidney-type glutaminase (KGA) inhibitors. Bioorg Med Chem 2019;29:1673–6.

55. Engman L, Al-Maharik N, McNaughton M, Birminggham A, Powis G. Thioredoxin reductase and cancer cell growth inhibition by organotellurium compounds that could be selectively incorporated into tumor cells. Bioorg Med Chem 2003;11:5091–100.

56. Smith WE, Franklin DV, Goutierrez KL, Fronczek FR, Mautner FA, Junk T. Organotellurium chemistry: synthesis and properties of 2-acylamino- and 2-arylamino-1,3-benzotellurazoles. Am J Heterocyclic Chem 2019;5:49–54.

57. Brodsky M, Yosef S, Galit R, Albeck M, Longo DL, Albeck A, et al. The synthetic tellurium compound, AS101, Is a novel inhibitor of IL-1β converting enzyme. J Interferon Cytokine Res 2007;27:453–62.

58. Vázquez-Tato MP, Mena-Menéndez A, Feás X, Seijas JA. Novel microwave-assisted synthesis of the immunomodulator organotellurium compound ammonium trichloro(dioxoethylene-O,O')tellurate (AS101). Int J Mol Sci 2014;15:3287–98.

59. Nobre PC, Borges EL, Silva CM, Casaril AM, Martinez DM, Lenardão EJ, et al. Organochalcogen compounds from glycerol: synthesis of new antioxidants. Bioorg Med Chem 2014;22:6242–9.

60. Xu Q, Huang X, Yuan J. Facile synthesis of β-organotellurobutenolides via electrophilic tellurolactonization of α-allenoic acids. J Org Chem 2005;70:6948–51.

61. Fonseca SF, Lima DB, Alves D, Jacob RG, Perin G, Lenardãoa EJ, et al. Synthesis, characterization and antioxidant activity of organoselenium and organotellurium compounds derivatives of chrysin. New J Chem 2015;39:3043–50.

62. Hu K, Wang W, Cheng H, Pan S, Ren J. Synthesis and cytotoxicity of novel chrysin derivatives. J Med Chem Res 2011;20:838–46.
63. Ericsson C, Engma L. Microwave-Assisted group-transfer cyclization of organotellurium compounds. J Org Chem 2004;69:5143–6.
64. Singh P, Chauhan AKS, Butcher RJ, Duthie A. Synthesis and structural aspects of 1-naphthyltellurium(IV) trichloride (1), bis(mesityl)tellurium(IV) dichloride (2) and *bis*(chlorobis(2-thiophenyl)tellurium)oxide (3). Polyhedron 2013;62:227–33.

Anjaly Das, Aparna Das and Bimal Krishna Banik*

9 Tellurium-based chemical sensors

Abstract: The various tellurium-based chemical sensors are described. This article focuses on four types of Tellurium sensors such as CdTe quantum dots-based sensor, Te thin films-based sensor, Te nanostructures or nanoparticles-based sensor, and TeO_2-based sensor.

Keywords: luminescence; nanostructures; semiconductors; sensors; tellurium.

9.1 Introduction

Tellurium (Te) is a semiconductor having low band gap. Te is practicable for wide range of applications including transistors, optical recording, strain sensitive devices, gas sensors, thermoelectric devices, and infrared detectors [1, 2]. Tellurium is tractable to the making of one-dimensional nanostructures because of the anisotropic crystal system, for example as nanotubes and nanowires.

Te-based materials have shown remarkable sensing properties and it is a good option for gas sensor application, operating at room temperature. Te gas sensors got considerable attention, as they are sensitive, compact, economical, and consume low power. In this article, we disclose the fabrication and sensing properties of four different Te-based sensors.

9.2 CdTe quantum dots-based sensor

The luminescent semiconductor nanocrystals quantum dots (QDs) have several excellent characteristics compared to the conventional fluorescent materials. The features include broad absorption, strong fluorescence, narrow and symmetric emission bands, feasibility for surface modification, large surface area, outstanding chemical stability, high photostability, high resistance to photo-bleaching, and biocompatibility.

***Corresponding author: Bimal Krishna Banik**, Department of Mathematics and Natural Sciences, College of Sciences and Human Studies, Prince Mohammad Bin Fahd University, Al Khobar 31952, Kingdom of Saudi Arabia, E-mail: bimalbanik10@gmail.com
Anjaly Das, National Institute of Electronics & Information Technology, Calicut 673601, Kerala, India
Aparna Das, Department of Mathematics and Natural Sciences, College of Sciences and Human Studies, Prince Mohammad Bin Fahd University, Al Khobar 31952, Kingdom of Saudi Arabia. https://orcid.org/0000-0002-2502-9446

As per De Gruyter's policy this article has previously been published in the journal Physical Sciences Reviews. Please cite as: A. Das, A. Das and B. K. Banik "Tellurium-based chemical sensors" *Physical Sciences Reviews* [Online] 2022. DOI: 10.1515/psr-2021-0116 | https://doi.org/10.1515/9783110735840-009

9.2.1 Ascorbic acid detection

Ascorbic acid (Vitamin C, AA) plays a critical role in several pathological and physiological procedures in the human [3, 4]. In the human body, it is a vitally necessary biomolecule. Ascorbic acid can be utilized as an enzyme complement, cofactor, or a combination of some biochemical actions to forbid human disorders [5], also ascorbic acid has good antioxidant properties. In addition, to raise the iron absorption in the human, it could cut down ferric ions into ferrous ions [6]. Thus, the deficiency of ascorbic acid can regulate biological operations and make several disorders such as cataracts, cancer, scurvy cardiovascular disease, aging and epidermal atrophy [7, 8]. Consequently, a quantitative ascorbic acid determination is very crucial for the rapid and early sensing of these diseases. Currently available methods for the spotting of ascorbic acid includes electrochemistry, chromatography [9], capillary electrophoresis [10], fluorescence [11, 12], and electrochemiluminescence [13]. A solid-state Electrochemiluminescence (ECL) sensor established on CdTe quantum dots and nitrogen-doped graphene (NG) was reported [14]. The sensor displayed long-time electrochemiluminescence stability along with high electrochemiluminescence intensity. Together with the electrochemiluminescence mechanism, the factors effecting on sensor's electrochemiluminescence emission were investigated in detail. The sensor was used for the detection of ascorbic acid and the detection was based on the electrochemiluminescence quenching of ascorbic acid. The study showed that due to the electrochemiluminescence quenching of the ascorbic acid, the electrochemiluminescence intensity of the sensor diminished in a gradual manner along with the increment of ascorbic acid concentrations (Figure 9.1A). In the range from 0.04 to 200 µM, the sensor showed a linear response as shown in Figure 9.1B. The detection limit was observed at 0.015 µM for ascorbic acid.

Ding et al. reported CdTe QDs capped with Mercaptopropionic acid (MPA-CdTe QDs) as an "on-off-on" ultrasensitive fluorescence probe for the detection of ascorbic acid through redox reaction [15]. The capped QDs were fabricated by hydrothermal method in an aqueous medium. For metal ions, the capped QDs fluorescence probe also could be applied as a successive sensor with an "on-off-on" process. The fluorescence of QDs was quenched by the addition of Fe^{3+}. After that, the fluorescence can be turned on in a sensitive manner by ascorbic acid to give an "on-off-on" fluorescence response in accordance with the oxidation-reduction among Fe^{3+} and ascorbic acid (Figure 9.2). Since Fe^{3+} sensitively reacted with CdTe QDs, there was a linear relationship between the Fe^{3+} concentration in the range from 2 to 10 µM and fluorescence intensity quenching value. For ascorbic acid, the linear detection scale was 0.1–1 µM and the detection limit was observed at 6.6 nM. This procedure was applied to detect the ascorbic acid in the human plasma sample in a successful manner.

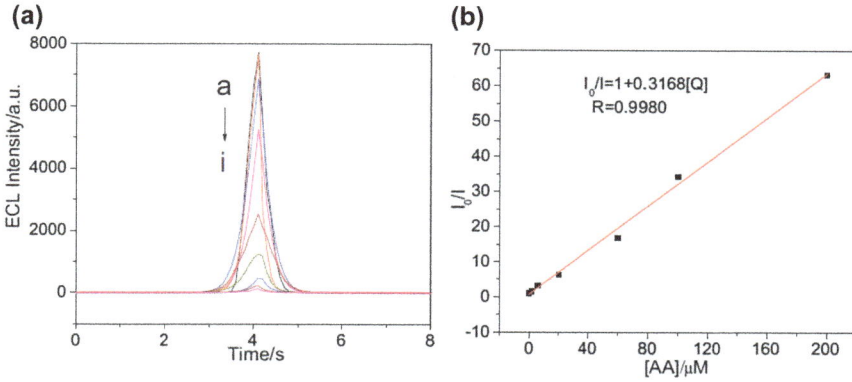

Figure 9.1: (A) Electrochemiluminescence intensity changes versus ascorbic acid concentrations: (a) 0; (b) 0.04 μM; (c) 0.2 μM; (d) 2 μM; (e) 6 μM; (f) 20 μM; (g) 60 μM; (h) 100 μM; (i) 200 μM; (B) calibration curve for the detection of ascorbic acid. Adapted with permission from [14].

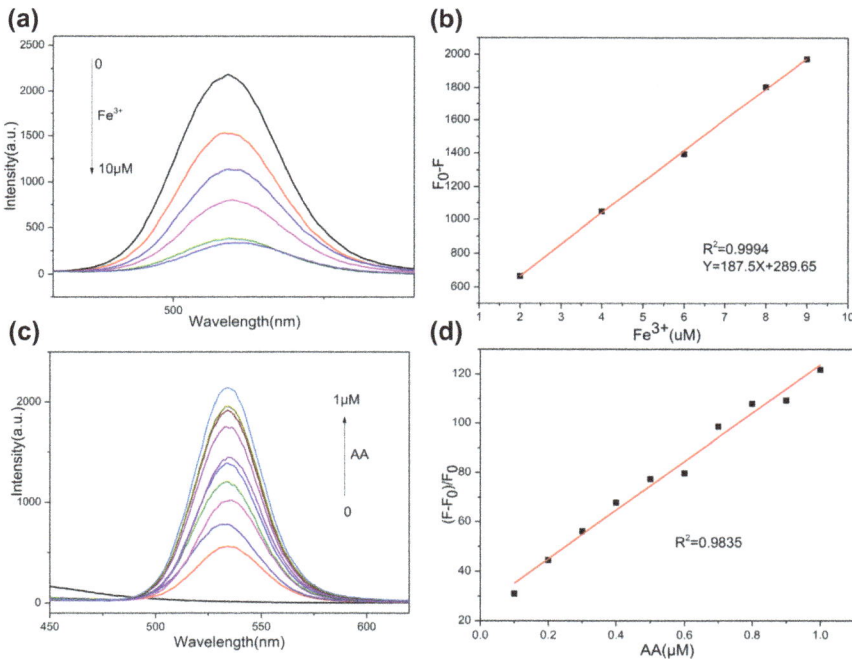

Figure 9.2: (a) Fluorescence spectra of Fe^{3+}@MPA-CdTe QDs. (b) The linearity relationship among various concentrations of Fe^{3+} and a fluorescence intensity value of (F_0-F). (c) Fluorescence spectra with various concentrations of ascorbic acid. (d) The linearity relationship among different concentrations of ascorbic acid and fluorescence intensity value of $(F-F_0)/F_0$. Adapted with permission from [15].

9.2.2 Drugs detection

To forbid a diverse of diseases, Sulfadimidine (SM$_2$) is used in a great degree in livestock feeds and stock farming [16]. It is one of the important sulfonamide antibacterials. Through drinking water and animal food containing the residue of sulfadimidine, the sulfadimidine can get into the human body in a gradual manner. Studies have shown that the excess amount of sulfadimidine can make allergies in the human and destruct the microflora's normal ecological balance [17]. Several methods are available for the detection of SM$_2$, including the solid-phase extraction method, supercritical fluid extraction and high-performance liquid chromatography. However, these methods have several limitations including low efficiency, high cost and complexity of process, [18, 19].

For the detection of sulfadimidine, Zhou et al. investigated the fabrication of surface molecular imprinting polymer (MIP) on CdTe QDs coated with SiO$_2$ as a sensor [20]. In this fluorescent sensor, for the fluorescent signal readout CdTe quantum dots were employed, as the functional monomer 3-aminopropyltriethoxysilane (APTES) was used, and as cross-linking agent tetraethyloxysilane (TEOS) was used. In the range of 10–60 µmol/L, the relative fluorescent intensity was decreased with the raising of sulfadimidine concentration (Figure 9.3). The recoveries were at the range of 90–99% and the relative standard diversion was in the scale of 1.9–3.1%. The Stern-Volmer plot analysis with diverse types of sulfonamides is shown in Figure 9.4.

Tetracycline is a member of broad-spectrum antibiotics and it has advantageous actions toward pathogenic microorganisms. Tetracycline is widely used in food,

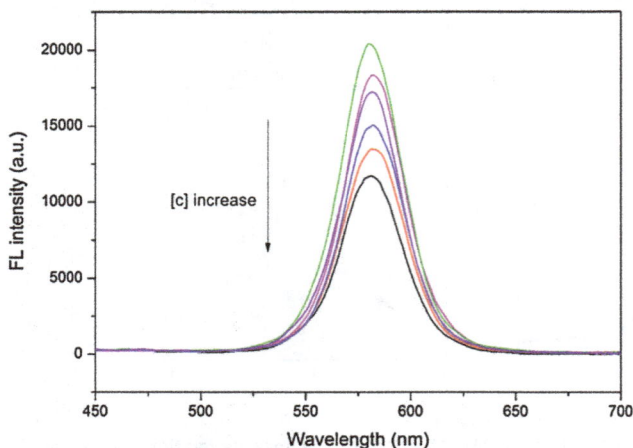

Figure 9.3: Fluorescence emission spectra of QDs@SiO$_2$-MIPs with different concentrations of sulfonamides (10, 20, 30, 40, 50, 60 µmol/L) in ethanol solution and water. Adapted with permission from [20].

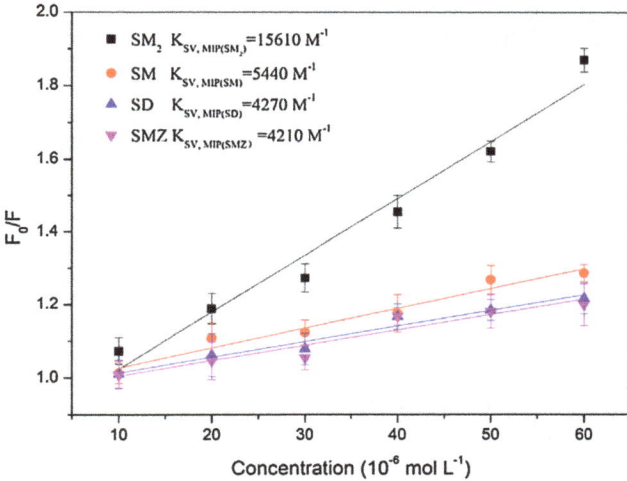

Figure 9.4: Stern-Volmer plots from QDs@SiO$_2$-MIPs with different types of SM$_2$. Adapted with permission from [20].

pharmaceuticals, and the environment because of its worths, such as hypotoxicity, proper oral absorption, and low cost [21, 22]. However, the residues in drinking water and animal food induced by the overuse of Tetracycline and other antibiotics guide to cell resistance and several chronic disorders in human. Thus, it is important to make an easy-to-use, fast, on-site or real-time visualization process to detect the Tetracycline and other related antibiotics. For colorimetric and visual detection of tetracycline, a dual-response ratiometric fluorescent sensor was reported recently by using europium-doped CdTe QDs [23].

With the addition of tetracycline, the fluorescence of the probe can be observed from green to yellow and finally to red, by the naked eye. It exhibits a broad-chromatic and dosage-sensitive sensing scheme for tetracycline. The fluorescence intensity ratio of the sensor showed a proper linear relation to concentrations of tetracycline in the range of 0–80 μM and the detection limit was 2.2 nM. The changes in the fluorescence spectra with concentrations of tetracycline are shown in Figure 9.5. The fluorescence of europium ions (616 nm) is raised with the increase of tetracycline, while the CdTe quantum dot's fluorescence (512 nm) is diminished gradually.

Amoxicillin (AMX) is another widely employed antibiotic for the treatment of several infections [24–26]. Wong et al. described an electrochemical platform using CdTe quantum dots and other inexpensive nanomaterials to detect the AMX in various matrices [27]. The device is based on an integration of Printex 6L Carbon and CdTe QDs within a poly(3,4-ethylenedioxythiophene) polystyrene sulfonate film. The sensor depicted excellent selectivity and sensitivity for AMX detection, with a linear scale of 0.9–69 μmol L^{-1}. It showed a low limit of detection of about 50 nmol L^{-1} (Figure 9.6).

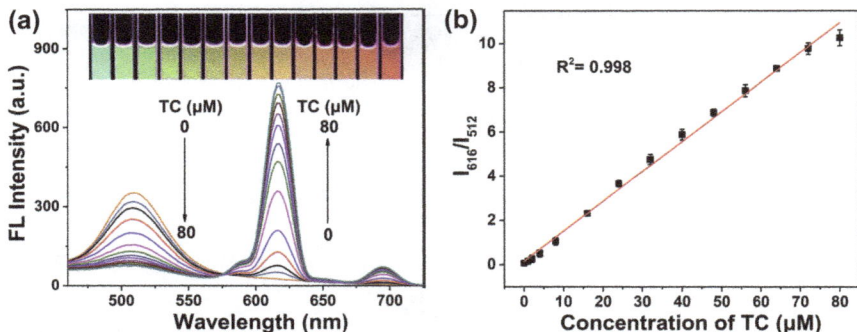

Figure 9.5: (a) Fluorescence spectral changes of Eu/CdTe quantum dots. With Tetracycline concentrations of 0–80 μM. (b) The ratiometric calibration plot of Eu/CdTe quantum dots. Adapted with permission from [23].

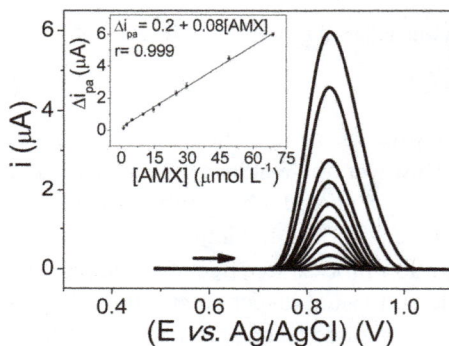

Figure 9.6: Square wave voltammetry (SWV) voltammograms for sensor, with the optimized parameters. The amoxicillin concentrations in μmol L^{-1} were in a range of 0.9–69 μmol L^{-1}. Linear dependence of the peaks current with amoxicillin concentrations is shown in inset. Adapted with permission from [27].

From possible biological interferences such as paracetamol, uric acid, caffeine and ascorbic acid, no substantial interference was detected (Figure 9.7). The study also showed that applying an uncomplicated measurement device and without any sample pre-treatment, the device can constitute an alternative way not only for the analysis of clinical samples and pharmaceutical products but also for analysis of food products.

Gemcitabine hydrochloride (GEM) is a chemotherapeutic agent used to treat several tumors [28, 29]. Together with its excellent effects, it has lots of side effects. To improve drug efficacy and reduce toxicity effects monitoring the GEM concentration in biological samples is important [30]. Normally it is determined by gas chromatography [31], liquid chromatography [32], the electroanalytical and spectroscopic analysis on the actions with DNA [33, 34]. At the same time, these methods require expensive instrumentation, are time-consuming, and suffered from low sensitivity [35].

Najafi et al. demonstrated, a procedure for the sensitive and selective quantitative finding of GEM in biological material by evaluating the fluorescence quenching of

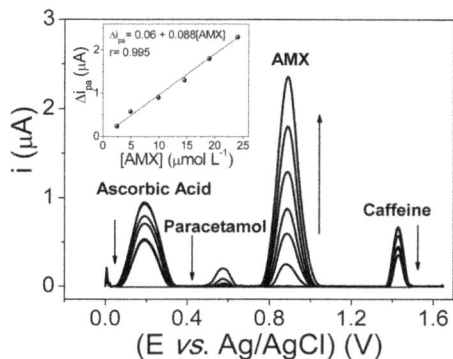

Figure 9.7: Determination by square wave voltammetry of amoxicillin in presence of caffeine ascorbic and acid. Amoxicillin at various concentrations (in a range of 0.9–24 µmol L^{-1}), 20 µmol L^{-1} paracetamol, 20 µmol L^{-1} ascorbic acid, and 10 µmol L^{-1} caffeine in 0.1 mol L^{-1} phosphate buffer solution. Adapted with permission from [27].

functionalized Au doped CdTe quantum dots [36]. The thioglycolic acid-capped Au doped CdTe QDs synthesized via hydrothermal methods showed good photostability and excellent photoluminescence intensity. The application of gemcitabine anticancer drugs quenches the fluorescence of the QDs *via* photoinduced charge transfer. For Gemcitabine in aqueous media, the limits of sensing were 0.1 µM. The sensor was also can be used to detect the drug in human urine and plasma with high sensitivity, broad linear range, excellent selectivity and small detection limit.

DES (Diethylstilbestrol) is an exogenous estrogen. It is widely applied in aquaculture and animal husbandry. DES is also used as a growth promoter. Studies showed that Diethylstilbestrol can enter to the human food chain via milk, meat, and water. Also it can stay in the human body for a longer duration [37]. It can create several health issues [38]. For the detection of DES, several traditional methods such as chemiluminescence assay [39], gas chromatography together with mass spectrometry [40], liquid chromatography together with mass spectrometry [41], electrochemistry [42], and immunoassay [43, 44] are available. However, these methods have several drawbacks. Zhao et al. demonstrated an electro-chemiluminescence solid-state imprinted sensor using graphene/CdTe@ZnS QDs as luminescent probes [45]. The ultrasensing detection of diethylstilbestrol was possible with the sensor.

Glyphosate (N-Phosphomethylglycine or GP) is one of the most widely used herbicides worldwide [46, 47]. It has outstanding weed control capability. Even though it makes low noxiousness to the living organism, its residue cumulates both in water and soil because of its high degree solubility in water and leakage. Because of the irreversible inhibition mechanism on acetylcholinesterase, such residue can cause unplayful damaging effects on the central nervous system. Thus, developing effective methods for the detection of glyphosate is extremely worthy.

For the detection of glyphosate, several methods have been accounted in the literature which includes ion-exchange chromatography [48], mass spectrometry [49], colorimetric [50], capillary electrophoresis [51], ion chromatography [52], gas chromatography [53], and enzyme-linked immune sorbent assay [54]. However, most of these processes are costly and demand advanced instrumentation.

Bera et al. reported the ultrasensitive sensing of GP via efficient photoelectron transfer among chitosan-derived carbon quantum dot (CQD) and CdTe [55]. Because of the effective photoelectron transfer from CdTe to chitosan-derived carbon quantum dot, the photoluminescence spectrum showed weakened fluorescence at 619 nm. Addition of GP into CdTe-CQD pair at pH 8 disintegrated the pair and it suppresses the photoelectron transfer process. Because of that the emission intensity of the probe was regained at 619 nm. To get the operational pH range of the sensor, the outcome of variations in pH on the intensity enhancement of the emission band was assessed with a constant glyphosate concentration (Figure 9.8). For glyphosate, the fluorescence assay handles a linear concentration scale of 0–1000 nM. The limit of detection was observed at 2 pM. The selectivity of the probe for glyphosate was analyzed in presence of various interfering ions and pesticides (Figure 9.9).

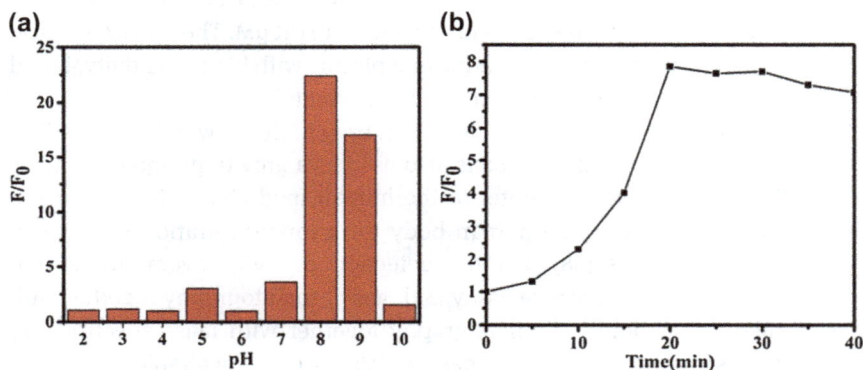

Figure 9.8: (a) Effect of pH on sensing, (b) time on sensing. Adapted with permission from [55].

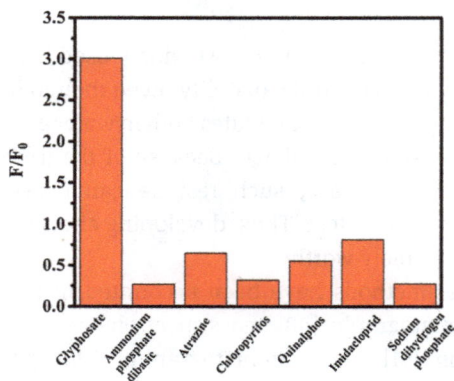

Figure 9.9: Comparison of selectivity toward glyphosate and other intervening species. Adapted with permission from [55].

9.2.3 Mercury (II) ions detection

Mercury (II) ions (Hg^{2+}) is one of the most notorious contaminants that are widely circulated with both anthropologic and natural sources. Mercury exposure can make dangerous side effects on the digestive system, immune system, central nervous system, kidney, and brain, even with a very low concentration [56–58]. Thus, the fabrication of efficient analytical ways for the specific and accurate detection of Hg^{2+} is a pressing goal.

Hallaj et al. covered the fabrication and characterization of Hg^{2+} fluorescence sensor based on bithiazolidine derivatives-capped CdTe/CdS QDs [59]. The study demonstrated that under optimal conditions, Hg^{2+} can be observed with a sensing limit of 0.08 nM in a linear scale from 0.3 to 21 nM. The study demonstrated that with the concentration of Mercury ions, the intensity of the fluorescence diminished gradually in the scale of 0.3–21 nM (Figure 9.10). Even in the presence of the several interposing metal ions, the sensor showed high selectivity for Hg^{2+}. This approach had appealing rewards including high selectivity and sensitivity, wide linear range, accuracy, and simple operation.

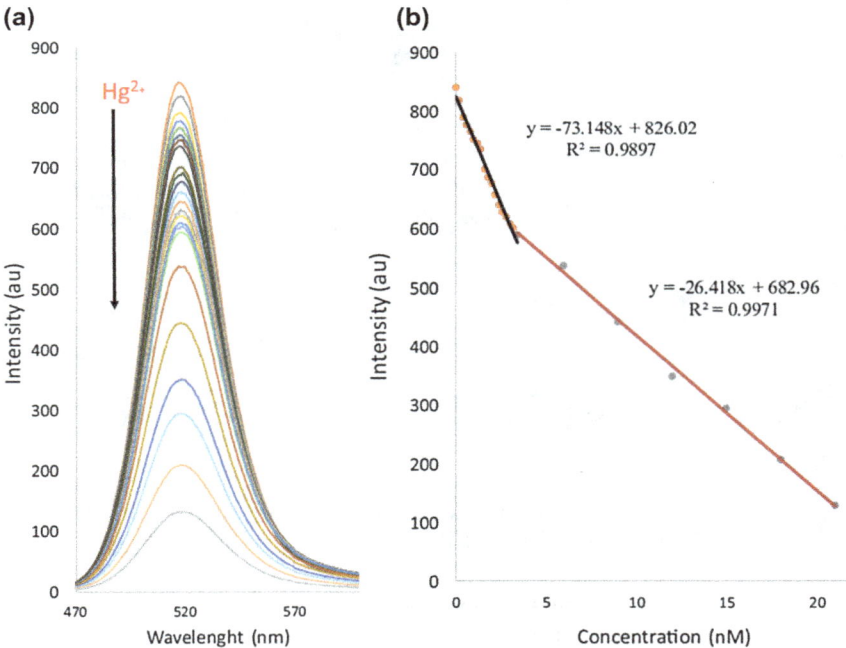

Figure 9.10: (a) Fluorescence responses of the fabricated sensor in various concentrations of Mercury ions in 0.1 M pH 6, (b) calibration curve for Mercury ion detection. Adapted with permission from [59].

9.2.4 Nitrogen dioxide (NO$_2$) gas detection

For room-temperature gas sensing applications, molybdenum disulfide (MoS$_2$) can show promising results. However, it has numerous limitations. Incomplete recovery, the limitation of fabrication methods, and selectivity at room temperature are the main disadvantages of this technique. Jaiswal et al. reported the synthesis of molybdenum disulfide nanoworms (NWs) thin film and CdTe QDs decorated molybdenum disulfide nanoworms hybrid heterostructure thin film [60]. The thin films were fabricated using the sputtering technique on the p-Silicon substrate and examined for room temperature nitrogen dioxide sensing applications. Figure 9.11 shows the FESEM images of the thin films.

The hybrid heterostructure thin-film sensor showed a sensor response of about 40%, complete recovery (114 s) and a fast response time (16 s). The results are summarized in Figure 9.12. Also, the sensor was highly selective toward 10 ppm nitrogen dioxide at room temperature compared to pristine molybdenum disulfide nanoworms thin-film sensor which showed a response of about 26%, recovery/response time nearly 23 s, and the recovery was incomplete.

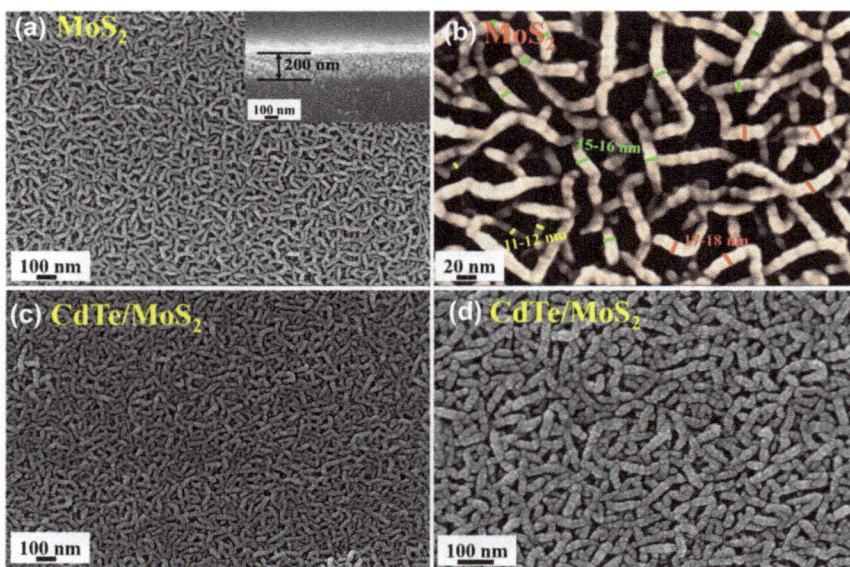

Figure 9.11: (a) FESEM surface micrographs at low magnification and (b) at high magnification of the molybdenum disulfide nanoworms thin film. (c), (d) FESEM images of CdTe quantum dots decorated molybdenum disulfide nanoworms hybrid heterostructure thin film at various ranges. Adapted with permission from [60].

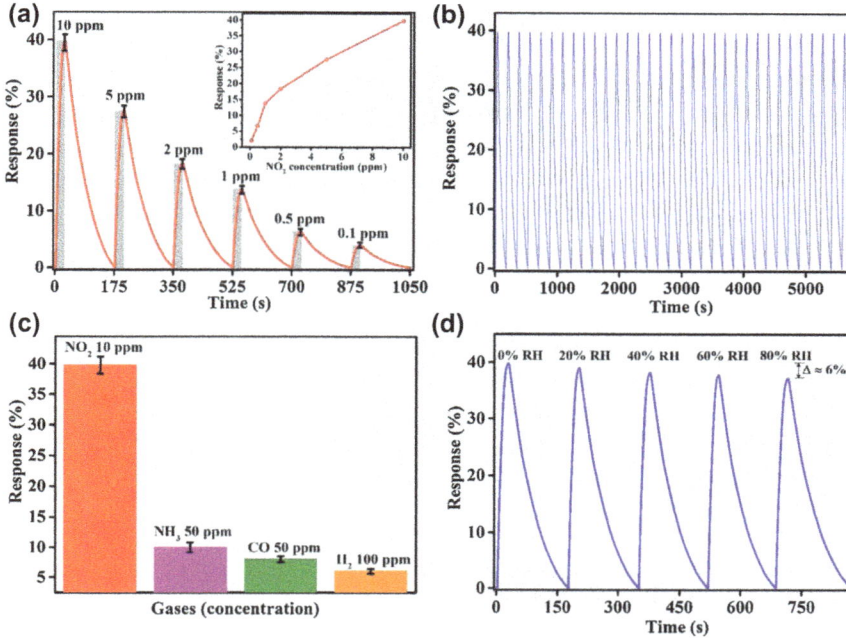

Figure 9.12: (a) The transient relative response of the CdTe/molybdenum disulfide hybrid heterostructure thin-film sensor with various nitrogen dioxide concentrations at room temperature in dry air. (b) Cyclic test of sensor toward 10 ppm of nitrogen dioxide at room temperature. (c) Selectivity histogram of sensor at room temperature. (d) The gas response curve of sensor toward 10 ppm nitrogen dioxide in various relative humidity conditions (0–80%) at room temperature. Adapted with permission from [60].

9.3 Te thin films-based sensor

For the room temperature sensing of unfavorable gases, Te-based films can be used [61, 62]. It is reported that Tellurium thin films display excellent sensitivity to NO_2 [63], ammonia [2], hydrogen sulfide [64], carbon oxide and propylamine [65], oxygen, nitrogen, and water vapors [66].

9.3.1 Nitrogen dioxide (NO_2) gas detection

The sensitivity of the Te thin film depends on many factors such as substrate micro-structure and deposition temperature of film [67, 68], operating temperature and post-deposition heating [67, 69]. The sensor sensitivity and conductivity also depend on the thickness of the film. For example, the step-down of a Te film thickness to 40 nm from 120 nm resulted in enhancing sensitivity toward nitrogen dioxide by 10 times more [69]. The sensitivity of the sensor as a function of the thickness of the film is shown in Figure 9.13.

Tsiulyanu et al. reported the fabrication, characteristics, and NO_2 sensing properties of ultrathin compact tellurium films [70]. The weakening of sensitivity caused by high gas concentration in Te films was reported. The sensitivity of Te film (30 nm thick) was decreased linearly with concentration increment in the range of 0.15–0.5 ppm of NO_2. Figure 9.14 shows the tellurium ultrathin sensor sensitivity versus the NO_2 concentration at room temperature diluted in ambient air.

Figure 9.15 demonstrates the transient characteristics of current induced by the gas in a tellurium ultrathin film by exposure to diverse NO_2 concentrations.

Te sputtered thin films were demonstrated as a nitrogen dioxide gas sensor [68]. By using RF sputtering system, the thin films were deposited onto alumina and glass substrates. The gas-sensing characteristics were calculated by analyzing the resistance changes of the films as a function of gas concentration and operating temperature. It was observed that the sensitivity was mostly regulated by the Te film's surface

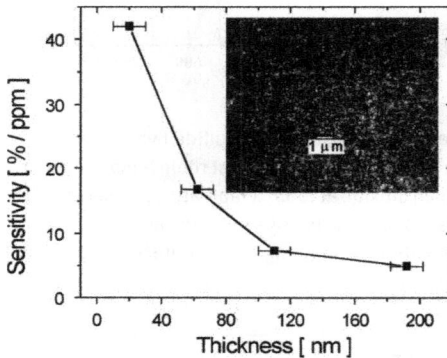

Figure 9.13: Outcome of thickness on sensitivity at room temperature toward 1.5 ppm of nitrogen dioxide. SEM micrograph of an as-grown film is shown in inset. Adapted with permission from [69].

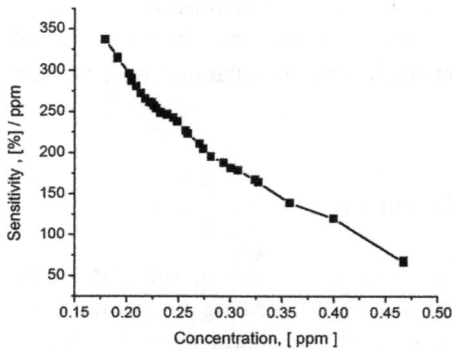

Figure 9.14: The ultrathin Te sensor sensitivity versus concentration of NO_2 gas. Adapted with permission from [70].

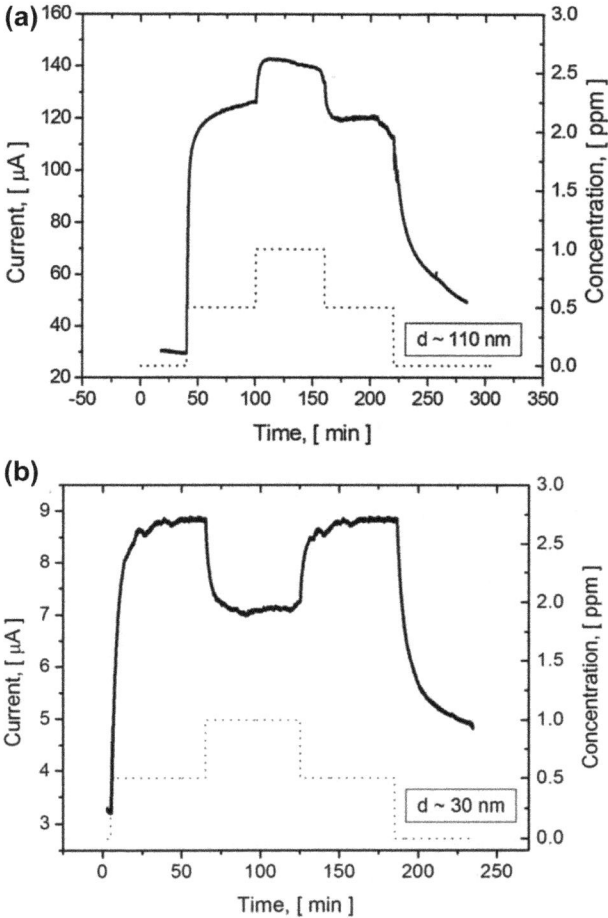

Figure 9.15: Transient characteristics of current induced by the gas by exposure of tellurium films to different NO_2 concentrations according to the profile shown in dotted lines of the bottom. The film thickness: (a) 110 nm and (b) 30 nm. Adapted with permission from [70].

morphology. At room temperature, the films deposited on a glass substrate depicted the greatest response to nitrogen dioxide.

Figure 9.16 demonstrate the responses of the sensor versus the operating temperature in the range from 18 to 140 °C for four Te thin films deposited on alumina and glass substrates and disclosed to nitrogen dioxide (40 ppm) in dry air. The room temperature sensor response of a Tellurium sample as a function of nitrogen dioxide concentration is shown in Figure 9.17.

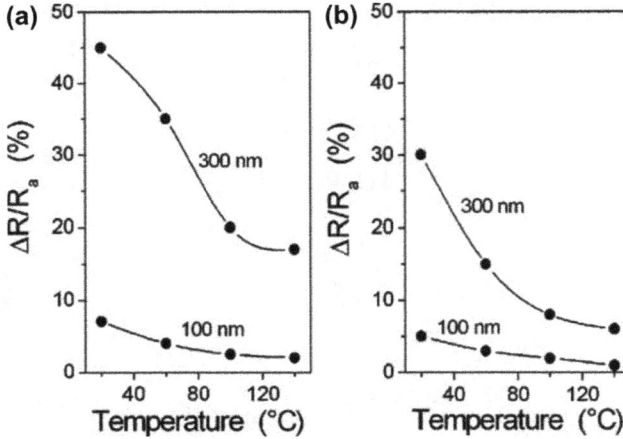

Figure 9.16: Sensor response versus the operating temperature for tellurium films with thicknesses of 100 and 300 nm deposited on (a) glass and (b) alumina substrates exposed to nitrogen dioxide (40 ppm) in dry air. Adapted with permission from [68].

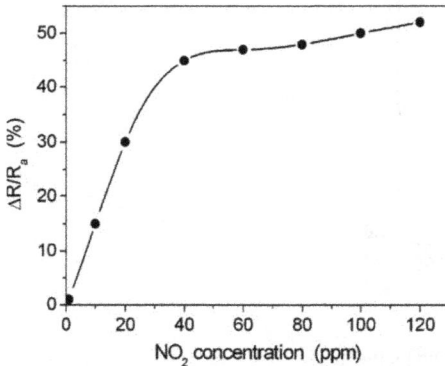

Figure 9.17: Room temperature sensor response of a tellurium sample as a function of nitrogen dioxide concentration. Adapted with permission from [68].

9.3.2 Hydrogen sulfide (H₂S) gas detection

The influence of the Te thin film thickness and film temperature on observing hydrogen sulfide (H_2S) gas was studied by Manouchehrian et al. [71]. Also, the Ultra Violet radiation influence during evaluating the gas was analyzed. By using thermal evaporation, the Te thin film was deposited on Al_2O_3 substrates. The study depicted that with the thickness, the sensor sensitivity drop-offs (Figure 9.18). Also, the recovery and response times were increased with thickness. The recovery and response times were observed to be strongly decreased with UV radiation.

The Tellurium sensor's sensitivity with diverse gases, such as H_2S, NH_3, NO_2, and CH_3OH at room temperature is shown in Figure 9.19(a). The response kinetics of a

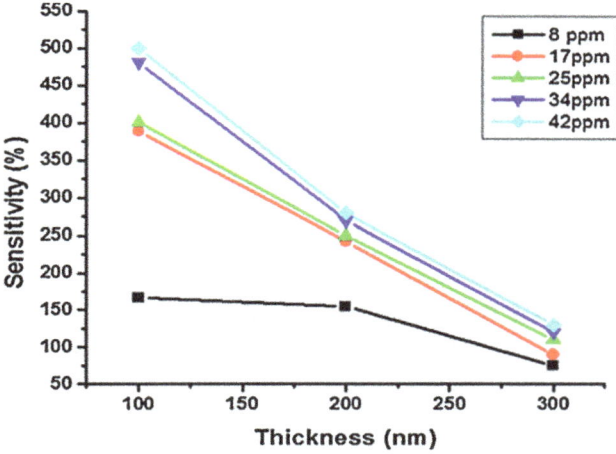

Figure 9.18: The variations on sensitivity of H_2S with thickness, at room temperature. Adapted with permission from [71].

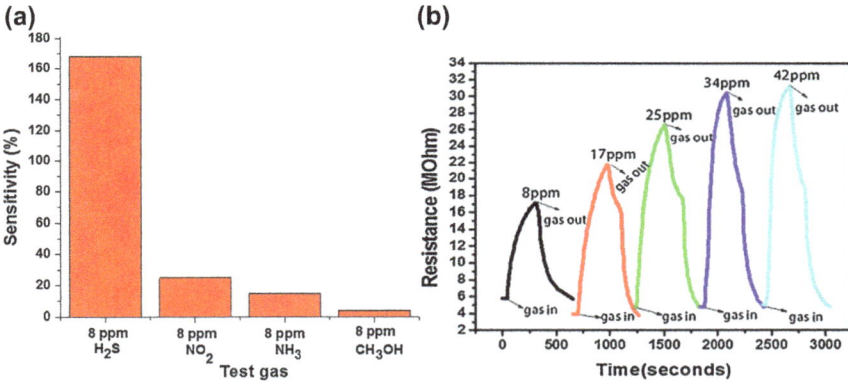

Figure 9.19: (a) Tellurium sensor sensitivity with diverse gases at the concentration of 8 ppm. (b) room temperature response curve of 100 nm tellurium film after exposure to H_2S. Adapted with permission from [71].

Tellurium film (100 nm thick) at diverse concentrations of H_2S gas is shown in Figure 9.19(b).

9.3.3 Ammonia (NH_3) gas detection

Room temperature functioning ammonia (NH_3) gas sensor using Te thin films was reported by Sen et al. [2]. On exposure to ammonia, the Te films depicted a reversible increase in resistance. Also, the response was linear in the range from 0 to 100 ppm. Figure 9.20(a) depicts the film sensitivity as a function of NH_3 concentration. The study showed that NH_3 cuts down tellurium oxide on the grain boundary region and surface of the film to Te. Thus, in the film the majority carrier density decreases, and because of that the film's conductivity decreases. To find the optimum working temperature, the film's sensitivity was also evaluated as the function of temperature at a constant ammonia concentration (Figure 9.20(b)).

9.4 Te nanostructures or nanoparticles-based sensors

The usage of Tellurium nanostructures as gas sensors also has been explored. The following section discusses Te nanostructures or Te nanoparticles-based sensors.

9.4.1 Chlorine (Cl_2) gas detection

When emitted into the environment chlorine is very unfavorable and it is used in a great degree in several industrial procedures. Thus, it is important to use sophisticated techniques for the detection of chlorine.

Chlorine gas sensors based on one-dimensional Te nanostructures were demonstrated by Sen et al. [72]. Under an inert environment under vacuum conditions

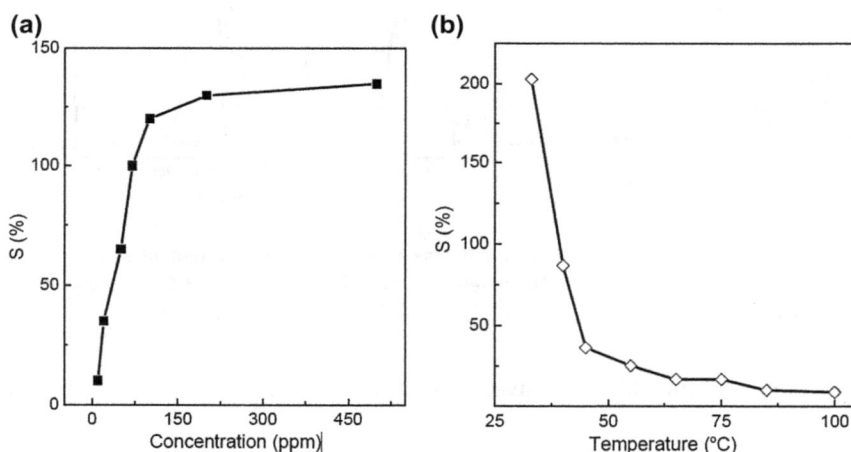

Figure 9.20: (a) Room temperature sensitivity (S) of Te films as a function of NH_3 concentration. (b) Sensitivity of tellurium films calculated after exposure to 100 ppm of ammonia as a function of temperature. Adapted with permission from [2].

as well as at atmospheric pressure, the Tellurium nanotubes have been grown by physical vapor deposition. Figure 9.21 presents the SEM images of the Te nanotubes. Films synthesized using both kinds of Te nanotubes were tested for sensitivity to reducing and oxidizing gases. The study demonstrated that the comparative response of the samples to gases relies on the microstructure. The samples made at atmospheric pressure (of argon) depicted better selectivity and high sensitivity to Cl_2 gas. The spectroscopic analysis pointed that the response to Cl_2 is mostly brought by grain boundaries, hence heightened for nanotubes synthesized under argon atmosphere.

Figure 9.22(a) shows the response of the sensor to Cl_2 at a concentration of 2 ppm. The recovery time was 2 h and the response time was 30 s. Figure 9.22(b) shows the response of the sensor to chlorine at diverse concentrations. Response of the sensor with different reducing and oxidizing gases is presented in Figure 9.22(c). It was mentioned that the samples were completely selective toward Cl_2.

Figure 9.21: SEM images of typical (a) tellurium nanotubes grown in the furnace, the inset shows the TEM of the nanotube (b) tellurium micro-rods, and (c) tellurium nanotubes on silicon. (d) TEM and HRTEM images of tellurium nanotubes on silicon. The inset of (a) shows the TEM of a type-I nanotube. Adapted with permission from [72].

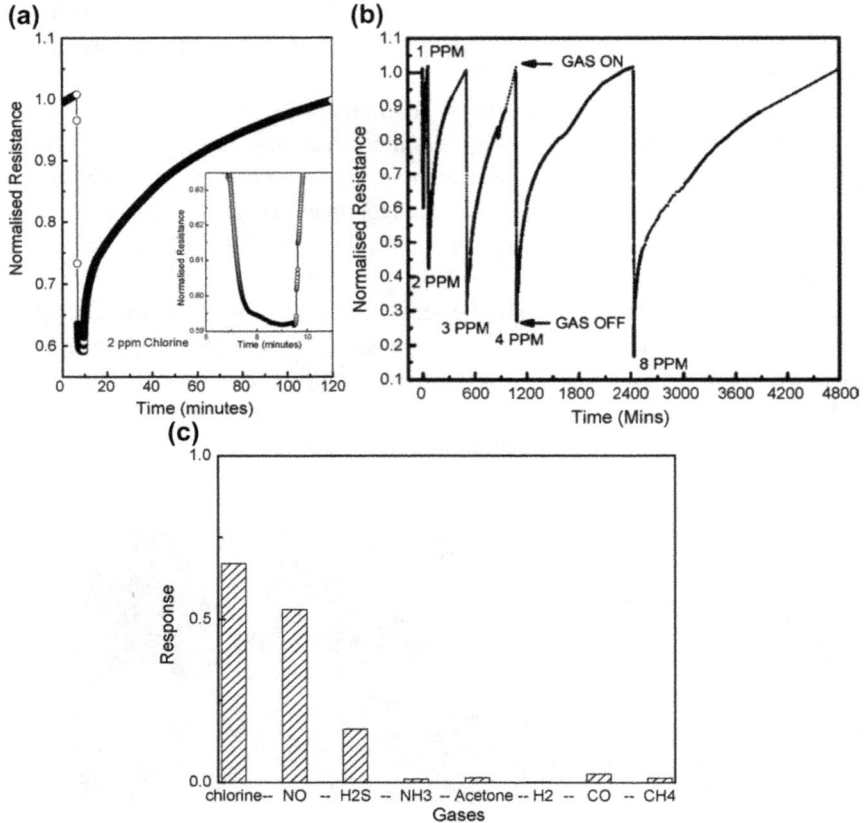

Figure 9.22: (a) Response of a type-I sensor to 2 ppm Cl_2. (b) Response and recovery characteristics for various concentrations of chlorine gas for another type-I sample. (c) Response of the sensor to Cl_2, H_2S, NO, acetone, CH_4, CO, and NH_3 gases at 4 ppm concentration and H_2 at 2% concentration. Adapted with permission from [72].

9.4.2 Ammonia (NH_3) and propylamine detection

The room temperature sensing properties of Tellurium nanoparticles grown on silicon nanowires (SiNWs) substrate was analyzed using propylamine and ammonia as probe molecules [73]. Through a solution method, Te nanoparticles were synthesized on SiNWs, the particles were having an average diameter of 5 nm. The sensitivity of the sensor was evaluated at a constant concentration of ammonia (100 ppm) or propylamine (10 ppm), in the temperature range from 35 to 80 °C to find the optimum working temperature (Figure 9.23). The detection ranges of propylamine and ammonia

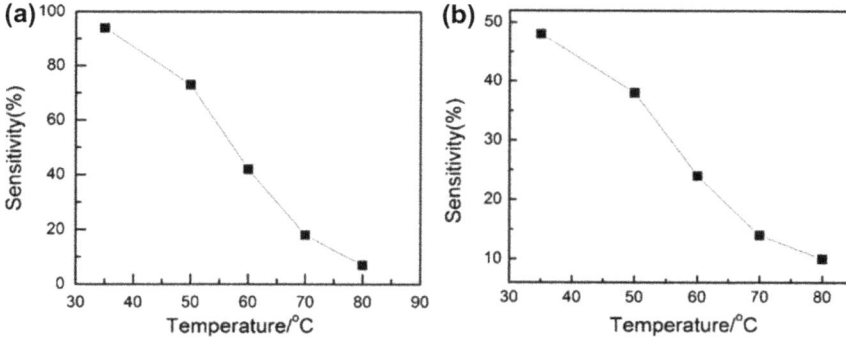

Figure 9.23: The responses of the sensor device at various temperatures upon exposure to (a) 100 ppm ammonia, and (b) 10 ppm propylamine. Adapted with permission from [73].

were 5–25 ppm and 10–400 ppm, and the detection limits were 174 and 196 ppb, respectively. The times of response during to the exposure to propylamine and ammonia were 15 and 5 s and the times of the recovery were 6 and 8 s, respectively. At the optimal working temperature, when the samples were exposed to 25 ppm propylamine or 400 ppm ammonia, the response of the sensor was 164% or 208%. The fast response and outstanding sensitivity might be attributed to the small size of Te nanoparticles.

The response curves of the sensor when cycled by raising ammonia or propylamine concentrations at the optimal working temperature of 35 °C are shown in Figure 9.24. Figure 9.25 displays the room temperature response sensitivities of the sensors during exposure to ammonia and propylamine with enhancing concentration.

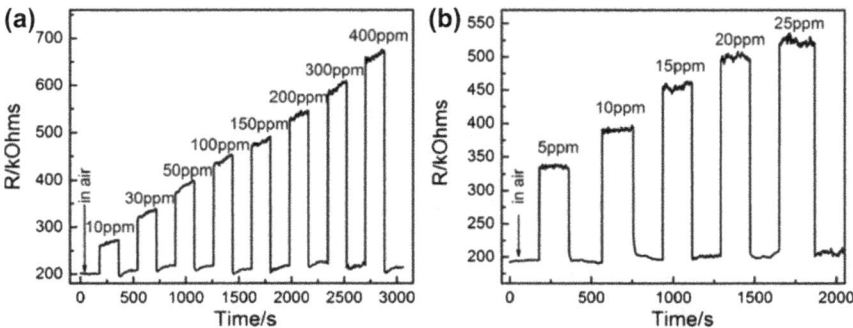

Figure 9.24: The room temperature response curves of resistance with time during the exposure to the gases of (a) ammonia and (b) propylamine from air to diverse concentrations. Adapted with permission from [73].

(a)

$S = 0.122 \cdot x^{0.469}$

Response (%)

Concentration(ppm)

(b)

$S = 0.319 \cdot x^{0.505}$

Response (%)

Concentration(ppm)

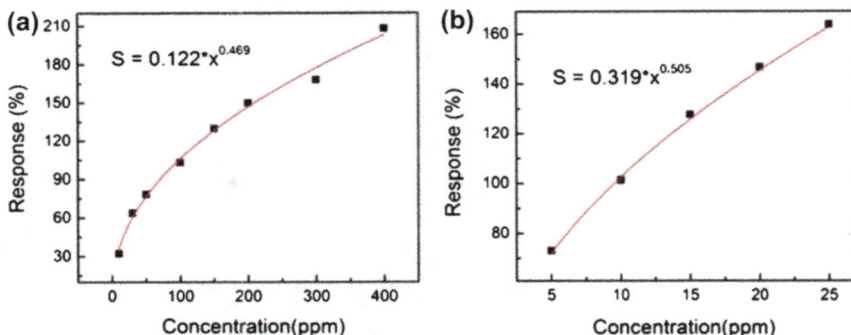

Figure 9.25: Sensor response as a function of (a) ammonia, and (b) propylamine concentration. Adapted with permission from [73].

9.4.3 Nitrogen dioxide (NO$_2$) gas detection

A chemiresistive sensor using tellurium nanotube for the room-temperature ultrasensitive sensing of nitrogen dioxide was demonstrated by Guan et al. [74]. The Tellurium nanotubes were fabricated by employing the microwave reflux method. The constructed chemiresistive sensor exhibited impressive selectivity and sensitivity to detect the amount of nitrogen dioxide at room temperature (Figure 9.26). The limit of detection was at 500 ppt. The study showed that the gas sensor response is completely reversible with the aid of ultra violet radiation.

To analyses the operation of the Tellurium sensor in a real environment, the sensor was also screened in an air environment (Figure 9.27). For the analysis, rather than with pure nitrogen gas, various quantities of nitrogen dioxide were mixed with dry air. The high crystallinity and large surface-to-volume ratio of the Tellurium nanotubes contributed to the better functioning of the tellurium-based gas sensor.

9.4.4 Hydrogen peroxide detection

Manikandan et al. demonstrated a superior non-enzymatic and supercapacitor hydrogen peroxide sensor using Te nanoparticles [75]. Tellurium nanoparticles were fabricated by the wet chemical process. To make the electrodes for biosensor applications and supercapacitors, the Te nanoparticles were coated on Glassy carbon electrodes and graphite foil. The SEM and TEM images of the NPs are shown in Figure 9.28.

The supercapacitor performance of Te NPs was evaluated in the electrolyte by both galvanostatic charge-discharge and Cyclic Voltammetry method. The impedance plot and frequency-dependent specific capacitance of Tellurium nanoparticles are shown

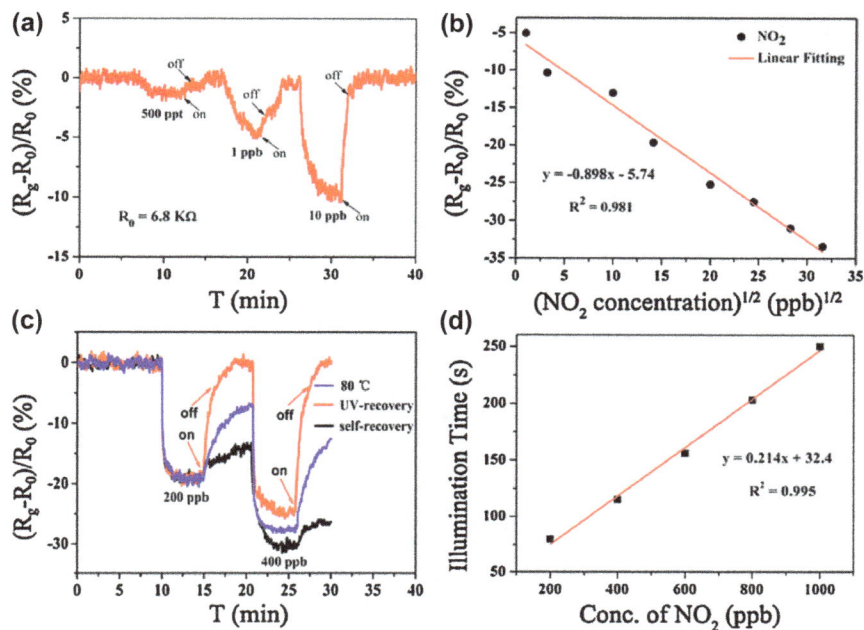

Figure 9.26: (a) Real-time sensor response to nitrogen dioxide gas, (b) the changes in the response of gas sensor on the nitrogen dioxide concentration, (c) recovery actions under different configurations, and (d) the required illumination time versus the analyte concentration. Adapted with permission from [74].

in Figure 9.29. Using CV and Chronoamperometry in phosphate buffer solution, the hydrogen peroxide sensor operation of Tellurium nanoparticles altered glassy carbon electrode was analyzed (Figure 9.30). In the linear range of nearly 0.6–8 µM of hydrogen peroxide, tellurium nanoparticles showed a good sensitivity of 0.83 mAmM^{-1} cm^{-2}, and the correlation coefficient was 0.99. The detection limit was 0.3 µM and the time of response was less than 5 s.

A low-cost, non-enzymatic sensor using TeNPs for the analytical detection of hydrogen peroxide was reported [76]. For the fabrication of the sensor, thin films of Tellurium nanoparticles were deposited on fluorine-doped tin oxide substrate. The sensor showed well stability and selectivity with an impressive amperometric time of response (5 s). The analysis also evidenced that this Te nanoparticle-based sample is a promising sensing material for the finding of hydrogen peroxide. Figure 9.31 shows the FE-SEM images of pristine fluorine-doped tin oxide and electrochemically deposited Tellurium nanoparticles on fluorine-doped tin oxide at a diverse employed potential.

The sensitivity of TeNPs/FTO sensors is illustrated in Figure 9.32(a) and Figure 9.32(b) shows the selectivity of the sensor toward hydrogen peroxide.

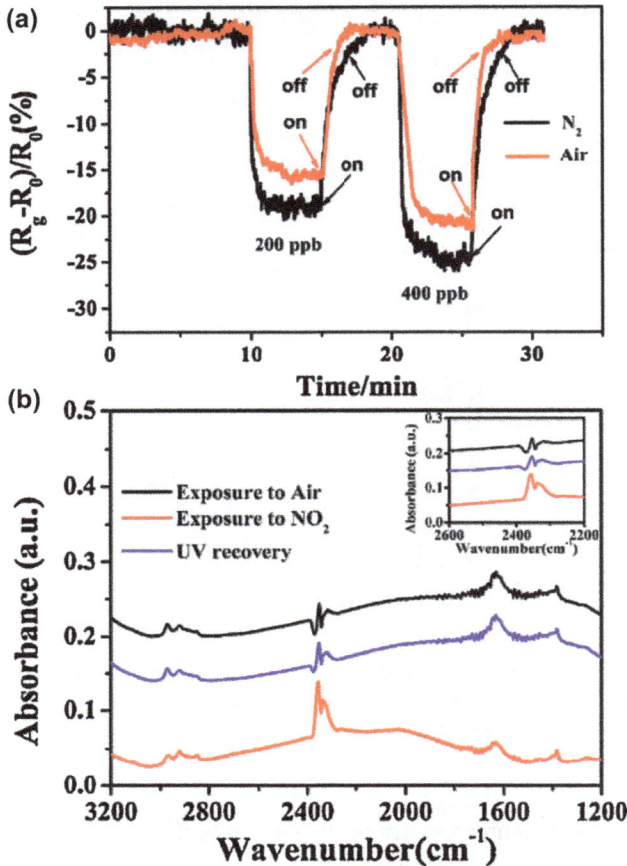

Figure 9.27: (a) Real-time response of the sensor to nitrogen dioxide in air and pure nitrogen gas environment, and (b) IR spectra of the tellurium nanotubes before and after exposure to nitrogen dioxide. Adapted with permission from [74].

9.5 Tellurium dioxide (TeO₂)-based sensor

Tellurium dioxide is a versatile wide bandgap oxide semiconductor. Because of its unique chemical and physical characteristics, TeO_2 is a promising and important functional material. TeO_2 is widely used for several electrical and electronic devices including laser devices [77], modulators [78], deflectors [79], solar cells [80], optical storage devices [81], tunable filters [82], and sensors [83–90].

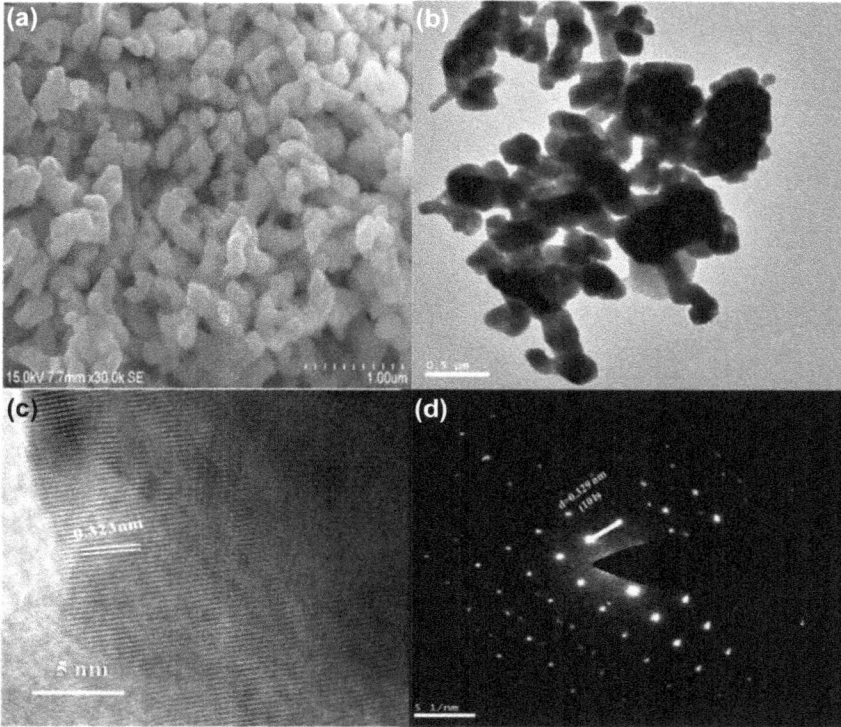

Figure 9.28: (a) FESEM (b) TEM (c) lattice-resolved TEM (d) SAED pattern of TEM image for tellurium nanoparticles. Adapted with permission from [75].

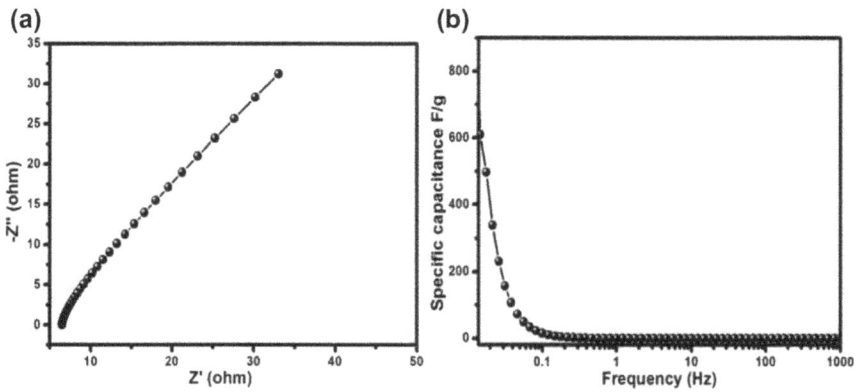

Figure 9.29: (a) Impedance plot of tellurium nanoparticles, (b) frequency-dependent specific capacitance of tellurium nanoparticles. Adapted with permission from [75].

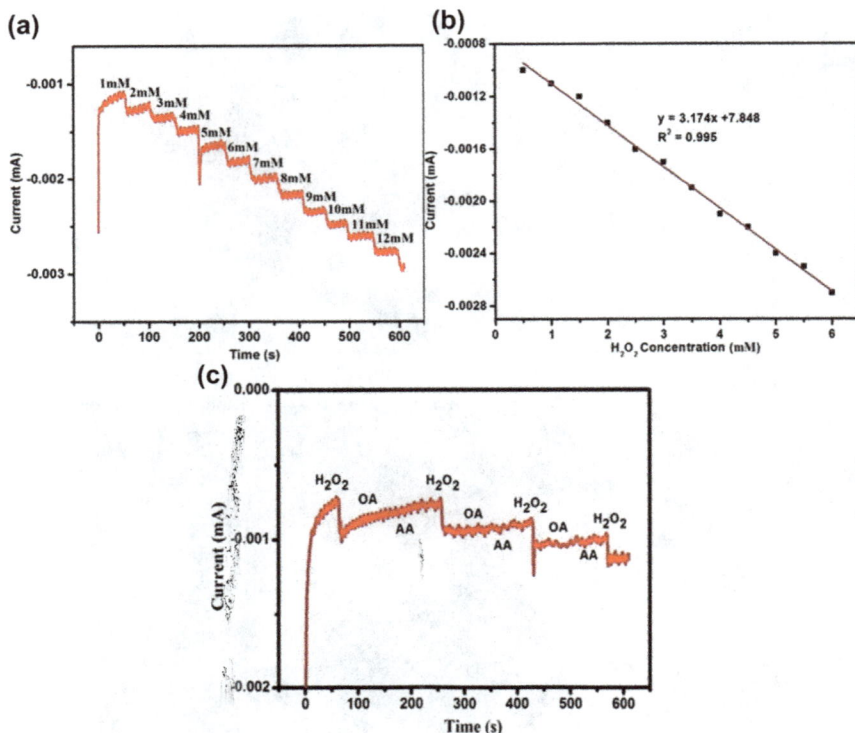

Figure 9.30: (a) Chronoamperometric analysis for diverse hydrogen peroxide concentrations, (b) linear relation between current and hydrogen peroxide concentrations, (c) chronoamperometric analysis with the exposure of hydrogen peroxide, AA and OA. Adapted with permission from [75].

9.5.1 Nitrogen dioxide (NO$_2$) gas detection

Room temperature nitrogen dioxide sensing properties of reactively sputtered tellurium dioxide thin films was reported [89]. By using sputtering method, the Tellurium dioxide thin films were deposited on quartz substrates. Even though the as-deposited films were amorphous, it became crystalline after thermal annealing. The analysis of functioning of the Tellurium dioxide thin films for room temperature nitrogen dioxide gas sensing showed that the as-deposited films have low sensitivity to nitrogen dioxide gas. At the same time, films prepared by thermal annealing demonstrated a promising response and sensitivity. Figure 9.33 shows the plot for absorption coefficient $(\alpha h v)^2$ against the photon energy (hv).

Figure 9.34(a) presents the dynamic electrical conductance of a tellurium dioxide thin film for various nitrogen dioxide concentrations at room temperature. The changes in the conductance with the concentration of nitrogen dioxide is shown in Figure 9.34(b).

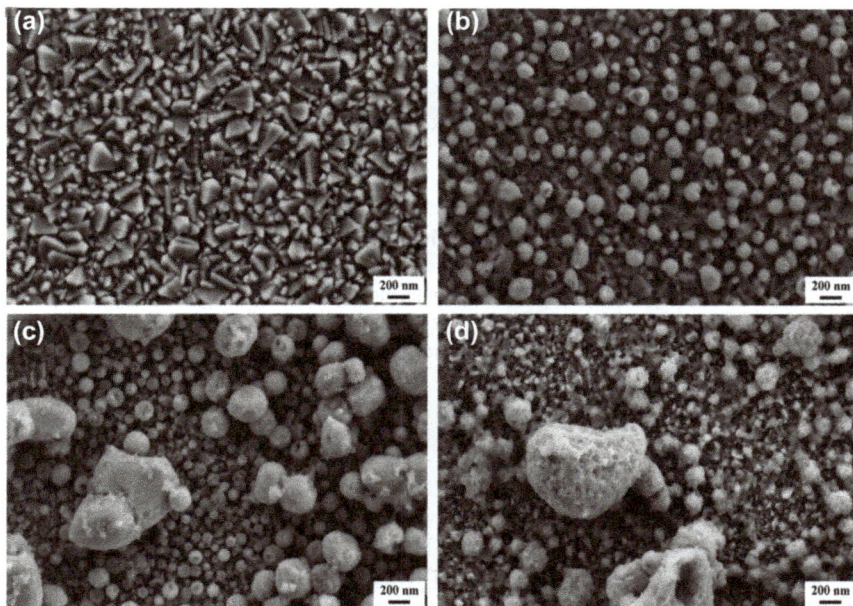

Figure 9.31: FE-SEM images of tellurium nanoparticles electrochemically deposited at diverse used potential onto fluorine-doped tin oxide substrate at 90 °C. (a) Pristine fluorine-doped tin oxide (b) −1.30 V (c) −1.40 V (d) −1.50 V. Adapted with permission from [76].

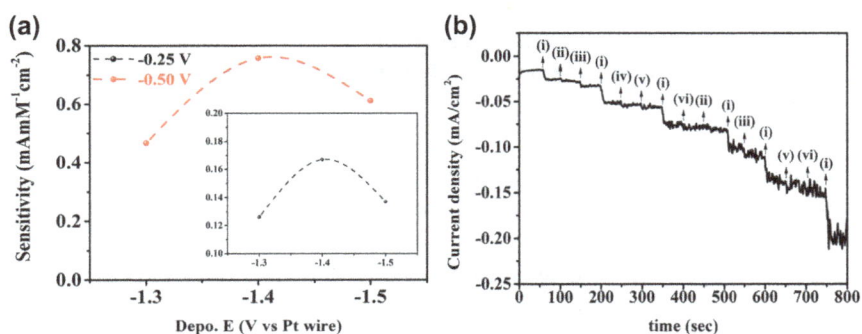

Figure 9.32: (a) Sensitivity of the sensor. (b) Selectivity of the sensor (i) hydrogen peroxide (ii) D-glucose (iii) D-fructose (iv) sucrose (v) lactic acid (vi) ascorbic acid. Adapted with permission from [76].

Kim et al. fabricated one-dimensional structures of Tellurium dioxide by heating of Te powders and investigated the NO$_2$ sensing property [91]. The received Tellurium dioxide products were crystalline with a tetragonal structure and displayed a high transmission rate. The room-temperature optical transmittance and Raman spectrum

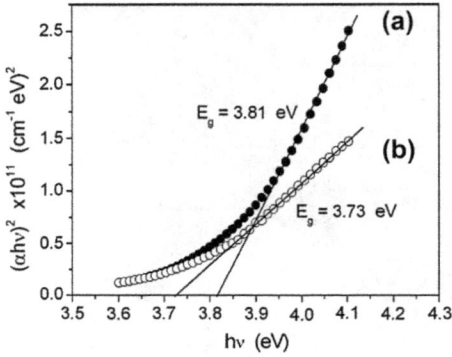

Figure 9.33: Plot of absorption coefficient versus photon energy for (a) as-deposited tellurium dioxide thin films and (b) heat-treated tellurium dioxide thin films. Adapted with permission from [89].

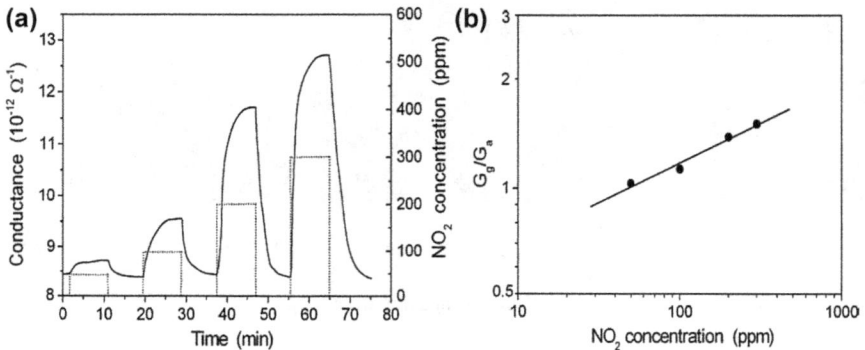

Figure 9.34: (a) Room temperature gas-sensing response of the tellurium dioxide thin films to diverse nitrogen dioxide gas concentration. (b) Log-log plot of G_g/G_a versus NO_2 concentration. Adapted with permission from [89].

of the TeO_2 nanowire films are shown in Figure 9.35. The sensing study demonstrated a linear relationship among the nitrogen dioxide gas concentration and sensitivity (Figure 9.36).

Porous TeO_2 microtubes were demonstrated as room-temperature NO_2 gas sensors [83]. The sensors were synthesized by the thermal annealing of Tellurium microtubes for 2 h at 420 °C in an oxygen atmosphere. Figure 9.37 shows the schematic diagram of the proposed sensor created with Tellurium dioxide microtubes.

The study showed that the sensitivity of the sensor enhanced with NO_2 concentration in the range from 5 to 200 ppm and saturated at higher levels. Changes in the electrical conductance during exposure to diverse gas concentrations are shown in Figure 9.38(a). The time of response decreased and the time of recovery increased with NO_2 concentration. At the same time, at higher concentrations the recovery and response times were stable (Figure 9.38(b)).

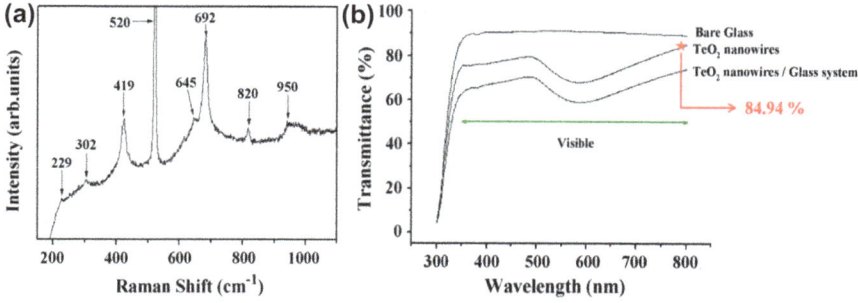

Figure 9.35: (a) Room-temperature Raman spectrum of tellurium dioxide nanowires. (b) Optical transmittance spectrum. Adapted with permission from [91].

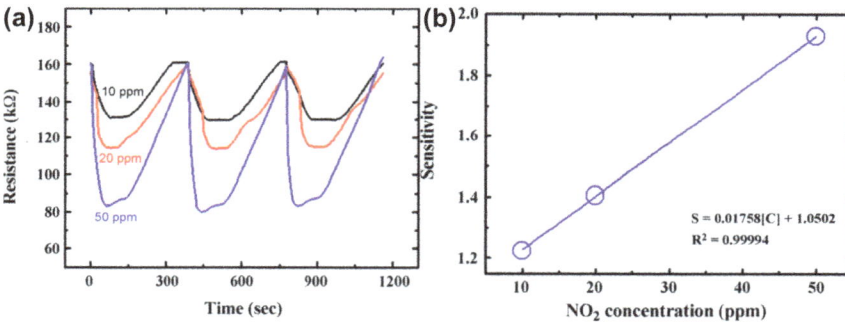

Figure 9.36: (a) Dynamic response of a tellurium dioxide nanowire sensor to nitrogen dioxide gas. (b) Changes in the sensitivity with varying nitrogen dioxide concentration. Adapted with permission from [91].

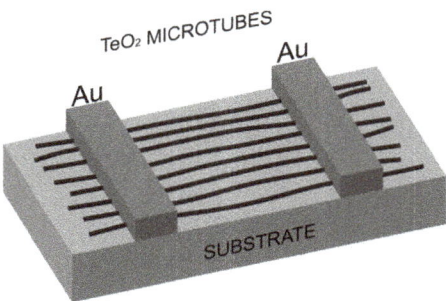

Figure 9.37: Schematic diagram of the proposed gas sensor. Adapted with permission from [83].

Wu et al. reported the sensing properties of p-porous silicon (substrate)/ p-Tellurium dioxide (nanowires) sensor for nitrogen dioxide sensing [92]. The composite structure was fabricated using porous silicon as a growth substrate and

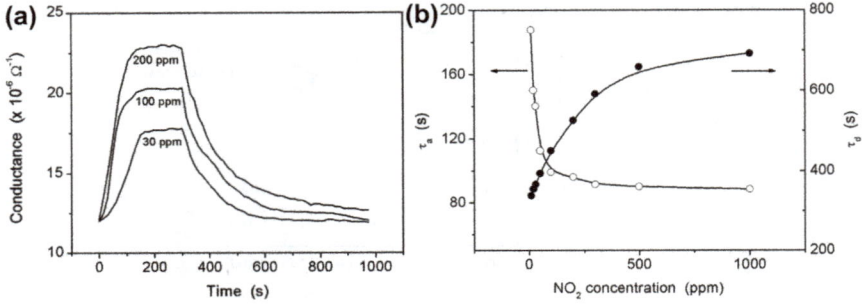

Figure 9.38: (a) Variations in the electrical conductance upon exposure to diverse gas concentrations. (b) Response (τ_a) and recovery (τ_d) times of tellurium dioxide sensor as a function of nitrogen dioxide concentration. Adapted with permission from [83].

Tellurium powder as source materials by the employing thermal evaporation method. Gas sensing characteristics of both composite structure sensor and pristine porous silicon sensor were analysed with nitrogen dioxide concentration in the range of 0.05–3 ppm at different working temperatures. The response of both the sensors by sensing 1 ppm nitrogen dioxide in the temperature range from 26 to 150 °C is presented in Figure 9.39.

Figure 9.40(a) and (b) displays the dynamic responses curves. Figure 9.40(c) depicts the received profiles of gas sensor responses as a function of nitrogen dioxide concentrations. Figure 9.41 compares the responses of the sensors for various gases at room temperature. The data disclosed that the composite structure sensor displayed excellent repeatability, good selectivity and high response to nitrogen dioxide at room temperature.

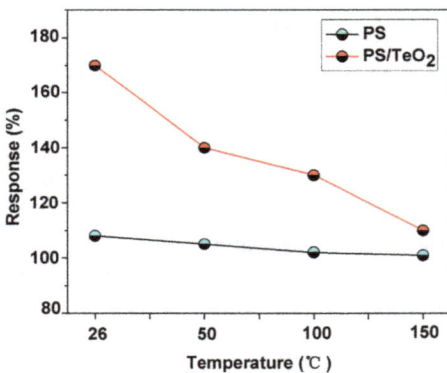

Figure 9.39: The sensors response to 1 ppm nitrogen dioxide at various temperatures. Adapted with permission from [92].

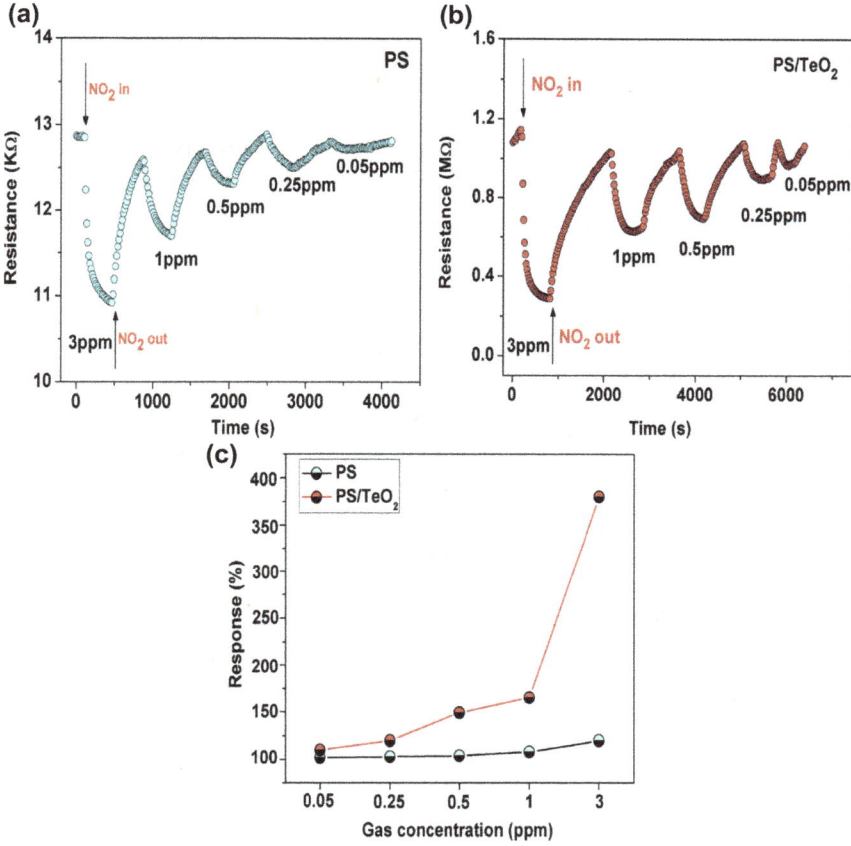

Figure 9.40: Room temperature responses curves of (a) PS and (b) PS/TeO$_2$ nanowire gas sensor (c) the responses of PS/TeO$_2$ and PS nanowire gas sensor as a function of nitrogen dioxide concentrations. Adapted with permission from [92].

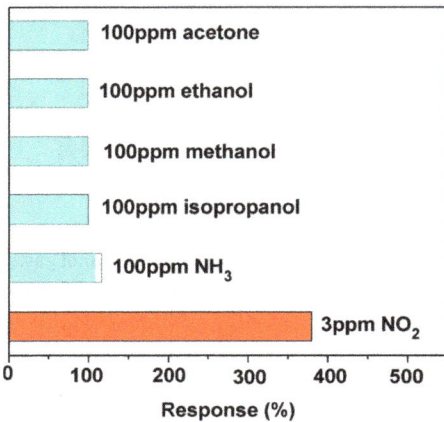

Figure 9.41: Room temperature selectivity histogram of PS/TeO$_2$ nanowire sensor to diverse gases. Adapted with permission from [92].

9.5.2 Ethanol detection

Shen et al. reported the ethanol detection characteristics of Tellurium dioxide thin films synthesized by the non-hydrolytic sol-gel process [93]. Microstructural characterizations depicted that the thin films were described as tetragonal in the Tellurium dioxide structure. The films having porous cotton-shaped nanostructures were gathered by several tellurium dioxide nanostrips. Gas sensing measurements evidenced that at a working temperature of 200 °C the sensors showed the greatest response to ethanol gas (Figure 9.42(a)). The sensors presented quick and reversible responses to ethanol gas at diverse concentrations of ethanol (Figure 9.42(b)) and operating temperatures. The continuous crystalline network and high porosity of TeO_2 thin films were the main factors for their impressive functioning in the sensing of ethanol gas. The sensing mechanism of the sensor is illustrated in Figure 9.43.

A very selective and sensitive room temperature alcohol gas sensor using tellurium dioxide nanowires was demonstrated by Shen et al. [94]. The nanowires were fabricated using Tellurium powders by thermal evaporation at ambient pressure in the air. The synthesized TeO_2 nanowires were about 70–200 nm in diameter and approximately hundreds of micrometers to 2 mm in length. Gas sensing analysis showed that at room temperature the tellurium dioxide nanowires with n-type conduction evidenced a reversible and quick response to ethanol gas (Figure 9.44(a)). They also demonstrated that the sensor response enhanced in the order of methanol>ethanol>propanol under the same conditions and with raising concentration of alcohol, as shown in Figure 9.44(b). Figure 9.45 illustrates the sensing mechanism of tellurium dioxide nanowires to ethanol gas.

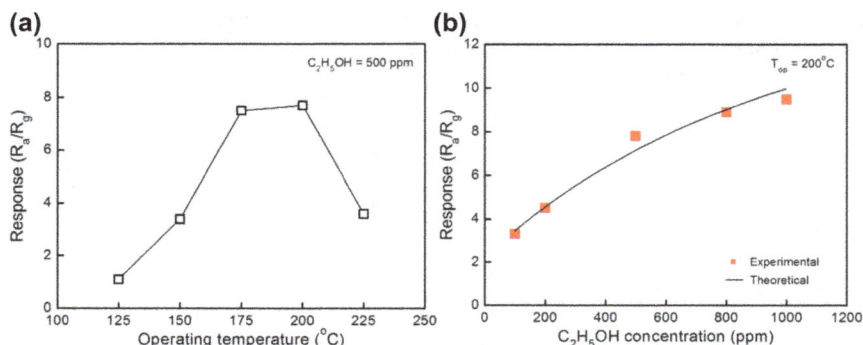

Figure 9.42: (a) Dependence of the response of tellurium dioxide thin films to 500 ppm ethanol gas. (b) Relationship among the response of tellurium dioxide thin films and ethanol gas concentration at a temperature of 200 °C. Adapted with permission from [93].

C$_2$H$_5$OH in   2CO$_2$ + 3H$_2$O

electron depletion layer

O^{2-}

O$_2^-$

e$^-$

e$^-$ e$^-$ O$^-$

e$^-$

O$^-$ e$^-$ e$^-$

e$^-$ O$_2^-$

O^{2-}

conduction band

e$^-$

e$^-$

O$_2$

C$_2$H$_5$OH out

electron depletion layer

conduction band

TeO$_2$ thin film

Al$_2$O$_3$ substrate

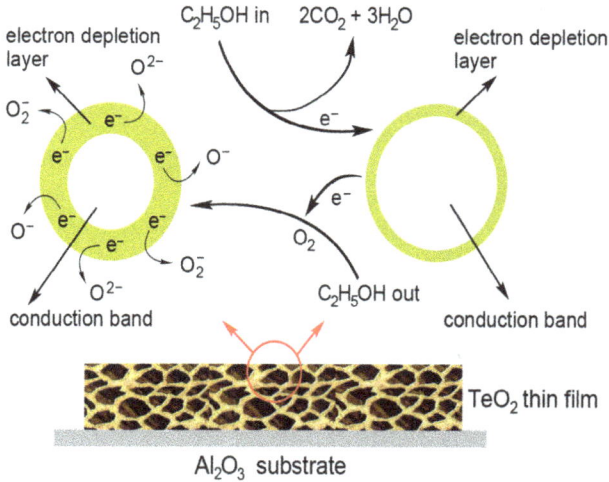

Figure 9.43: Sensing mechanism of sensor during the exposure to ethanol gas. Adapted with permission from [93].

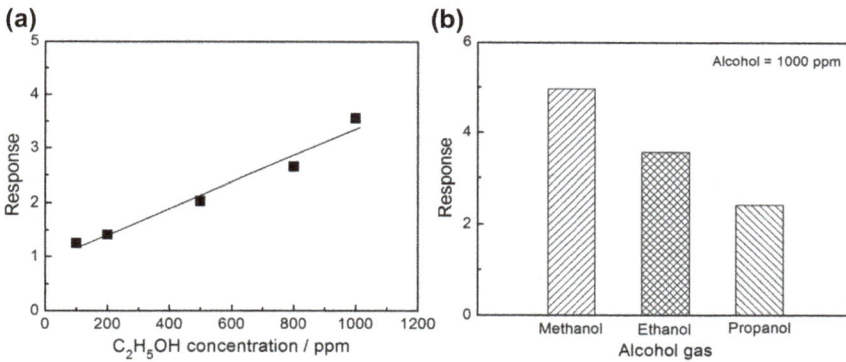

(a)

Response

C$_2$H$_5$OH concentration / ppm

(b)

Alcohol = 1000 ppm

Response

Methanol   Ethanol   Propanol

Alcohol gas

Figure 9.44: (a) Relationship between the room temperature response of tellurium dioxide nanowire gas sensor and ethanol gas concentration. (b) Room temperature responses of tellurium dioxide nanowire gas sensor during exposure to 1000 ppm methanol, propanol, and ethanol gases. Adapted with permission from [94].

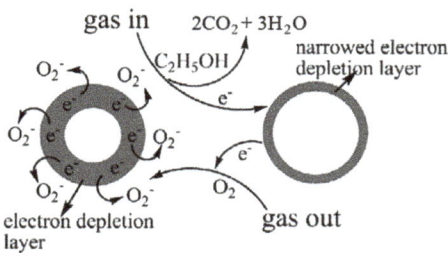

gas in   2CO$_2$ + 3H$_2$O

C$_2$H$_5$OH

O$_2^-$ O$_2^-$

narrowed electron depletion layer

e$^-$

O$_2^-$ e$^-$ e$^-$ O$_2^-$

e$^-$

e$^-$ e$^-$

O$_2^-$ O$_2^-$ O$_2$

electron depletion layer

gas out

Figure 9.45: Sensing mechanism of tellurium dioxide nanowires upon exposure to ethanol gas. Adapted with permission from [94].

9.6 Other Te-based sensors

Same as CdTe nanostructures NiTe nanostructures also can be used for sensing applications. An electrochemical sensor based on carbon paste electrode modified with nickel telluride (CPE/Ni$_{3-x}$Te$_2$) was demonstrated for the sensing of the neurotransmitters dopamine and adrenaline simultaneously in phosphate buffer solution (pH 7.0) [95]. Dopamine (DA) and adrenalin (AD) are the neurotransmitters that stimulate the lungs breathing, the heart beating, and the stomach digestion. In health care medicine, they are clinically important molecules. DA and AD are used for bronchial asthma, hypertension, myocardial infection and cardiac surgery. Several neurological disorders, drug addiction problems and immune-suppressing diseases are associated to low levels of these biomolecules [96, 97].

The CPE/Ni$_{3-x}$Te$_2$ electrode evidenced electrocatalytic characteristics toward dopamine and adrenalin with two distinguished peaks. The 4–31 µmol/L was the linear response range for the finding of both dopamine and adrenalin and the limits of detection were 0.15 µmol/L and 0.35 µmol/L for dopamine and adrenalin, respectively. The important results are depicted in Figure 9.46.

Room temperature ac operating gas sensors using quaternary chalcogenides (As–Ge–S–Te) was reported [98]. Impedance spectra of quaternary chalcogenides-based alloys were analyzed in both dry synthetic air and mixture with NO$_2$ to check the sensing property at room temperature. To analyze the action of Te, the quaternary compositions As$_2$Te$_{130}$Ge$_8$S$_3$ and As$_2$Te$_{13}$Ge$_8$S$_3$, with enhancing concentration of Tellurium was viewed along with pure Te films. At room temperature, the frequency-dependent sensitivity of the films toward nitrogen dioxide is shown in Figure 9.47(a). Figure 9.47(b) displays the effect of humidity on the As$_2$Te$_{13}$Ge$_8$S$_3$-based film to nitrogen dioxide sensing, using impedance measurements.

L-cysteine (L-Cys) is one of the sulfur-containing α-amino acids [99, 100]. As L-cysteine has significant role in biological systems, it is required for phospholipid metabolism in the liver and for cell reduction process [101, 102]. L-cysteine can shield hepatocytes from harm and boost the retrieval of liver function. It is primarily utilized for radiopharmaceutical poisoning, tincture poisoning, and heavy metal poisoning. It can also be employed for serum disease, hepatitis, and toxic hepatitis. Other than these, L-cysteine deficiency can cause many clinical conditions, including early atherosclerosis, diabetes, liver disease, Alzheimer's disease, and Parkinson's disease [103–106]. Different methods are available for the sensing of L-cysteine, such as capillary zone electrophoresis [107], colorimetric assay, spectrophotometry [108], electrochemical techniques [109], and high-performance liquid chromatography [110, 111]. However, most of these processes have drawbacks and are normally less sensitive, complicated, and particular for the sensing of target analytes [112, 113].

Yang et al. demonstrated inner filter effect sensors using Ag nanoparticles (NPs) and CdTeS quantum dots for the sensing of L-cysteine [114]. For the sensor, the ternary alloyed CdTeS quantum dots were fabricated in two steps, employing

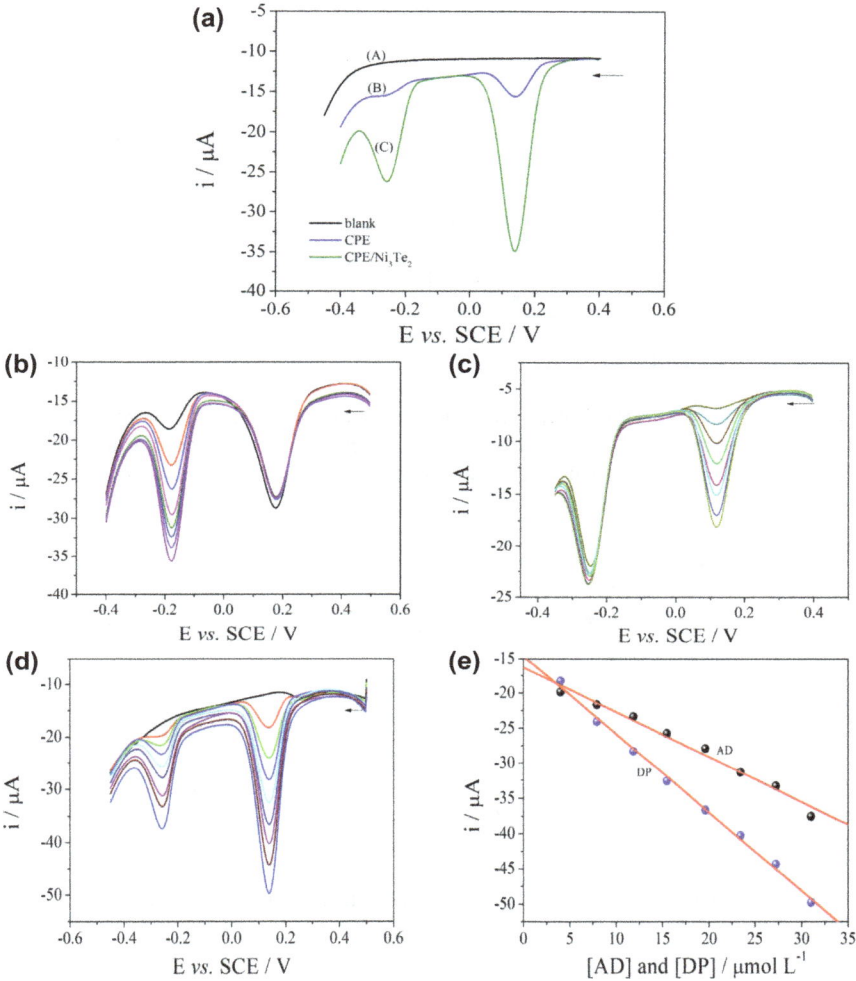

Figure 9.46: (a) Square wave voltammograms of 23 µmol/L dopamine and adrenalin in PBS buffer (A) blank; (B) CPE; (C) CPE/Ni$_{3-x}$Te$_2$. (b) A fixed concentration of 23 µmol/L adrenalin with successive additions of dopamine. (c) A fixed concentration of 23 µmol/L dopamine with successive additions of adrenalin. (d) Square wave voltammograms for the simultaneous determination of dopamine and adrenalin using the CPE/Ni$_{3-x}$Te$_2$ in PBS buffer. (e) Calibration curves for the simultaneous dopamine and adrenalin reduction process. Adapted with permission from [95].

3-mercaptopropionic acid as a stabilizer. The polyvinylpyrrolidone stabilized Ag nanoparticles were made by the crystal-seed method. The CdTeS QDs/Ag NPs (QNs) emitting at 580 nm were employed as sensors for L-cysteine. Figure 9.48 shows the schematic diagram of the CdTeS QDs/Ag NPs for the sensing of L-cysteine. The QNs fluorescence was enhanced with the addition of L-cysteine. The intensity of

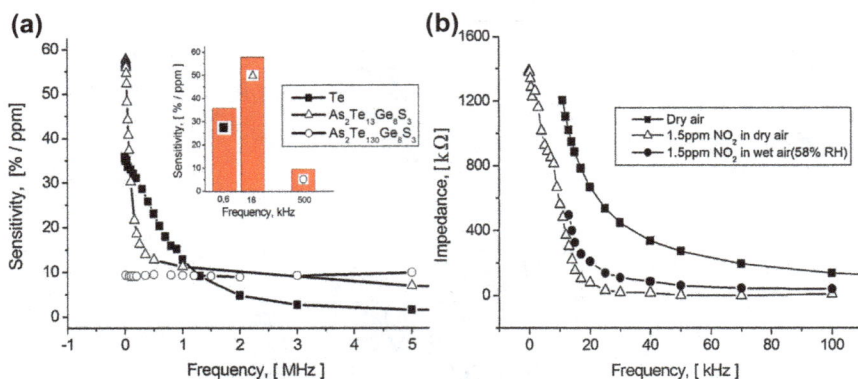

Figure 9.47: (a) Sensitivity spectra of quaternary chalcogenide thin films to 1.5 ppm nitrogen dioxide in dry air at room temperature. (b) Effect of humidity on the frequency-dependent impedance of $As_2Te_{13}Ge_8S_3$-based film to nitrogen dioxide sensing. Adapted with permission from [98].

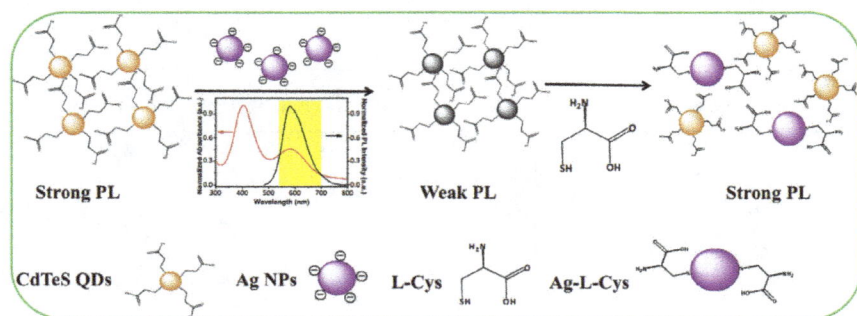

Figure 9.48: The schematic diagram of the CdTeS QDs/Ag NPs for the sensing of L-cysteine. Adapted with permission from [114].

fluorescence was linearly proportional to the L-cysteine concentrations (20–400 µM) as shown in Figure 9.49. The coefficient of correlation was 0.9942 and the limit of detection was observed at 0.025 µM.

Even though many gas sensors based on chalcogenide materials were fabricated for monitoring pollutants in ambient air. Most attractive in this respect were sensors based on pure tellurium or its alloys. For example, Te-based sensors showed remarkable sensing properties to NO_2 even at room temperature, with high sensitivity and low response time [69]. Considering sensors based on semiconductor oxides, for example In_2O_3 and WO_3, most of the sensors have been used for the detection of gases (for example chlorine gas) but most of these sensors require elevated temperatures for operation. Te-based sensor can overcome these issues [72]. Thus, Te is an auspicious candidate for sensor application.

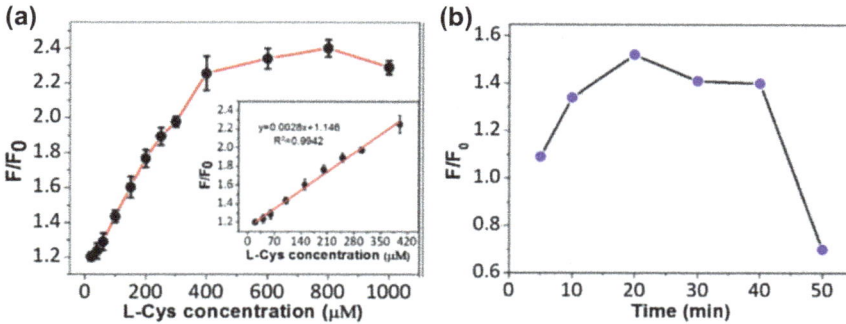

Figure 9.49: (a) Relationship between fluorescence intensity F/F_0 and L-cysteine concentrations. (b) The effect of the response time of CdTeS QDs/Ag NPs to L-cysteine on detection results. Adapted with permission from [114].

9.7 Conclusions

Because of its unique physical and chemical properties, Tellurium is a promising and important functional material. This article has discussed the synthesis and sensing properties of Te-based sensors. Te-based sensors have shown promising results. The unique properties and morphologies of these sensors may also contribute to the great improvement in the gas sensing field.

Acknowledgments: AD and BKB are grateful to Prince Mohammad Bin Fahd University for support.

References

1. Ikari T, Berger H, Levy F. Electrical properties of vapour grown tellurium single crystals. Mater Res Bull 1986;21:99–105.
2. Sen S, Muthe K, Joshi N, Gadkari SC, Gupta SK, Jagannath J, et al. Room temperature operating ammonia sensor based on tellurium thin films. Sensor Actuator B Chem 2004;98:154–9.
3. Li Y, Lin X. Simultaneous electroanalysis of dopamine, ascorbic acid and uric acid by poly (vinyl alcohol) covalently modified glassy carbon electrode. Sensor Actuator B Chem 2006;115:134–9.
4. Huang Z-N, Zou J, Teng J, Liu Q, Yuan M-M, Jiao F-P, et al. A novel electrochemical sensor based on self-assembled platinum nanochains—multi-walled carbon nanotubes-graphene nanoparticles composite for simultaneous determination of dopamine and ascorbic acid. Ecotoxicol Environ Saf 2019;172:167–75.
5. Chen Q, Espey MG, Sun AY, Pooput C, Kirk KL, Krishna MC, et al. Pharmacologic doses of ascorbate act as a prooxidant and decrease growth of aggressive tumor xenografts in mice. Proc Natl Acad Sci U S A 2008;105:11105–9.

6. Li L, Wang C, Luo J, Guo Q, Liu K, Liu K, et al. Fe^{3+}-functionalized carbon quantum dots: a facile preparation strategy and detection for ascorbic acid in rat brain microdialysates. Talanta 2015; 144:1301–7.
7. Clark Huff J, Weston WL, Tonnesen MG. Erythema multiforme: a critical review of characteristics, diagnostic criteria, and causes. J Am Acad Dermatol 1983;8:763–75.
8. Lee W, Hamernyik P, Hutchinson M, Raisys VA, Labbé RF. Ascorbic acid in lymphocytes: cell preparation and liquid-chromatographic assay. Clin Chem 1982;28:2165–9.
9. Etsu K, Yuko N, Shosuke K. Specific determination of ascorbic acid with chemical derivatization and high-performance liquid chromatography. Anal Chem 1992;64:1505–7.
10. Peng Y, Zhang Y, Ye J. Determination of phenolic compounds and ascorbic acid in different fractions of tomato by capillary electrophoresis with electrochemical detection. J Agric Food Chem 2008;56:1838–44.
11. Shamsipur M, Molaei K, Molaabasi F, Alipour M, Alizadeh N, Hosseinkhani S, et al. Facile preparation and characterization of new green emitting carbon dots for sensitive and selective off/on detection of Fe3+ ion and ascorbic acid in water and urine samples and intracellular imaging in living cells. Talanta 2018;183:122–30.
12. Na W, Qu Z, Chen X, Su X. A turn-on fluorescent probe for sensitive detection of sulfide anions and ascorbic acid by using sulfanilic acid and glutathione functionalized graphene quantum dots. Sensor Actuator B Chem 2018;256:48–54.
13. Wang H, Pu G, Devaramani S, Wang Y, Yang Z, Li L, et al. Bimodal electrochemiluminescence of G-CNQDs in the presence of double coreactants for ascorbic acid detection. Anal Chem 2018;90: 4871–7.
14. Zhang C, Wang L, Wang A, Zhang S, Mao C, Song J, et al. A novel electrochemiluminescence sensor based on nitrogen-doped graphene/CdTe quantum dots composite. Appl Surf Sci 2014;315:22–7.
15. Ding M, Wang K, Fang M, Zhu W, Du L, Li C. MPA-CdTe quantum dots as "on-off-on" sensitive fluorescence probe to detect ascorbic acid via redox reaction. Spectrochim Acta Mol Biomol Spectrosc 2020;234:118249.
16. Tansakul N, Niedorf F, Kietzmann M. A sulfadimidine model to evaluate pharmacokinetics and residues at various concentrations in laying hen. Food Addit Contam 2007;24:598–604. Taylor & Francis.
17. Teixidó M, Hurtado C, Pignatello JJ, Beltrán JL, Granados M, Peccia J. Predicting contaminant adsorption in black carbon (Biochar)-Amended soil for the veterinary antimicrobial sulfamethazine. Environ Sci Technol 2013;47:6197–205.
18. Zhang Y, Xu J, Zhong Z, Guo C, Li L, He Y, et al. Degradation of sulfonamides antibiotics in lake water and sediment. Environ Sci Pollut Res 2013;20:2372–80.
19. Ji K, Kim S, Han S, Seo J, Lee S, Park Y, et al. Risk assessment of chlortetracycline, oxytetracycline, sulfamethazine, sulfathiazole, and erythromycin in aquatic environment: are the current environmental concentrations safe? Ecotoxicology 2012;21:2031–50.
20. Zhou Z, Ying H, Liu Y, Xu W, Yang Y, Luan Y, et al. Synthesis of surface molecular imprinting polymer on SiO_2-coated CdTe quantum dots as sensor for selective detection of sulfadimidine. Appl Surf Sci 2017;404:188–96.
21. Li C, Zeng C, Chen Z, Jiang Y, Yao H, Yang Y, et al. Luminescent lanthanide metal-organic framework test strip for immediate detection of tetracycline antibiotics in water. J Hazard Mater 2020;384:121498.
22. Han Q, Wang R, Xing B, Chi H, Wu D, Wei Q. Label-free photoelectrochemical aptasensor for tetracycline detection based on cerium doped CdS sensitized BiYWO6. Biosens Bioelectron 2018; 106:7–13.

23. Han S, Yang L, Wen Z, Chu S, Wang M, Wang Z, et al. A dual-response ratiometric fluorescent sensor by europium-doped CdTe quantum dots for visual and colorimetric detection of tetracycline. J Hazard Mater 2020;398:122894.
24. Isla A, Trocóniz IF, Canut A, Labora A, Martín-Herrero JE, Pedraz JL, et al. Pharmacokinetic/ pharmacodynamic evaluation of amoxicillin, amoxicillin/clavulanate and ceftriaxone in the treatment of paediatric acute otitis media in Spain. Enferm Infecc Microbiol Clín 2011;29:167–73.
25. Fernández J, Ribeiro IAC, Martin V, Martija OL, Zuza E, Bettencourt AF, et al. Release mechanisms of urinary tract antibiotics when mixed with bioabsorbable polyesters. Mater Sci Eng C 2018;93: 529–38.
26. Mathur S, Fuchs A, Bielicki J, Van Den Anker J, Sharland M. Antibiotic use for community-acquired pneumonia in neonates and children: WHO evidence review. Paediatr Int Child Health 2018;38: S66–75. Taylor & Francis.
27. Wong A, Santos AM, Cincotto FH, Moraes FC, Fatibello-Filho O, Sotomayor MDPT. A new electrochemical platform based on low cost nanomaterials for sensitive detection of the amoxicillin antibiotic in different matrices. Talanta 2020;206:120252.
28. Sun Y, Zhen L, Peng Y, Wang J, Fei F, Aa L, et al. Simultaneous determination of gemcitabine prodrug, gemcitabine and its major metabolite 2′,2′-difluorodeoxyuridine in rat plasma by UFLC-MS/MS. J Chromatogr B 2018;1084:4–13.
29. Yam C, Murthy RK, Valero V, Szklaruk J, Shroff GS, Stalzer CJ, et al. A phase II study of tipifarnib and gemcitabine in metastatic breast cancer. Invest New Drugs 2018;36:299–306.
30. Malekigorji M, Alfahad M, Lin PKT, Jones S, Curtis A, Hoskins C. Thermally triggered theranostics for pancreatic cancer therapy. Nanoscale 2017;9:12735–45. The Royal Society of Chemistry.
31. Sun Y, Tang D, Chen H, Zhang F, Fan B, Zhang B, et al. Determination of gemcitabine and its metabolite in extracellular fluid of rat brain tumor by ultra performance liquid chromatography– tandem mass spectrometry using microdialysis sampling after intralesional chemotherapy. J Chromatogr B 2013;919–920:10–9.
32. Raju N, Rao J, Prakash K, Khagga M. The estimation of gemcitabine hydrochloride in the parenteral dosage forms by gradient RP-HPLC. Orient J Chem 2008;24:135–8.
33. Florea A, Guo Z, Cristea C, Bessueille F, Vocanson F, Goutaland F, et al. Anticancer drug detection using a highly sensitive molecularly imprinted electrochemical sensor based on an electropolymerized microporous metal organic framework. Talanta 2015;138:71–6.
34. Naik KM, Nandibewoor ST. Electro-oxidation and determination of gemcitabine hydrochloride, an anticancer drug at gold electrode. J Ind Eng Chem 2013;19:1933–8.
35. Wang F, Cao M, Wang N, Muhammad N, Wu S, Zhu Y. Simple coupled ultrahigh performance liquid chromatography and ion chromatography technique for simultaneous determination of folic acid and inorganic anions in folic acid tablets. Food Chem 2018;239:62–7.
36. Najafi S, Amani S, Shahlaei M. Rapid determination of the anti-cancer agent Gemcitabine in biological samples by fluorescence sensor based on Au-doped CdTe. J Mol Liq 2018;266:514–21.
37. Bai J, Zhang X, Peng Y, Hong X, Liu Y, Jiang S, et al. Ultrasensitive sensing of diethylstilbestrol based on AuNPs/MWCNTs-CS composites coupling with sol-gel molecularly imprinted polymer as a recognition element of an electrochemical sensor. Sensor Actuator B Chem 2017;238:420–6.
38. Jiang W, Yan Y, Ma M, Wang D, Luo Q, Wang Z, et al. Assessment of source water contamination by estrogenic disrupting compounds in China. J Environ Sci (China) 2012;24:320–8.
39. Zhang Q-L, Li J, Ma T-T, Zhang Z-T. Chemiluminescence screening assay for diethylstilbestrol in meat. Food Chem 2008;111:498–502.
40. Sanfilippo K, Pinto B, Colombini MP, Bartolucci U, Reali D. Determination of trace endocrine disruptors in ultrapure water for laboratory use by the yeast estrogen screen (YES) and chemical analysis (GC/MS). J Chromatogr B 2010;878:1190–4.

41. Qin Y, Zhang J, Li Y, Han Y, Zou N, Jiang Y, et al. Multiplug filtration cleanup method with multi-walled carbon nanotubes for the analysis of malachite green, diethylstilbestrol residues, and their metabolites in aquatic products by liquid chromatography–tandem mass spectrometry. Anal Bioanal Chem 2016;408:5801–9.

42. Wang Y. Electrochemical determination of diethylstilbestrol in animal food using a poly polylysine/graphene modified electrode. Int J Electrochem Sci 2019:5763–76. https://doi.org/10.20964/2019.06.35.

43. Liu S, Lin Q, Zhang X, He X, Xing X, Lian W, et al. Electrochemical immunosensor based on mesoporous nanocomposites and HRP-functionalized nanoparticles bioconjugates for sensitivity enhanced detection of diethylstilbestrol. Sensor Actuator B Chem 2012;166–167:562–8.

44. Tang J, Xiang L, Zhao F, Pan F, Wang S, Zhan X. Development of an up-conversion homogenous immunoassay for the determination of diethylstilbestrol in water. Anal Lett 2015;48:796–808. Taylor & Francis.

45. Zhao W-R, Kang T-F, Xu Y-H, Zhang X, Liu H, Ming A-J, et al. Electrochemiluminescence solid-state imprinted sensor based on graphene/CdTe@ZnS quantum dots as luminescent probes for low-cost ultrasensing of diethylstilbestrol. Sensor Actuator B Chem 2020;306:127563.

46. Dill GM. Glyphosate-resistant crops: history, status and future. Pest Manag Sci 2005;61:219–24.

47. González-Martínez MÁ, Brun EM, Puchades R, Maquieira Á, Ramsey K, Rubio F. Glyphosate immunosensor. Application for water and soil analysis. Anal Chem. American Chemical Society 2005;77:4219–27.

48. Marc J, Mulner-Lorillon O, Bellé R. Glyphosate-based pesticides affect cell cycle regulation. Biol Cell 2004;96:245–9.

49. Accinelli C, Koskinen WC, Seebinger JD, Vicari A, Sadowsky MJ. Effects of incorporated corn residues on glyphosate mineralization and sorption in soil. J Agric Food Chem. American Chemical Society 2005;53:4110–7.

50. Werth JA, Preston C, Taylor IN, Charles GW, Roberts GN, Baker J. Managing the risk of glyphosate resistance in Australian glyphosate- resistant cotton production systems. Pest Manag Sci 2008;64:417–21.

51. Walsh LP, McCormick C, Martin C, Stocco DM. Roundup inhibits steroidogenesis by disrupting steroidogenic acute regulatory (StAR) protein expression. Environ Health Perspect 2000;108:769–76.

52. Arregui MC, Lenardón A, Sanchez D, Maitre MI, Scotta R, Enrique S. Monitoring glyphosate residues in transgenic glyphosate-resistant soybean. Pest Manag Sci 2004;60:163–6.

53. Schönherr J, Schreiber L. Interactions of calcium ions with weakly acidic active ingredients slow cuticular penetration: a case study with glyphosate. J Agric Food Chem 2004;52:6546–51.

54. Kleter GA, Harris C, Stephenson G, Unsworth J. Comparison of herbicide regimes and the associated potential environmental effects of glyphosate-resistant crops versus what they replace in Europe. Pest Manag Sci 2008;64:479–88.

55. Bera MK, Mohapatra S. Ultrasensitive detection of glyphosate through effective photoelectron transfer between CdTe and chitosan derived carbon dot. Colloids Surf A Physicochem Eng Asp 2020;596:124710.

56. Campbell L, Dixon DG, Hecky RE. A review of mercury in Lake Victoria, East Africa: implications for human and ecosystem health. J Toxicol Environ Health B Crit Rev 2003;6:325–56.

57. Wren CD. A review of metal accumulation and toxicity in wild mammals: I. Mercury. Environ Res 1986;40:210–44.

58. Driscoll CT, Mason RP, Chan HM, Jacob DJ, Pirrone N. Mercury as a global pollutant: sources, pathways, and effects. Environ Sci Technol. American Chemical Society 2013;47:4967–83.

59. Hallaj R, Hosseinchi Z, Babamiri B, Zandi S. Synthesis and characterization of novel bithiazolidine derivatives-capped CdTe/CdS quantum dots used as a novel Hg^{2+} fluorescence sensor. Spectrochim Acta A Mol Biomol Spectrosc 2019;216:418–23.
60. Jaiswal J, Sanger A, Tiwari P, Chandra R. MoS2 hybrid heterostructure thin film decorated with CdTe quantum dots for room temperature NO2 gas sensor. Sensor Actuator B Chem 2020;305: 127437.
61. Reithmaier JP, Paunovic P, Kulisch W, Popov C, Petkov P. Nanotechnological Basis for Advanced Sensors. Springer; 2011.
62. Aswal DK, Gupta SK. Science and Technology of Chemiresistor Gas Sensors. Nova Science Publishers; 2007.
63. Tsiulyanu D, Marian S, Miron V, Liess H-D. High sensitive tellurium based NO 2 gas sensor. Sensor Actuator B Chem 2001;1:35–9.
64. Sen S, Bhandarkar V, Muthe K, Roy M, Deshpande S, Aiyer R, et al. Highly sensitive hydrogen sulphide sensors operable at room temperature. Sensor Actuator B Chem 2006;115:270–5.
65. Tsiulyanu D, Marian S, Liess H-D. Sensing properties of tellurium based thin films to propylamine and carbon oxide. Sensor Actuator B Chem 2002;85:232–8.
66. Tsiulyanu D, Stratan I, Tsiulyanu A, Liess H-D, Eisele I. Investigation of the oxygen, nitrogen and water vapour cross-sensitivity to NO 2 of tellurium-based thin films. Sens Actuators B: Chem 2007;121:406–13.
67. Bhandarkar V, Sen S, Muthe KP, Kaur M, Kumar MS, Deshpande SK, et al. Effect of deposition conditions on the microstructure and gas-sensing characteristics of Te thin films. Mater Sci Eng B 2006;1:156–61.
68. Siciliano T, Di Giulio M, Tepore M, Filippo E, Micocci G, Tepore A. Tellurium sputtered thin films as NO2 gas sensors. Sensor Actuator B Chem 2008;135:250–4.
69. Tsiulyanu D, Tsiulyanu A, Liess H-D, Eisele I. Characterization of tellurium-based films for NO2 detection. Thin Solid Films 2005;1–2:252–6.
70. Tsiulyanu D, Mocreac O. Concentration induced damping of gas sensitivity in ultrathin tellurium films. Sensor Actuator B Chem 2013;177:1128–33.
71. Manouchehrian M, Larijani MM, Elahi SM. Thickness and UV irradiation effects on the gas sensing properties of Te thin films. Mater Res Bull 2015;62:177–83.
72. Sen S, Sharma M, Kumar V, Muthe KP, Satyam PV, Bhatta UM, et al. Chlorine gas sensors using one-dimensional tellurium nanostructures. Talanta 2009;77:1567–72.
73. Yang L, Lin H, Zhang Z, Cheng L, Ye S, Shao M. Gas sensing of tellurium-modified silicon nanowires to ammonia and propylamine. Sensor Actuator B Chem 2013;177:260–4.
74. Guan L, Wang S, Gu W, Zhuang J, Jin H, Zhang W, et al. Ultrasensitive room-temperature detection of NO_2 with tellurium nanotube based chemiresistive sensor. Sensor Actuator B Chem 2014;196: 321–7.
75. Manikandan M, Dhanuskodi S, Maheswari N, Muralidharan G, Revathi C, Rajendra Kumar RT, et al. High performance supercapacitor and non-enzymatic hydrogen peroxide sensor based on tellurium nanoparticles. Sens Bio-Sens Res 2017;13:40–8.
76. Waldiya M, Bhagat D, RN, Singh S, Kumar A, Ray A, et al. Development of highly sensitive H2O2 redox sensor from electrodeposited tellurium nanoparticles using ionic liquid. Biosens Bioelectron 2019;132:319–25.
77. Venkataiah G, Jayasankar CK, Venkata Krishnaiah K, Dharmaiah P, Vijaya N. Concentration dependent luminescence properties of SM^{3+}-ions in tellurite–tungsten–zirconium glasses. Opt Mater 2015;40:26–35.
78. Antonov SN, Proklov VV, Rezvov YG, Chesnokov LN, Chesnokov VN. Improving the efficiency of an acoustooptic modulator with a two-lobe directivity pattern by correcting the two-frequency

electric signal. Technical Physics [Internet]; 2006. [cited 2021 Aug 14];51. Available from: https://link.springer.com/epdf/10.1134/s1063784206010099.

79. Shoham S, O'Connor DH, Sarkisov DV, Wang SS-H. Rapid neurotransmitter uncaging in spatially defined patterns. Nat Methods 2005;2:837–43.

80. Pi X-X, Cao X-H, Fu Z-X, Zhang L, Han P-D, Wang L-X, et al. Application of Te-based glass in silicon solar cells. Acta Metall Sin 2015;28:223–9.

81. Hodgson SNB, Weng L. Sol-gel processing of tellurium oxide and suboxide thin films with potential for optical data storage application. J Sol Gel Sci Technol 2000;18:145–58.

82. Mantsevich S, Korablev O, Kalinnikov YK, Ivanov A, Kiselev AV. Examination of temperature influence on wide-angle paratellurite crystal acousto-optic filters operation. Acta Phys Pol, A 2015;127:43–5.

83. Siciliano T, Genga A, Micocci G, Siciliano M, Tepore M, Tepore A. Porous tellurium oxide microtubes for room-temperature NO_2 gas sensors. Sensor Actuator B Chem 2014;201:138–43.

84. Park S, An S, Ko H, Lee C. Enhanced ethanol sensing properties of TeO_2 nanorods functionalized with Co_3O_4 nanoparticles. J Nanosci Nanotechnol 2015;15:439–44.

85. Park S, An S, Ko H, Jin C, Lee C. Enhancement of ethanol sensing of TeO_2 nanorods by Ag functionalization. Curr Appl Phys 2013;13:576–80.

86. Shen Y, Fan A, Wei D, Gao S, Liu W, Han C, et al. A low-temperature n-propanol gas sensor based on TeO_2 nanowires as the sensing layer. RSC Adv 2015;5:29126–30.

87. Liu Z, Yamazaki T, Shen Y, Kikuta T, Nakatani N. Synthesis and characterization of TeO_2 nanowires. Jpn J Appl Phys 2008;47:771.

88. Liu Z, Yamazaki T, Shen Y, Kikuta T, Nakatani N, Kawabata T. Room temperature gas sensing of p-type TeO_2 nanowires. Appl Phys Lett 2007;90:173119.

89. Siciliano T, Di Giulio M, Tepore M, Filippo E, Micocci G, Tepore A. Room temperature NO_2 sensing properties of reactively sputtered TeO_2 thin films. Sensor Actuator B Chem 2009;137:644–8.

90. Siciliano T, Di Giulio M, Tepore M, Filippo E, Micocci G, Tepore A. Ammonia sensitivity of rf sputtered tellurium oxide thin films. Sensor Actuator B Chem 2009;138:550–5.

91. Kim SS, Park JY, Choi S-W, Na HG, Yang JC, Kwak DS, et al. Drastic change in shape of tetragonal TeO_2 nanowires and their application to transparent chemical gas sensors. Appl Surf Sci 2011; 258:501–6.

92. Wu Y, Hu M, Qin Y, Wei X, Ma S, Yan D. Enhanced response characteristics of p-porous silicon (substrate)/p-TeO_2 (nanowires) sensor for NO_2 detection. Sensor Actuator B Chem 2014;195: 181–8.

93. Shen Y, Yan X, Zhao S, Chen X, Wei D, Gao S, et al. Ethanol sensing properties of TeO_2 thin films prepared by non-hydrolytic sol–gel process. Sensor Actuator B Chem 2016;230:667–72.

94. Shen Y, Zhao S, Ma J, Chen X, Wang W, Wei D, et al. Highly sensitive and selective room temperature alcohol gas sensors based on TeO_2 nanowires. J Alloys Compd 2016;664:229–34.

95. de Fatima Ulbrich K, Winiarski JP, Jost CL, Maduro de Campos CE. Mechanochemical synthesis of a $Ni_{3-x}Te_2$ nanocrystalline composite and its application for simultaneous electrochemical detection of dopamine and adrenaline. Compos B Eng 2020;183:107649.

96. Si B, Song E Recent Advances in the Detection of Neurotransmitters. Chemosensors 2020;6:1.

97. Pradhan S, Das R, Biswas S, Das DK, Bhar R, Bandyopadhyay R, et al. Chemical synthesis of nanoparticles of nickel telluride and cobalt telluride and its electrochemical applications for determination of uric acid and adenine. Electrochim Acta 2017;238:185–93.

98. Tsiulyanu D, Ciobanu M. Room temperature A.C. operating gas sensors based on quaternary chalcogenides. Sensor Actuator B Chem 2016;223:95–100.

99. Nie L, Ma H, Sun M, Li X, Su M, Liang S. Direct chemiluminescence determination of cysteine in human serum using quinine–Ce(IV) system. Talanta 2003;59:959–64.

100. Deng C, Chen J, Chen X, Wang M, Nie Z, Yao S. Electrochemical detection of l-cysteine using a boron-doped carbon nanotube-modified electrode. Electrochim Acta 2009;54:3298–302.
101. Huang H, Liu X, Hu T, Chu PK. Ultra-sensitive detection of cysteine by gold nanorod assembly. Biosens Bioelectron 2010;25:2078–83.
102. Sattarahmady N, Heli H. An electrocatalytic transducer for l-cysteine detection based on cobalt hexacyanoferrate nanoparticles with a core–shell structure. Anal Biochem 2011;409:74–80.
103. Lin W, Long L, Yuan L, Cao Z, Chen B, Tan W. A ratiometric fluorescent probe for cysteine and homocysteine displaying a large emission shift. Org Lett 2008;10:5577–80.
104. Yan F, Shi D, Zheng T, Yun K, Zhou X, Chen L. Carbon dots as nanosensor for sensitive and selective detection of Hg^{2+} and l-cysteine by means of fluorescence "Off–On" switching. Sensor Actuator B Chem 2016;224:926–35.
105. Hou C, Fan S, Lang Q, Liu A. Biofuel cell based self-powered sensing platform for l-cysteine detection. Anal Chem 2015;87:3382–7.
106. Zong J, Yang X, Trinchi A, Hardin S, Cole I, Zhu Y, et al. Carbon dots as fluorescent probes for "off–on" detection of Cu^{2+} and l-cysteine in aqueous solution. Biosens Bioelectron 2014;51:330–5.
107. Jin W, Chen H. Theory concerning the current for an end-column amperometric detector with a disk working electrode in capillary zone electrophoresis. J Chromatogr A 1997;765:307–14.
108. Lunar ML, Rubio S, Pérez-Bendito D, Carreto ML, McLeod CW. Hexadecylpyridinium chloride micelles for the simultaneous kinetic determination of cysteine and cystine by their induction of the iodine-azide reaction. Anal Chim Acta 1997;337:341–9.
109. Salimi A, Pourbeyram S. Renewable sol–gel carbon ceramic electrodes modified with a Ru-complex for the amperometric detection of l-cysteine and glutathione. Talanta 2003;60:205–14.
110. Kuśmierek K, Głowacki R, Bald E. Analysis of urine for cysteine, cysteinylglycine, and homocysteine by high-performance liquid chromatography. Anal Bioanal Chem 2006;385: 855–60.
111. Tcherkas YV, Denisenko AD. Simultaneous determination of several amino acids, including homocysteine, cysteine and glutamic acid, in human plasma by isocratic reversed-phase high-performance liquid chromatography with fluorimetric detection. J Chromatogr A 2001;913: 309–13.
112. Wei X, Qi L, Tan J, Liu R, Wang F. A colorimetric sensor for determination of cysteine by carboxymethyl cellulose-functionalized gold nanoparticles. Anal Chim Acta 2010;671:80–4.
113. Huang S, Xiao Q, Li R, Guan H-L, Liu J, Liu X-R, et al. A simple and sensitive method for l-cysteine detection based on the fluorescence intensity increment of quantum dots. Anal Chim Acta 2009; 645:73–8.
114. Yang M, Yan Y, Shi H, Wang C, Liu E, Hu X, et al. Efficient inner filter effect sensors based on CdTeS quantum dots and Ag nanoparticles for sensitive detection of l-cysteine. J Alloys Compd 2019;781: 1021–7.

Priya Rose Thankamani* and Sheenu Thomas

10 Tellurium based materials for nonlinear optical applications

Abstract: Materials having broadband nonlinear optical responses find applications in photonics and optoelectronics devices. Novel materials with improved nonlinear optical properties are necessary for realizing effective all-optical switches, modulators etc. Tellurium (Te) and novel low-dimensional derivatives of Te offer intriguing nonlinear optical responses, making them promising candidates for design of various photonic devices.

Keywords: 2-D material; nonlinear optics; tellurium.

10.1 Introduction

Tellurium is a well-known p-type semiconductor belonging to the chalcogenide group elements (group-VI) with a bandgap of 0.35 eV. It has been extensively investigated due to the interesting properties such as piezoelectricity [1], thermoelectricity [2] and photoconductivity [3]. These fascinating properties possessed by monoelement Te makes it a wonderful material to be used for sensors and optoelectronic applications.

Recently nanomaterials based on Te such as nanoclusters and nanotubes were synthesized and investigated for their applications in ultrafast switching and nonlinear optical responses. Another important development is the successful synthesis of 2D Tellurium (Tellurene) by several groups using methods such as liquid phase exfoliation. After the successful fabrication of graphene and identification of its interesting properties such as high conductivity and broadband optical transitivity, there has been a growing interest in developing various 2-dimensional materials. Black phosphorous and transition metal dichalcogenides (TMDCs) have gained attention due to their ability to possess 2-dimensional phase with properties which are otherwise difficult to obtain. But these materials also suffer from certain challenges due to their inherent properties. Graphene is a zero-bandgap material [4, 5] whereas black phosphorous is quite unstable under ambient conditions [6]. TMDCs suffer from low current mobility [7].

*Corresponding author: Priya Rose Thankamani, International School of Photonics, Cochin University of Science and Technology, Cochin 682022, Kerala, India; and Inter University Center for Nanomaterials and Devices (IUCND), Cochin University of Science and Technology, Cochin 682022, Kerala, India, E-mail: priyarose@cusat.ac.in
Sheenu Thomas, International School of Photonics, Cochin University of Science and Technology, Cochin 682022, Kerala, India

As per De Gruyter's policy this article has previously been published in the journal Physical Sciences Reviews. Please cite as: P. R. Thankamani and S. Thomas "Tellurium based materials for nonlinear optical applications" *Physical Sciences Reviews* [Online] 2022. DOI: 10.1515/psr-2021-0117 | https://doi.org/10.1515/9783110735840-010

Recent studies based on first principle density function theory reveals that 2D Te can form a stable state with very high carrier mobility [8]. 2D form of tellurium, tellurene was successfully synthesized recently using liquid-phase exfoliation [9]. Apart from their intriguing optoelectronic properties, 2D tellurium also shows excellent nonlinear optical response.

10.2 Nonlinear optics

The advent of lasers paved the way to extensive investigations of interaction of matter with intense light. When the field strength of the optical field is comparable to that of the atomic and interatomic fields, we start to see the intensity dependent optical properties exhibited by materials. At such large intensities, the relationship between the material polarization and the electric field will no longer be linear. This leads to some interesting effects, which can be harnessed for a wide variety of applications such as optical switches, modulators etc. [10]. This is especially important for achieving all-optical photonic devices.

When placed in an electric field, a dielectric medium will be polarized with all constituent molecules acting like dipoles. The dipole moment vector per unit volume can be written as

$$P = \sum_i P_i \tag{10.1}$$

where the summation is over all the dipoles in the unit volume.

The above polarization depends both on the strength of the electric field and the properties of the medium. So it can be written as

$$P = \epsilon_0 \chi E \tag{10.2}$$

where χ is the polarizability or dielectric susceptibility of the medium.

The above relation is valid for light coming from conventional sources. When we use high-intensity sources such as lasers, with associated electric fields between 10^7 to 10^{10} V/cm, the relationship between polarization and field become a more general one with nonlinear terms as follows because of the perturbed atomic and interatomic fields:

$$P = \epsilon_0\left(\chi^{(1)}E + \chi^{(2)}E^2 + \chi^{(3)}E^3 + \dots\right), \tag{10.3}$$

where $\chi^{(1)}$ is the linear optical susceptibility, $\chi^{(2)}$ is the second-order nonlinear optical susceptibility, $\chi^{(3)}$ is the third-order nonlinear optical susceptibility etc.

If we consider an optical field whose electric field is given by $E = E_0 \cos \omega t$, then equation (10.3) becomes:

$$P = \epsilon_0 E_0 \left(\chi^{(1)} \cos \omega t + \chi^{(2)} E_0 \cos^2 \omega t + \chi^{(3)} E_0^2 \cos^3 \omega t + \ldots \right). \tag{10.4}$$

Equation (10.4) can be re-written using trigonometric relations as:

$$P = \frac{1}{2}\epsilon_0 \chi^{(2)} E_0^2 + \epsilon_0 \left(\chi^{(1)} + \frac{3}{4}\chi^{(3)} E_0^2 \right) E_0 \cos \omega t + \frac{1}{2}\epsilon_0 \chi^{(2)} E_0^2 \cos 2 \omega t$$

$$+ \frac{1}{4}\epsilon_0 \chi^{(3)} E_0^3 \cos 3 \omega t + \ldots. \tag{10.5}$$

This indicates the appearance new frequencies such as 2ω, 3ω etc. as a result of nonlinear polarization.

If we look at each term in equation (10.5), the first term is just a DC field contribution whereas the second term has an oscillatory part with frequency ω. The third and fourth terms will give oscillatory contributions at frequencies 2ω and 3ω. This means that, when the contribution of nonlinear part is significant, we can obtain higher-order frequencies in the output- referred and harmonic generation. Apart from harmonic generation, other useful and interesting phenomena also take place under nonlinear regime. Nonlinear optical effects occur in the medium where the interaction can be described by a coupled wave equation with the nonlinear susceptibility as the coupling coefficient.

10.3 Second-order nonlinear optical effects

Optical second harmonic generation first demonstrated by Franken et al. [11] in quartz crystal irradiated with ruby laser was a starting point for the field of nonlinear optics. As the second-order susceptibility is zero for media with inversion symmetry under electric dipole approximation, not all crystals will have a non-zero value of $\chi^{(2)}$[12].

In order to describe the second-order NLO processes, we consider the second-order polarization from equation (10.3) omitting all higher-order terms for the time being. That is,

$$P_{NL} = P^{(2)} = \epsilon_0 \chi^{(2)} E^2. \tag{10.6}$$

This term leads to sum frequency generation (SFG), difference frequency generation (DFG) and second harmonic generation (SHG) if we consider a more general case of two electric fields corresponding to two different frequencies coherently interacting. For such a case, the ith component of complex polarization can be expressed as:

$$P^{(2)}(\omega_3) = \epsilon_0 \sum_{jk} \chi^{(2)}_{(ijk)} E(\omega_1) E(\omega_2), \tag{10.7}$$

where $\chi^{(2)}_{(ijk)}$ is the second-order susceptibility in tensor notations. Here we can think of three waves having frequencies ω_1, ω_2 and ω_3 being coupled through $\chi^{(2)}_{(ijk)}$.

10.3.1 Sum frequency generation (SFG) and different frequency generation (DFG)

For SFG, $\omega_3 = \omega_1 + \omega_2$, the second-order polarization can be expressed as

$$P^{(2)}(\omega_3 = \omega_1 + \omega_2) = \epsilon_0 \sum_{jk} \chi^{(2)}_{(ijk)}(\omega_3 = \omega_1 + \omega_2)E(\omega_1)E(\omega_2). \tag{10.8}$$

Similarly for DFG, $\omega_2 = \omega_3 - \omega_1$

$$P^{(2)}(\omega_2 = \omega_3 - \omega_1) = \epsilon_0 \sum_{jk} \chi^{(2)}_{(ijk)}(\omega_2 = \omega_3 - \omega_1)E(\omega_3)E(\omega_1). \tag{10.9}$$

The above equations represent fields oscillating at frequencies which are combinations of input frequencies.

10.3.2 Second harmonic generation (SHG)

This can be considered as a special case of sum frequency generation where both the interacting fields have the same frequency, i.e., $\omega_1 = \omega_2$.

In this case the polarization can be expressed as

$$P^{(2)}(2\omega) = \epsilon_0 \sum_{jk} \chi^{(2)}_{(ijk)}(2\omega)E(\omega)E(\omega), \tag{10.10}$$

which represents a field oscillating at frequency 2ω.

10.3.3 Second-order nonlinear optical processes in tellurium-based materials

Investigations for efficient NLO materials capable of showing second-order nonlinear optical properties is important for realization of frequency doublers, optical parametric amplifiers etc. Crystals having non-centrosymmetry such as $LiNbO_3$ and BBO have been used for this purpose. Recently, materials of lower dimensions such as nanoparticles, nanotubes etc. have also been getting attention as efficient second-order nonlinear materials.

Trigonal Tellurium has been proved to exhibit second harmonic generation at 28.0 μm [13]. In this study a bulk Te crystal was used for generating the second harmonic output with phase matching condition. Recent investigations have made use of lower dimensional structures based on Te. Londoño-Calderon et al. [14] recently investigated chiral nature of tellurium nanowires (Te-NW). The chiral nature of Te lattice leads to linear and nonlinear optical dichroism (DC). The nonlinear optical dichroism is highly sensitive to structural asymmetry for lower dimensional structures. The DC properties were investigated by second harmonic generation which strongly depends on the handedness of optical excitation.

Microwave-assisted solution-based synthesis method was used to obtain the Te NWs which are 5.7 ± 1.1 nm in diameter. Structural characterization of the nanowires was done using TEM and four-dimensional scanning transmission electron microscopy (4D-STEM). Detailed analysis using these techniques revealed the intrinsic helical twist of about 25° in a length of 84 nm in the Te NWs. Nonlinear circular dichroism measurement was used to confirm the chirality in NWs. It was observed that the intensity of the second harmonic output was dependent on the handedness of the excitation beam. Left circularly polarized (LCP) and right circularly (RCP) input generated second harmonic output of different intensities.

The second harmonic generation-circular dichroism (SHG-CD) is given by:

$$SHG - CD = \frac{I_{\text{LCP}} - I_{\text{RCP}}}{(I_{\text{LCP}} + I_{\text{RCP}})/2}. \tag{10.11}$$

This quantity was measured to be up to 0.23 for certain angles of incidence, which is two orders of magnitude larger than linear CD. On the contrary, 2D-Te nanoplates did not show any significant SHG-CD which may be due to the minimal contribution to chirality from morphology unlike the Te-NWs.

A systematic analysis of second harmonic generation from one-dimensional tellurium and selenium based on first-principle calculations was carried out by Cheng et al. using DFT [15]. According to their calculations, 1D tellurium chains exhibited superior SHG properties compared to semiconductor materials with similar bandgaps. 1D tellurium is shown to have second-order nonlinear susceptibility almost five times more than that of GaN. These analyses prove the potential of tellurium nanostructures as effective NLO materials. They have also investigated the bulk photovoltaic properties of one-dimensional tellurium. Enhanced NLO response of 1D tellurium can be attributed to its high degree of anisotropy and one-dimensional structure.

Another promising report on second-order properties demonstrated by 2D Te (tellurene) was recently published by Deckoff-Jones et al. [16]. They measured the second harmonic generation from solution derived tellurene flakes using multiphoton spectroscopy. They also investigated the change in the refractive index of tellurene when the angle between applied electric field and (0001) direction of tellurene changes. They also demonstrated the use of tellurene for mid-IR detector as well as modulator applications by integrating it with a chalcogenide glass platform.

10.4 Higher-order nonlinear optical effects

10.4.1 Nonlinear index of refraction

Most of the fascinating applications of nonlinear optical materials are based on the nonlinear index of refraction. There are several complex factors contributing to the occurrence of nonlinear index of refraction. Generally, if the refractive index or its spatial distribution is changed by the application of an optical field, then the medium is said to have a nonlinear index of refraction [17]. A medium might possess a nonlinear refractive index due to optical Kerr effect in which the change in refractive index is proportional to the square of the electric field of the light. Another effect contributing to nonlinear index of refraction is the absorption of the light by the medium which in turn causes an intensity dependent change of refractive index. There are also other effects like heating of the medium due to light which might also cause the change in refractive index with intensity and fluence of the incident light.

The nonlinear refractive index finds a wide variety of applications in optical switching, realization of optical logic gates, passive mode-locking and modulators.

The general dependence of nonlinear refractive index on intensity is given by [17]:

$$n(r,t) = n_0(r,t) + \Delta n[I(r,t)], \tag{10.12}$$

where $n_0(r,t)$ is the lower-intensity value of the refractive index and Δn is the change in refractive index which depends on the intensity $I(r,t)$. The actual form of Δn will depend upon the nature of the process contributing to the change.

10.4.2 Nonlinear optical absorption (NLA)

Nonlinear optical absorption happens when the optical absorption of a sample depends on the incident light intensity. There are several processes which can induce nonlinear optical absorption in materials such as multi-photon absorption, excited state absorption etc. Materials having nonlinear optical absorption find applications as optical limiters and in several areas of laser spectroscopy.

10.4.2.1 Two-photon absorption (TPA/2PA)

When a system is excited to a higher lying energy level from the ground state by the simultaneous absorption of two photons, it is referred to as two-photon absorption. The selection rules for this process are different from those of single photon absorption. In the case of single beam two photon absorption, where two photons from the same optical field are absorbed simultaneously by the material,the following differential equation describes the optical loss:

$$\frac{dI}{dz} = -\alpha I - \beta I^2, \tag{10.13}$$

where α is the linear absorption coefficient and β is the two-photon absorption (TPA) coefficient.

10.4.2.2 Multiphoton absorption

The discussions on TPA can be extended to understand the simultaneous absorption of three or more numbers of photons by a material. For a process in which $(n + 1)$ photons are absorbed from a single beam, we can write:

$$\frac{dI}{dz} = -(\alpha + \gamma^{(n+1)}I^n)I, \tag{10.14}$$

where $\gamma^{(n+1)}$ is the $(n + 1)$-photon absorption coefficient.

10.4.2.3 Excited state absorption

When the incident light intensity is well above the saturation intensity, excited states can also become populated. When this occurs in materials such as semiconductors and large molecular systems in which the density of states near the state involved in the transition is large, the excited electron may fall into one of these states before it decays back to the ground state. So the electrons in the excited state also contribute to the absorption of light by the material which is referred to as excited state absorption.

If the absorption cross-section of the excited state is smaller compared to the ground state, the material shows an increased transmittance with increased light intensity. This is called saturable absorption (SA). On the other hand, if the excited state possesses a large absorption cross-section compared to the ground state, the material will exhibit a lower transmittance at higher intensities. This is called reverse saturable absorption (RSA).

10.4.3 Higher-order nonlinear optical processes in tellurium-based materials

In a recent report, Wu et al. demonstrated the self-phase modulation properties of 2D tellurium [18]. They explored the possibility of realizing a nonlinear photonic diode using 2D Te.

When the medium possesses intensity dependent refractive index, the incident laser light will lead to refractive index gradient within the medium depending on the transverse intensity profile of the beam. This change in refractive index will in turn result in a phase change of the incident light beam. Basically, the phase of the beam is modulated by its own intensity. This is known as self-phase modulation (SPM). The light waves at different transverse positions will experience different phase changes and thus result in spatial self-phase modulation (SSPM). SSPM can be used effectively to study the nonlinear optical response of materials. There are several studies where the nonlinear refractive index of materials such as graphene [19] and black phosphorus [20] were estimated using SSPM.

In the above study, the 2D Te samples fabricated by liquid phase exfoliation were characterized using TEM and AFM. Te nanosheets of 130–210 nm were observed. They also confirmed the crystal features of 2D Te using electron diffraction pattern. The number of layers of 2D Te were estimated using AFM.

In order to investigate the nonlinear optical response of 2D Te, three CW laser sources having wavelengths of 457 nm (blue light), 532 nm (green light) and 671 nm (red light) were employed. Tellurene dispersion was taken in a 10 mm cuvette and was allowed to interact with the focused (with an objective lens of f = 200 mm) laser beam. Diffraction pattern was observed on a screen placed behind the sample. The schematic diagram of the SSPM experiment is shown in Figure 10.1(a). After a given time, on interacting with the dispersion, the light beam will expand to form complete diffraction rings. The time taken to get the complete pattern depends on the wavelength of the laser used, shorter wavelengths taking shorter time to reach the complete pattern. As time progresses, the diffraction pattern gets distorted due to the thermal convection created by the laser heating.

As shown in Figure 10.2(c), the number of diffraction rings appearing depends upon the incident laser intensity. Fitting this curve yields the value of dN/dI, which gives a

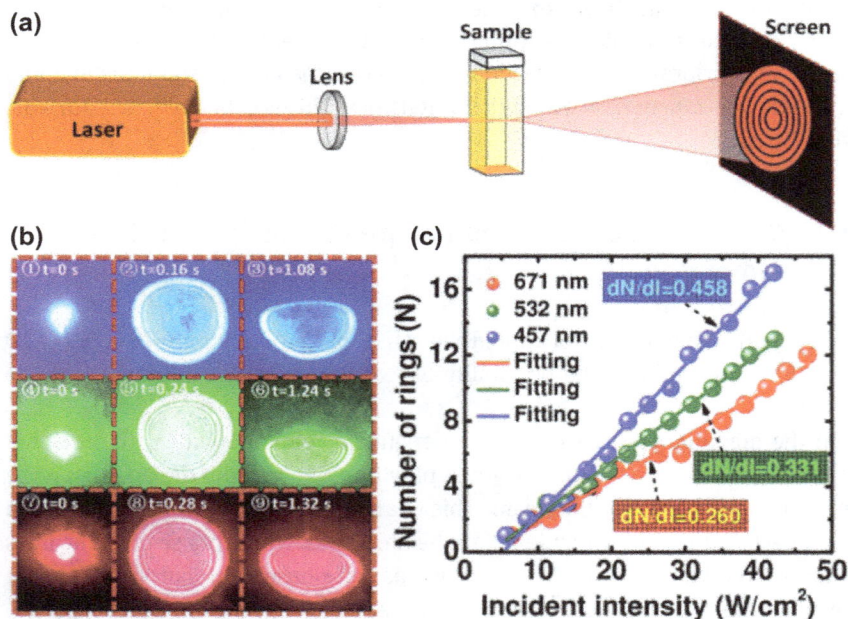

Figure 10.1: (a) Experimental setup of the 2D Te Ns-based SSPM. (b) Images of the diffraction rings received by a screen of 2D Te Ns dispersions at λ = 457, 532, and 457 nm, respectively. (c) The number of the diffraction rings changes with the increase of incident intensity at λ = 457, 532, and 671 nm, respectively. Reproduced with permission from ref. [18].

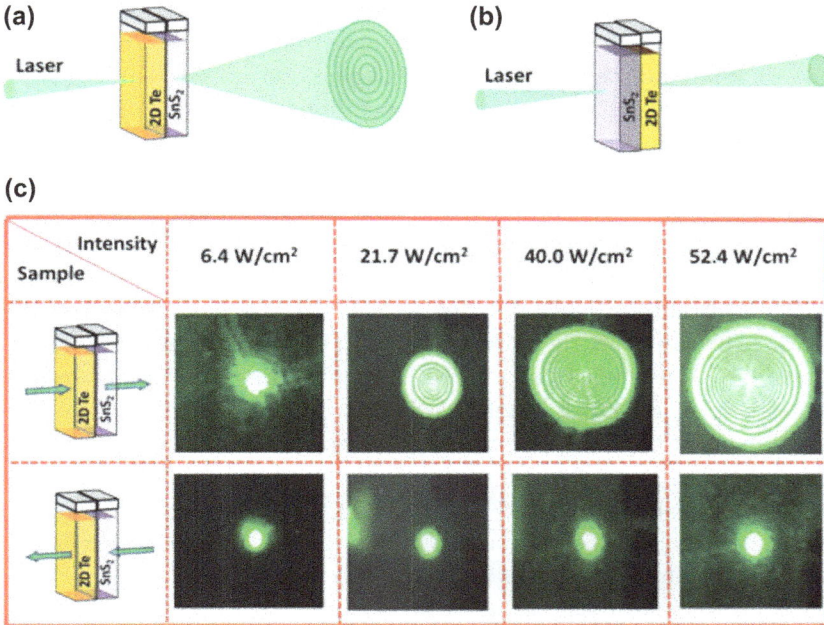

Figure 10.2: (a and b) Experimental setup of the 2D Te/SnS$_2$-based nonlinear photonic diode. (c) Images of the diffraction rings obtained from the experiment under different incident intensity for the hybrid structure of 2D Te and SnS2 at λ = 532 nm when the laser beam passes from the forward (2D Te/SnS2) and reverse (SnS2/2D Te) directions of the hybrid structure. Reproduced with permission from ref. [18].

measure of the optical nonlinearity of the sample. The dN/dI values for wavelengths 457, 532 and 671 nm were 0.458, 0.331, and 0.260 cm^2 W^{-1} respectively, implying the higher nonlinear optical response invoked by 457 nm laser input. They also calculated the nonlinear refractive indices of 2D Te at these wavelengths as 6.148 × 10^{-5} cm^2 W^{-1} (at 671 nm), 6.202 × 10^{-5} cm^2 W^{-1} (at 532 nm) and 7.37 × 10^{-5} cm^2 W^{-1} (at 457 nm).

An all-optical diode was envisaged by combining the strong Kerr-nonlinearity of the 2D Te sample with the reverse saturable absorption (RSA) behavior of SnS$_2$ [21] nanosheets. RSA property of SnS$_2$ will lead to increased absorption of the incident beam at higher intensities. On the other hand, 2D Te will excite more diffraction rings when the light intensity is increased. Combining these two different NLO properties will lead to a non-reciprocal propagation of light leading to the formation of an all-optical diode. In order to demonstrate this, two matching cuvettes were filled with 2D Te dispersion and SnS$_2$ dispersion and kept closely together in the path of the laser beam (Figure 10.2). When light passes through the hybrid structure of 2D Te/SnS$_2$ from the side of 2D Te, diffraction rings will be formed by the 2D Te dispersion which is subsequently passed through SnS$_2$. It was observed that the number of rings remain unchanged while the intensity of the rings get reduced as a result of increased absorption in the SnS$_2$ part.

On the other hand, when the beam is passed from the SnS_2 side, the transmittance reduces as the intensity increases due to the RSA nature of SnS_2. As the beam passes through 2D Te, no diffraction rings are formed because the intensity of the light is reduced below the threshold required for the pattern formation. So, depending upon from which side of the hybrid structure laser beam enters, a completely different transmission characteristic is obtained. They also demonstrated a light-modulated system utilizing the 2D Te exhibiting all optical switching properties. A strong pump beam was used to modulate a weaker probe beam to obtain ON (maximum light intensity) and OFF (minimum light intensity) conditions using the principle of cross-phase modulation.

Molecular beam epitaxy was successfully used by Huang et al. [22] to fabricate monolayer 2D Te on graphene. They also reported that the band gap of Te depends on its thickness (0.33 eV for bulk Te and 0.92 eV for monolayer Te). There has been a growing interest in investigating ultrafast switching properties of two-dimensional materials. Compared to other most commonly employed 2D materials like graphene or black phosphorous, 2D Te offers some advantages when it comes to ultrafast optical response. Intrinsic zero bandgap of graphene limits its applications in optical switching, while black phosphorous has limitations due to its instability and ease of getting oxidized in ambient conditions [23].

Guo et al. [23] fabricated 2D Te using facile liquid phase exfoliation and the exfoliated 2D Te is mixed with polyvinylpyrrolidone (PVP) to obtain a uniform 2D Te/PVP membrane. This structure is then characterized with various methods like TEM, HR-TEM, absorption spectroscopy and Raman spectroscopy as shown in Figure 10.3. Te NSs were found to have a lattice spacing of 0.22 nm, corresponding to the (1−210) plane of Te crystals.

The ultrafast response was estimated using the z-scan method. Then the 2D Te/PVP membrane was also used to produce ultrashort laser pulses in fiber lasers at 1556 and 1060 nm. This method offers some advantages over the solvent-dispersed 2D Te. As PVP is not having appreciable nonlinear optical response for the wavelengths used, contribution of solvent to the NLO properties is avoided. PVP also protects 2D Te from oxidation and optical bleaching.

The prepared 2D Te showed broadband absorption ranging between 200 and 2000 nm (solvent dependent). Even though there was degradation of Te over time, the suspensions showed the absorption peaks corresponding to Te even after one week of preparation. Incorporating 2D Te in the PVP matrix helped to alleviate the problem of degradation.

Open aperture z-scan measurements were employed for estimating the nonlinear optical properties of 2D Te/PVP membrane. The z-scan method [24, 25] can be used to measure both nonlinear index of refraction and nonlinear absorption coefficient. In this technique, the sample is translated across a focused laser beam. Focusing the laser beam gives a beam whose intensity is varied along the axis (z-direction). This enables the measurement of intensity dependent optical parameters. Placing an aperture before the

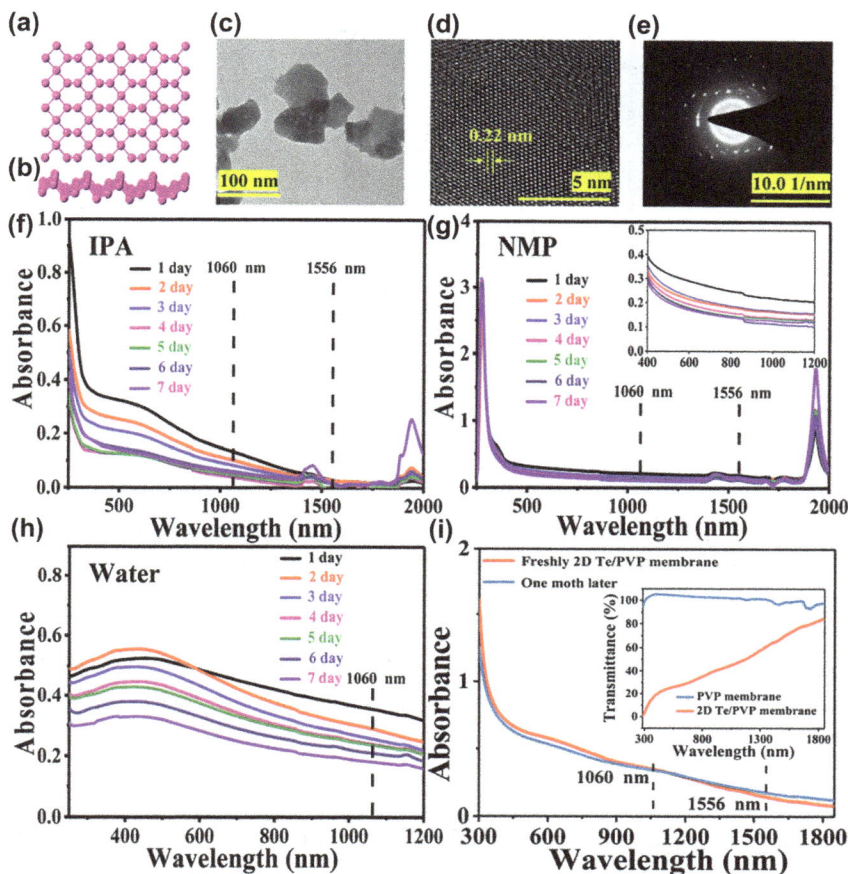

Figure 10.3: Characterization of 2D Te NSs and 2D Te/PVP membrane.
(a) Top and (b) side views of the atomic structure of monolayer Te. (c) TEM image of 2D Te NSs. (d) HR-TEM image illustrating the atomic arrangement of a 2D Te NSs. (e) Selected-area electron diffraction image. Typical absorption spectra of 2D Te NSs in (f) IPA, (g) NMP, and (h) water. (i) Absorption and transmittance (inset) of neat PVP and 2D Te/PVP membrane. Reproduced with permission from ref. [23].

detector (closed-aperture z-scan) will help to estimate the nonlinear refractive index. If no aperture is used, we get the nonlinear absorption coefficient (open-aperture z-scan).

In order to study the nonlinear optical absorption of 2D Te/PVP, z-scan measurements were done at three different wavelengths viz. 800, 1060, and 1550 nm. The z-scan experiments showed that the 75 μm-thick 2D Te/PVP membrane casted on a glass plate showed the behavior of a saturable absorber for which the normalized transmittance shows an increase when the incident intensity increases (Figure 10.4). Z-scan measurements were also performed with PVP alone to verify that the nonlinear optical response

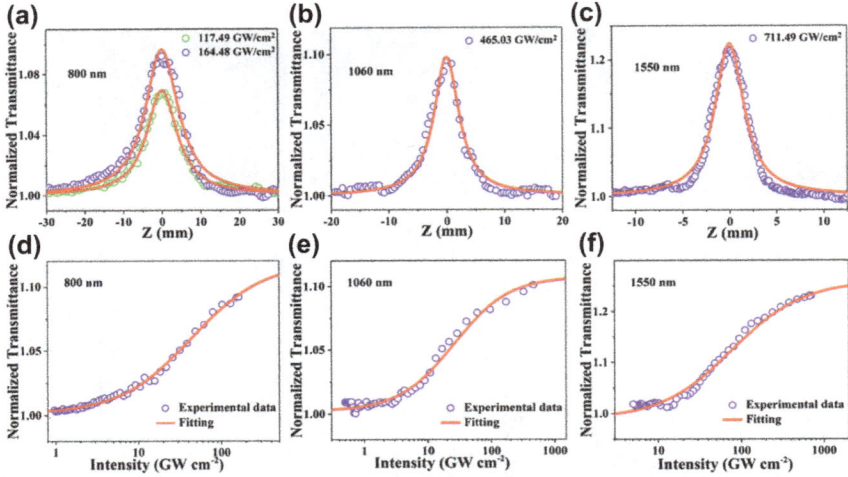

Figure 10.4: Open-aperture Z-scan measurements and fittings of a 2D Te/PVP membrane at (a) 800 nm, (b) 1060 nm, and (c) 1550 nm. Relationships between normalized transmittance and input peak intensity of the 2D Te/PVP membrane at (d) 800 nm, (e) 1060 nm, and (f) 1550 nm. Reproduced with permission from ref. [23].

comes solely from 2D Te. PVP did not show any nonlinear optical response. The scan data was fitted using a nonlinear absorption model to obtain the nonlinear optical parameters of the 2D Te.

$$T(z) = 1 - \beta I_0 L_{\text{eff}} / 2^{3/2} \left(1 + z^2/z_0^2\right), \tag{10.15}$$

where $T(z)$, β, I_0, L_{eff}, and z_0 are the normalized transmittance, nonlinear absorption coefficient, peak on-axis intensity at the focus ($Z = 0$), the effective length and the Rayleigh range, respectively.

β values were obtained by fitting the experimental data, as -3.56×10^{-1}, -1.13×10^{-1}, and -1.39×10^{-1} cm GW^{-1} at 800, 1060, and 1550 nm, respectively, which are comparable to the β values of other 2D materials like graphene and black phosphorous.

The saturable absorption in 2D Te happens due to the fact that when a very intense laser beam interacts with 2D Te, the energy levels near the band-edge get filled up easily if the energy of individual photons is slightly greater than the bandgap energy. As a result, the light will no longer be absorbed by the sample and the light passes through the sample without getting absorbed. This is reflected as an increase in the transmittance at higher intensities.

The absorption saturation property of the 2D Te/PVP membrane was employed to design a mode-locked erbium-doped fiber laser. The 2D Te/PVP was inserted in an erbium-doped fiber (EDF) ring cavity for mode-locking operation to obtain a highly stable mode-locked output. Autocorrelation measurement was used to estimate the pulse width which was calculated as 879 fs assuming sech^2 pulse. The calculated time-bandwidth product was 0.344 which is close to the theoretical limit indicating that the pulse obtained is almost transform limited (Figure 10.5).

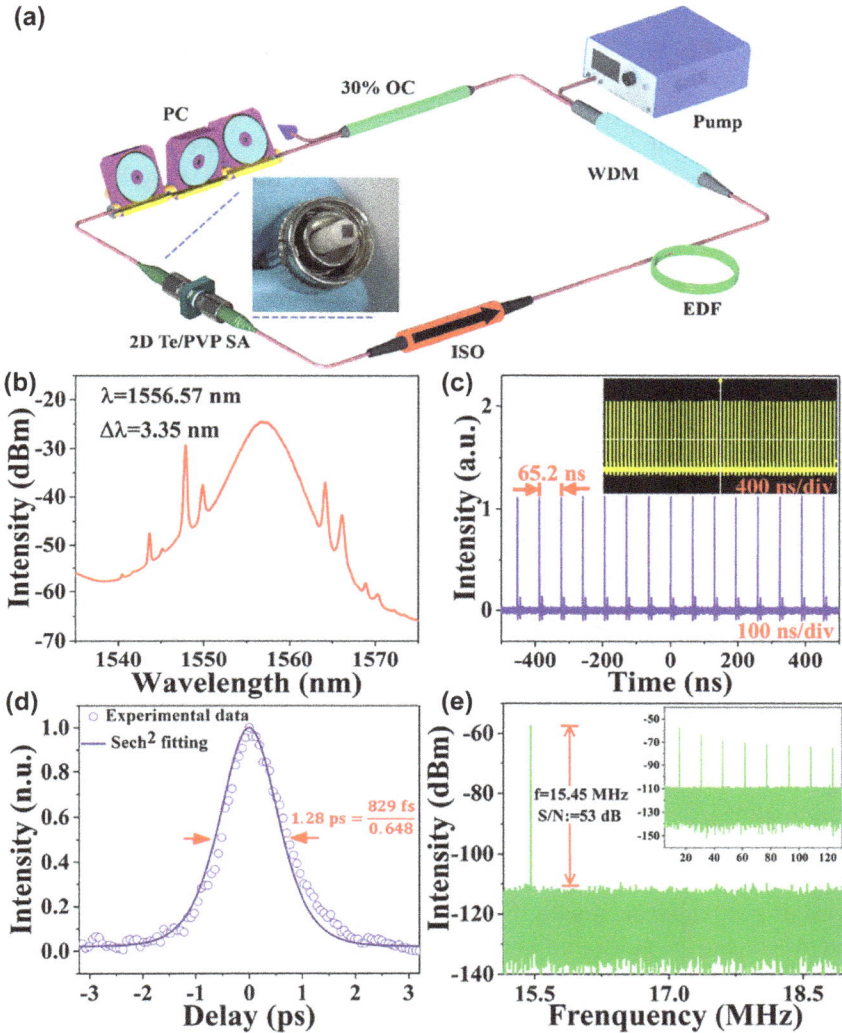

Figure 10.5: (a) Schematic of the 2D Te/PVP membrane-based mode-locking erbium-doped fiber (EDF; OFS EDF80, 50 cm) laser. WDM: wavelength division multiplexer. ISO: polarization independent isolator. 2D Te/PVP SA: 2D Te/PVP membrane-based SA. PC: polarization controller. OC:output coupler (30%). (b) Mode-locked optical spectrum centered at 1556.57 nm. (c) Pulse trains. Inset: 4 µs pulse trains. (d) Autocorrelation trace with a sech2 fitting, $\tau = 829/0.648$ fs. (e) Radio-frequency (RF) spectrum. SNR: signal-to-noise ratio. Inset: broadband RF spectrum. Reproduced with permission from [23].

Another investigation in the direction of realizing ultrafast laser action using 2D Te was done by Yang et al. [26]. The 2D Te was prepared by using liquid-phase exfoliation where Te powder with purity of 99.99% was dispersed in an alcohol suspension and ultrasonicated for 3 h to fabricate thin nanosheets. This suspension was centrifuged to remove aggregate powders and the obtained supernatant was transferred to a sapphire substrate by spin coating and dried under an infrared lamp. AFM was used to study the morphology of the Te nanosheets and the sheet-like 2D structure was confirmed. Te nanosheets of average thickness of 5 nm were observed corresponding to about 12 layers of Te nanosheets. XRD and Raman spectroscopy were used to analyze the crystal phase of the Te samples and the crystallinity was confirmed.

Nonlinear optical absorption of 2D Te samples were studied using open-aperture z-scan technique. Pulses from a femtosecond laser at 1030 nm with a pulse width of 350 fs and a repetition rate of 1 kHz were used for the measurements with pulse energy of 100 nJ. The normalized transmittance gradually increased as the sample was moved across the focused laser beam indicating saturable absorption behavior. The experimental data was fitted with saturable absorption model with

$$T(I) = 1 - \Delta R \times \exp\left(-\frac{I}{I_s}\right) - T_{ns}, \tag{10.16}$$

where $T(I)$ is the transmittance rate, I is the input intensity, and ΔR, I_s, and T_{ns} are the modulation depth, saturation intensity, and non-saturable loss, respectively.

The modulation depth and saturation intensity of the 2D Te sample were calculated to be 1.9% and 5.7 GW/cm² , respectively.

The 2D Te saturable absorber was then employed to design a mode-locked solid state laser with Nd:YVO$_4$ crystal as gain medium. A 1.495 m long W-type optimum resonator having a round trip time of 9.97 ns was designed based on the parameters of both the crystal and 2D Te saturable absorber (2D Te SA). The pump light was 808 nm and the oscillating light was 1064 nm. A stable continuous wave mode-locked (CWML) operation was obtained when the absorbed pump power increased to 6.44 W with an output power of 442 mW. The repetition rate of the laser was measured to be 00.3 MHz which was in good agreement with the cavity round trip time. Autocorrelation trace was used to estimate the pulse width of the laser using sech² fitting as 5.8 ps. The pulse also showed a spectral width (FWHM) of 0.30 nm with 1064 nm as the center wavelength. The maximum obtained pulse energy was calculated to be 8.48 nJ with a peak power of 1.46 kW. A time-bandwidth product of 0.455 was obtained, which was higher than the Fourier transform limit of 0.315 indicating that the pulse was not transform-limited. Optimization of the cavity along with dispersion compensation is required to further compress the pulse.

Photonic properties of elemental tellurium particles obtained from a culture of a tellurium-oxyanion respiring bacteria were extensively investigated by Wang et al. [27]. The bacterially synthesized nanocrystals were shown to exhibit photonic properties comparable to those of chemically synthesized nanomaterials, paving a way to an environment-friendly way of preparing these materials.

Wang et al. used the anaerobic bacterium *Bacillus selenitireducens* with Te (IV) as its electron acceptor to harvest elemental tellurium nanostructures. The nonlinear optical properties of these biologically synthesized tellurium (Bio-Te) nanostructures were investigated using open aperture z-scan technique at different pulse widths and wavelengths. The method of preparing the nanocrystals is shown in Figure 10.6a. Micro-pellets were obtained as a result of the aggregation of the bio-Te nanostructures. As these aggregates are not suitable for the optical characterization studies used, poly(*m*-phenylenevinylene)-co-2,5-dioctoxy-phenylenevinylene (PmPV) was used to disperse Te(0) to form nanocomposites. Lattice spacings of 0.32 and 0.59 nm were observed corresponding to the (101) and (001) crystal planes using HR-TEM (Figure 10.6d) implying the trigonal crystal geometry of the Bio-Te. Raman spectroscopy was also used to confirm the trigonal crystalline Te in both pristine Bio-Te and the Bio-Te-PmPV composite (Figure 10.6e). The Raman spectrum showed the characteristic peaks of Te (0) and PmPV. Some peaks corresponding to TeO_2 were also observed indicating the slight oxidation of Te (0) over time due to exposure to ambient air.

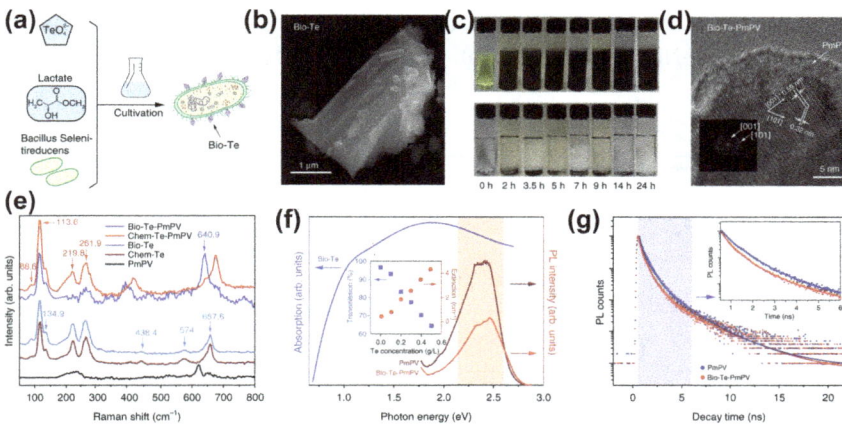

Figure 10.6: Synthesis and characterization of biological synthesized tellurium (Bio-Te). (a) Synthesis scheme of tellurium nanocrystals by anaerobic bacteria, *Bacillus selenitireducens*. (b) Image of Bio-Te crystalline nanoflake taken by scanning transmission electron microscope. (c) Dispersion of Bio-Te in PmPV/toluene (upper) and in toluene only (bottom) with increasing stirring times on the *X*-axis, showing the preparation of Bio-Te-PmPV composites. PmPV is abbreviation of poly(*m*-phenylenevinylene)-co-2,5-dioctoxy-phenylenevinylene. (d) Transmission electron microscopy image of a Bio-Te crystalline nanoflake wrapped by PmPV layers. Inset is the image after fast Fourier transformation (FFT). (e) Raman spectra of Bio-Te, Bio-Te-PmPV, chemically synthesized tellurium nanocrystals (Chem-Te), Chem-Te-PmPV, and PmPV. (f) Left: the absorption spectrum of Bio-Te. This curve was obtained by subtracting the absorption of PmPV (0.5 g/L in toluene) from that of Bio-Te-PmPV. Right: photoluminescence (PL) spectra of Bio-Te-PmPV and pure PmPV. Inset: optical linear transmission and extinction coefficient as functions of Te concentrations at 532 nm g The PL decay kinetics of PmPV and Bio-Te-PmPV at 528 nm excitation. The inset shows a blow up of the blue shaded region, showing the quenching effect caused by tellurium. Reproduced from ref. [27] under Creative Commons Attribution 4.0 (http://creativecommons.org/licenses/by/4.0/).

The linear optical absorption of Bio-Te-PmPV composite (after subtracting the effect of PmPV) shows a broad curve with peak around 1.9 eV (~650 nm) as shown in Figure 10.6f. Their experimental investigations also showed that the Bio-Te-PmPV shows a photoluminescence (PL) spectrum similar to that of PmPV. This indicates the lack of photoluminescence from Bio-Te. It was also observed that the PL of PmPV was quenched in the presence of Bio-Te as the PL intensity showed a reduction of one order of magnitude (Figure 10.6g).

In order to study the nonlinear optical effects in Bio-Te-PmPV, open aperture z-scan measurements with femtosecond (340 fs) laser pulses of three different wavelengths (515, 800 and 1030 nm) were used. Bio-Te-PmPV composite exhibited saturable absorption (SA) behavior under 800 nm excitation with pulse energy 200 nJ as can be seen in Figure 10.7a. The z-scan measurement was carried out on a sample of PmPV alone, which showed no SA response, indicating that the nonlinear optical response is coming solely from Bio-Te nanocrystals. Here again, the mechanism of saturable absorption can be attributed to the band filling at higher intensities and Pauli blocking. The excitation energy (1.55 eV) is much larger than the bandgap of Te (0.323 eV). When the excitation pulse energy was increased to 400 nJ, the nonlinear optical response changes to a combination of absorption saturation and nonlinear absorption (nonlinear extinction-NLE) (Figure 10.7b). This could be due to the multiphoton

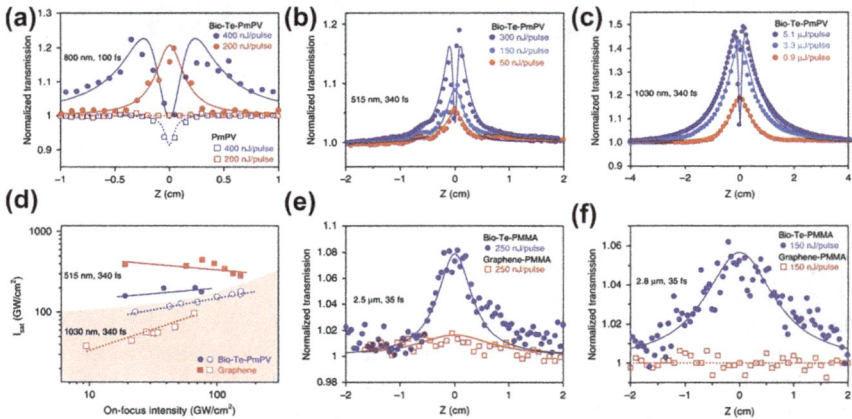

Figure 10.7: Open-aperture z-scan results of the biological synthesized tellurium samples. (a) Experimental results with fs pulses, 800 nm. Bio-Te-PmPV (solid circles); PmPV (hollow squares). Solid lines: fitted z-scan curves using equation (10.1); Dashed lines are for visual guide. (b and c) Experimental results with 340 fs laser at 515 nm (b) and 1030 nm (c). (d) Saturated intensity Isat of Bio-Te-PmPV and graphene dispersion as functions of the laser intensity at 1030 and 515 nm. (e and f) Mid-infrared open z-scans of Bio-Te and graphene polymethyl methacrylate (PMMA) films at 2.5 and 2.8 μm wavelengths, showing better saturable absorptive responses of Bio-Te than those of graphene. Reproduced from ref. [27] under Creative Commons Attribution 4.0 (http://creativecommons.org/licenses/by/4.0/).

absorptions taking place in PmPV at higher intensities. The nonlinear optical co-efficients of Bio-Te nanocrystals were calculated based on theoretical fitting of the experimental data. The following model was used for the fitting:

$$\frac{dI}{dz} = \frac{\alpha_0 I}{1 + \frac{I}{I_{sat}}} - \beta I^2,$$ (10.17)

where α_0 is the linear absorption coefficient, I is the photon intensity, z is the distance traveled by light in the NLO medium, I_{sat} is the saturated intensity contributed by Bio-Te, and β is the NLE coefficient contributed by PmPV.

At lower input pulse energies, the saturable absorption caused by Bio-Te dominates while at higher energies, the multi-photon absorption caused by PmPV is evident from the z-scan plots (Figure). They also studied the changes in I_{sat} as a function of on-focus beam intensity for these different wavelengths. The I_{sat} value of Bio-Te at 1030 nm was almost one third of black phosphorous [28] implying stronger saturation of absorption by Bio-Te at IR region.

Further investigations of absorption saturation by Bio-Te samples were carried out in mid-infrared (MIR) region with 35 fs pulses. In order to avoid the effects of two photon absorption of PmPV at this region, polymethyl methacrylate (PMMA) was used as a host material. The host material showed negligible nonlinear optical effects in the z-scan measurements done at 2.5 and 2.8 μm. Saturation of absorption was observed at these wavelengths as well for the Bio-Te/PMMA composite. This shows that Bio-Te can be used as a saturable absorber to obtain passive mode-locking in a large range of wavelengths.

This study also compared the saturable absorption behavior of Bio-Te with graphene, which is also a broad band saturable absorber [29–31]. It was shown that Bio-Te/PmPV shows better nonlinear optical response at 515 nm, comparable values at 800 nm and inferior response at 1030 nm. At the MIR region, Bio-Te exhibited much better SA compared to graphene at both 2.5 and 2.8 μm.

The z-scan measurements were also carried out with nanosecond (ns) pulses at 532 and 1064 nm (6 ns at 10 Hz Q-switched Nd:YAG laser). At this pulsewidth, z-scan plots were obtained for pulse energies from 52 to 199 μJ. Strong nonlinear scattering was exhibited by Bio-Te under these experimental conditions, which is attributed to the Mie scattering caused by the formation of microbubbles due to the absorption of laser by Te nanostructures. The z-scan traces obtained indicate strong nonlinear optical extinction (NLE), suppressing the saturation of absorption. The theoretical fit was obtained by incorporating the effects of scattering as well in the fitting equation as:

$$\frac{dI}{dz} = -(\alpha_0 + \beta_{NLE} I) I,$$ (10.18)

where β_{NLE} is the effective NLE coefficient. The broadband NLE nature can be utilized for realizing laser-protective equipment. This equation is an adaptation of equation

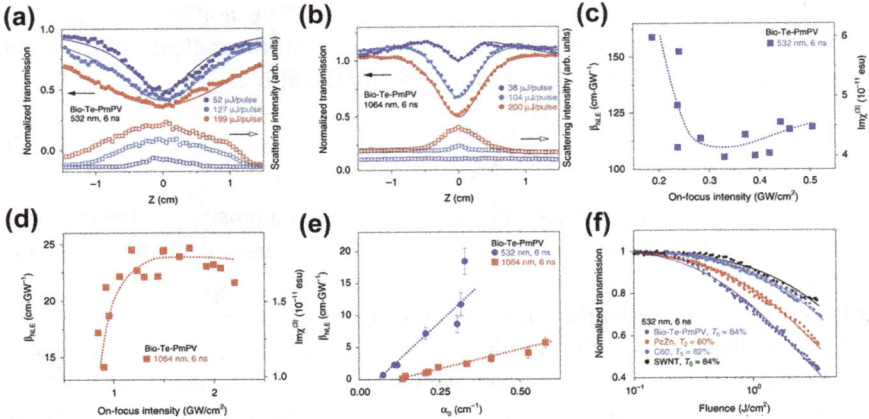

Figure 10.8: Nonlinear optical responses to ns pulses.
(a and b) Circles: normalized transmission of Bio-Te-PmPV as a function of z at 532 and 1064 nm; squares: light intensity scattered by the sample at 35° to the laser's direction; lines: Z-scan fitting results. (c and d) Effective nonlinear extinction (NLE) coefficient β_{NLE} and corresponding $Im\chi(3)$ as a function of on-focus intensity for ns pulses at 532/1064 nm. (e) Effective nonlinear extinction coefficient β_{NLE} as a function of linear absorption coefficient $\alpha 0$ with error bars indicating s.e.m. (f) Comparison of optical limiting performance of Bio-Te-PmPV, PcZn (t-Bu4PcZn), C60, and single-walled carbon nanotube (SWNT) dispersions at 532 nm, 6 ns irradiation. Reproduced from ref. [27] under Creative Commons Attribution 4.0 (http://creativecommons.org/licenses/by/4.0/).

(10.13). Here β_{NLE} incorporates the extinction due to nonlinear scattering as well. The z-scan plots are presented in Figure 10.8.

This work also demonstrated the use of the ultrafast SA response of the tellurium nanostructures for realizing passive modelocking and Q-switching at 1.55 μm. Bio-Te film was integrated with PMMA and was coated on the face of a fiber connector and inserted into the cavity of an erbium-doped fiber (EDF) laser. The SA nature of Bio-Te lead to the self-starting mode-locking at a pump power of 360 mW. The laser output was obtained with a central wavelength of 1.55 μm, whose autocorrelation trace gave a pulse width of 1.81 ps. They also demonstrated the Q-Switching properties of Bio-Te. Studies were done with pump powers from 0 to 110 mW. The observed the pulse width varied from 7 to 12 μs. Compared to the q-switching showed by black phosphorous [32] Bio-Te required lesser pump power to obtain a shorter pulse width.

An earlier work by the same group had compared the nanosecond nonlinear optical response of Bio-Te with C60 and other well-known nonlinear optical materials and showed that Bio-Te possessed superior optical limiting properties under the similar illumination conditions [33]. They also compared the NLO properties of Bio-Te with chemically derived Te(0)/PmPV composite and both were shown to have comparable nonlinear optical response with a dominance of nonlinear scattering as the major mechanism of optical limiting.

A very detailed analysis comparing the ultrafast nonlinear optical response of different Te-based low-dimensional structures were carried out by Xiao et al. [34] They investigated one-dimensional (1D) Te nanowires (NWs), quasi-1D Te nanorods (NRs), zero-dimensional (0D) Te nanodots (NDs) and two-dimensional (2D) Te nanosheets (NSs) prepared by electrochemical exfoliation and liquid phase exfoliation methods. Femtosecond z-scan measurements were used to compare the dimension dependent nonlinear optical response of these materials. They have carefully designed the nanomaterials to compare the effect of quantum confinement in each of these structures. They chose 1D and 0D Te nanomaterials in such a way that they both have the same diameter so that the effect of length of 1D structure can be studied. The thickness of the nanosheets were controlled to be less than the diameter of nanodots and nanowires.

Cathodic electrochemical exfoliation method was used to prepare nanomaterials with well-controlled dimensions. This method resulted in uniform NWs, NRs and NDs. In order to fabricate Te nanosheets, they used a modified liquid phase exfoliation method. High angled annular dark field (HAADF) STEM and bright field TEM revealed the ultralong morphology of the one-dimensional nanowires (Figure 10.9a) with an average diameter 9.4 ± 2.1 nm. The elemental mapping showed uniform distribution of Te with a negligible presence of O in the Te NWs, indicating the single crystal nature of

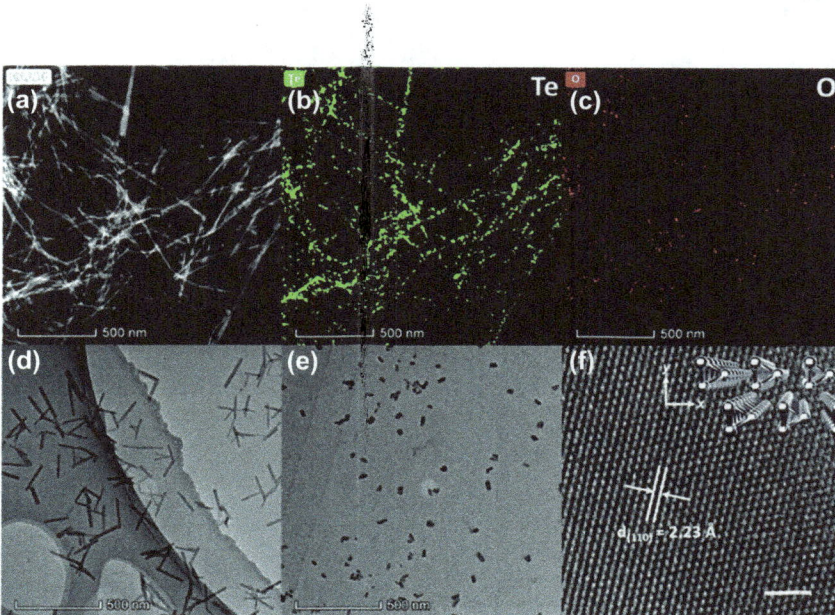

Figure 10.9: (a) High-angle annular dark-field (HAADF)-STEM image of Te NWs. (b and c) Elemental mapping of Te and O elements, respectively. (d) TEM image of Te NRs. (e) TEM image and (f) high-resolution TEM (HRTEM) image of Te NDs. Scale bar, 2 nm. Inset: the corresponding crystal structure. Reproduced with permission from ref. [34].

Te NWs. TEM images of quasi-1D Te nanorods (Te NRs) (Figure 10.9d) showed diameters of 12.1 ± 2.8 nm and length of around 200 nm. For these nanostructures as well, the elemental mapping indicated a negligible amount of oxygen with the majority of Te element. The 0D Te nanodots (NDs) were shown to have a uniform diameter of 12.0 ± 3.0 nm (Figure 10.9e). The HR-TEM images (Figure 10.9f) indicated an atomic matrix with three fold symmetry with a lattice spacing measured to be 2.23 Å corresponding to (110) plane of the trigonal Te.

The Te nanosheets (NSs) synthesized by the LPE method were also subjected to TEM analysis to reveal its 2D morphology (Figure 10.10a). The NSs were highly transparent to the electron beam indicating their ultrathin nature. The HR-TEM images showed a rectangle-like atomic matrix which is different from the Te NDs (Figure 10.10b). The measured lattice spacing was 1.98 Å. This corresponds to the (003) plane of the trigonal Te. A simulated crystal structure showed the atoms lying in the plane (Figure 10.10b).

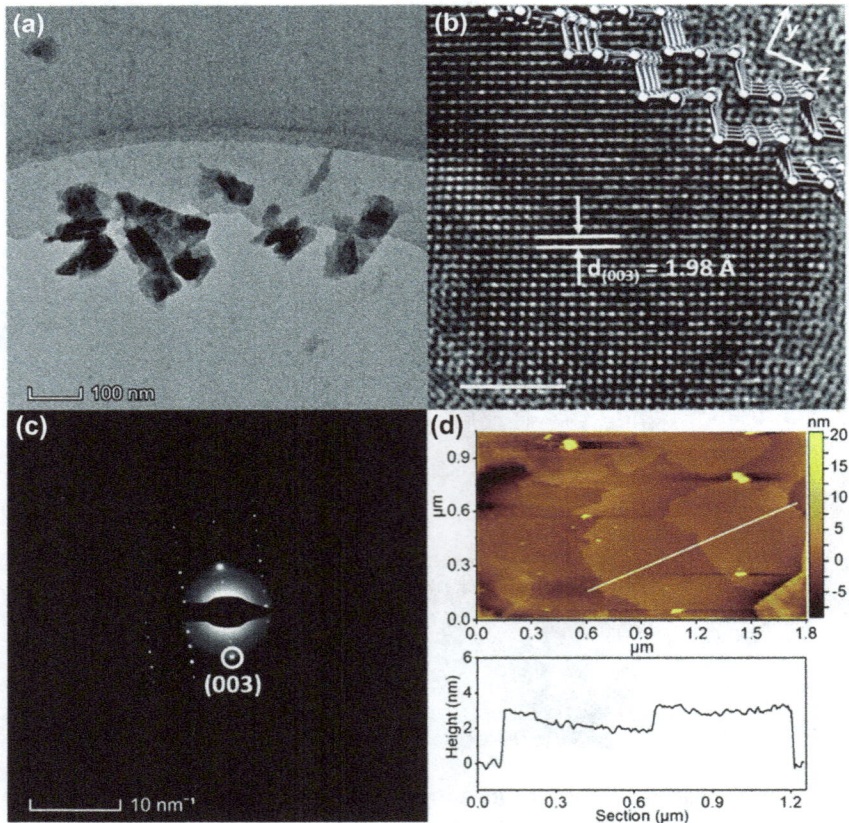

Figure 10.10: (a) TEM image and (b) HRTEM image of Te NSs. Inset: the simulated crystal structure. (c) The corresponding selected area electron diffraction (SAED) pattern. (d) Top: atomic force microscopy (AFM) image. Bottom: corresponding height profile of Te NSs. Scale bars, 2 nm (b); 10 nm^{-1} (c). Reproduced with permission from ref. [34].

Selected area electron diffraction (SAED) pattern also confirms the (003) plane observed HR-TEM (Figure 10.10c). The lateral size of the Te NSs was estimated using SEM analysis to be around 500 nm. AFM was employed to calculate the thickness of the nanosheets, which was found to be around 1.4–3.2 nm corresponding to 3–7 layers of Te monolayers (Figure 10.10d).

UV–Vis absorption spectroscopy of these nanostructures dispersed in NMP showed broad absorption ranging from 400 to 1000 nm. The absorption bands of these structures were at 695 nm for Te NWs, 622 nm for Te NRs and 573 nm for Te NDs showing a blue shift as the dimensionality of the structures change indicating a change in the bandgap energies. The corresponding bandgap energies were calculated to be 1.12, 1.28 and 1.3 eV respectively for Te NWs, Te NRs and Te NDs. The bandgap of 2D Te NSs were calculated to be 2 eV. Raman spectroscopy and XRD were used for further characterization of the nanostructures. High resolution X-ray photoelectron spectroscopy (XPS) was used to evaluate the percentage content of Te and oxygen in these nanostructures and analysis showed Te NSs having the highest percentage of oxygen content. The increased amount of oxidized states in the 2D Te NSs is attributed to the fact that in 2D layered structure, the ratio of surface atoms is increased.

Femtosecond open aperture z-scan was used to investigate the nonlinear optical properties of the Te nanostructures. The nanostructures were dispersed in NMP. NMP alone does not show any nonlinear optical effects. As discussed earlier, the z-scan traces give the normalized transmittance as a function of sample position along the z-direction in which the focused laser beam is providing a continuous variation in intensity.

Here the z-scan curves were fitted using the transmission equation (10.13). The samples were taken in a 2 mm quartz cuvette and were excited by a femtosecond laser beam having 600 nm wavelength which is close to the absorption band of the samples. Measurements were repeated for three different pulse energies viz. 280, 1050 and 1480 nJ.

Te nanowire sample showed a clear saturation of absorption in the z-scan measurement using 280 nJ laser pulse. For increased pulse energies the sample showed onset of reverse saturable absorption near $z = 0$. Repeatability of measurements using the same sample also confirmed that the Te NWs are not damaged by the exposure to high intensity laser beams. As discussed earlier, saturation of absorption is attributed to narrow bandgap and Pauli blocking. The reverse saturable absorption usually occurs due to multiphoton absorption processes (like two-photon absorption) or excited state absorption. In this case, the photon energy is larger than the bandgap energy of the samples. So the possibility of two-photon absorption can be omitted. Another investigation showed that depending upon the linear transmittance of the sample used, there were changes in the nonlinear optical properties. For this laser pulse and energy was fixed at 280 nJ and then three different samples of Te-NWs were made such that their linear transmittances were 0.70, 0.48 and 0.30 respectively. The absorption saturation behavior steadily improved when the transmittance was reduced. Solvent

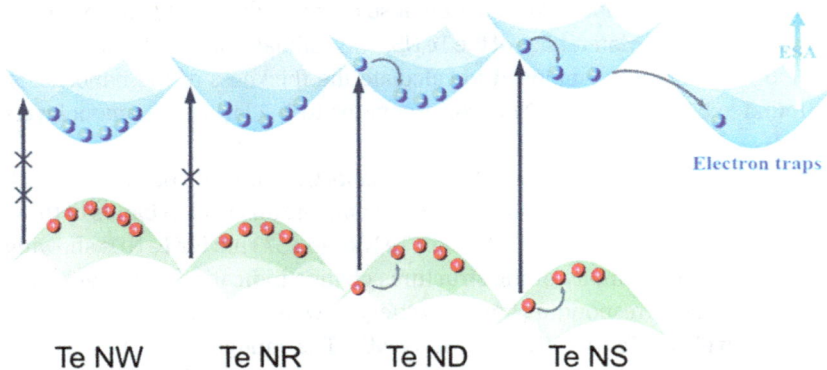

Figure 10.11: Schematic NLO mechanism of Te NWs, NRs, NDs and NSs, respectively. Reproduced with permission from ref. [34].

heating and evaporation effects also play a role in the nonlinear optical responses of the samples with higher transmittances as the amount of solvent is larger.

Z-scan investigations of other Te nanostructures also proved that the NLO behavior depends on the dimensionality of the material. For the same laser pulse energy and sample transmittance, different nanostructures showed different NLO properties. The NWs exhibited stronger SA behavior than the NRs and NDs showed a combination of SA and RSA behavior. In these studies, Te NSs showed RSA behavior. Similar behavior was also shown when the laser pulses of 532 nm wavelength (<110 fs) were used for z-scan measurements. They also investigated the NLO properties of these nanostructures embedded in a PMMA matrix to avoid the effect of solvent heating. In this case also dimension dependent changes in the NLO response were observed. They employed transient absorption measurements at femtosecond timescale to understand the carrier dynamics in these nanostructures. These observations as illustrated in Figure 10.11 also explain the saturation of absorption and reverse saturable absorption exhibited by these samples. This study also demonstrated the ultrafast RSA behavior of 2D Te unlike the earlier examples.

10.5 Conclusions

Novel materials based on elemental tellurium are gaining attention in optoelectronic and photonics research due to their interesting properties. Tellurium nanomaterials including nanoclusters, nanodots and nanorods show excellent nonlinear optical response. 2D Te is another promising material which shows ultrafast switching properties and can be used to design mode-locking in ultrafast lasers. Depending upon the morphology of the materials, incident laser wavelength, pulse width and pulse energy,

Te-based materials can achieve different nonlinear optical characteristics. These can be effectively used for photonic device fabrication.

References

1. Hermann JP, Quentin G, Thuillier JM. Determination of the d_{14} piezoelectric coefficient of tellurium. Solid State Commun 1969;7:161–3.
2. Peng H, Kioussis N, Snyder GJ. Elemental tellurium as a chiral p-type thermoelectric material. Phys Rev B 2014;89:195206.
3. Liu JW, Zhu JH, Zhang CL, Yu HW, Liang SH. Mesostructured assemblies of ultrathin superlong tellurium nanowires and their photoconductivity. J Am Chem Soc 2010;132:8945–52.
4. Castro Neto AH, Guinea F, Peres NMR, Novoselov KS, Geim AK. The electronic properties of graphene. Rev Mod Phys 2009;81:109–62.
5. Zhang H, Virally S, Bao Q, Ping LK, Massar S, Godbout N, et al. Z-scan measurement of the nonlinear refractive index of graphene. Opt Lett 2012;37:1856–8.
6. Lu SB, Miao LL, Guo ZN, Qi X, Zhao CJ, Zhang H, et al. Broadband nonlinear optical response in multi-layer black phosphorus: an emerging infrared and mid-infrared optical material. Opt Express 2015;23:11183–94.
7. Zhang H, Lu SB, Zheng J, Du J, Wen SC, Tang DY, et al. Molybdenum disulfide (MoS2) as a broadband saturable absorber for ultra-fast photonics. Opt Express 2014;22:7249–60.
8. Zhu Z, Cai X, Yi S, Chen J, Dai Y, Niu C, et al. Multivalency-driven formation of Te-based monolayer materials: a combined first-principles and experimental study. Phys Rev Lett 2017;119:106101.
9. Xie Z, Xing C, Huang W, Fan T, Li Z, Zhao J, et al. Ultrathin 2D nonlayered tellurium nanosheets: facile liquid-phase exfoliation, characterization, and photoresponse with high performance and enhanced stability. Adv Funct Mater 2018;28:1705833.
10. Laud BB. Lasers and nonlinear optics. India: New Age International Private Limited; 2011.
11. Franken PA, Hill AE, Peters CW, Weinreich G. Generation of optical harmonics. Phys Rev Lett 1961;7:118.
12. Shen YR. The principles of nonlinear optics. Hoboken, New Jersey, USA: Wiley; 2003.
13. Sherman GH, Coleman PD. Absolute measurement of the second-HarmonicGeneration nonlinear susceptibility of tellurium. IEEE J Quant Electron 1973;QE-9:403.
14. Londoño-Calderon A, Williams DJ, Schneider MM, Savitzky BH, Ophus CC, Ma S, et al. Evidence of intrinsic helical twist and chirality in ultrathin tellurium nanowires. Nanoscale 2021;13:9606–14.
15. Cheng M, Shi X, Wu S, Zhu Z-Z. Significant second-harmonic generation and bulk photovoltaic effect in trigonal selenium and tellurium chains. Phys Chem Chem Phys 2021;23:6823–31.
16. Deckoff-Jones S, Wang Y, Lin H, Wu W, Hu J. Tellurene: a multifunctional material for midinfrared optoelectronics. ACS Photonics 2019;6:1632–8.
17. Sutherland RL. Handbook of nonlinear optics. CRC Press; 2003.
18. Wu L, Huang W, Wang Y, Zhao J, Ma D, Xiang Y, et al. 2D tellurium based high-performance all-optical nonlinear photonic devices. Adv Funct Mater 2019;29:1806346.
19. Wu R, Zhang Y, Yan S, Bian F, Wang W, Bai X, et al. Purely coherent nonlinear optical response in solution dispersions of graphene sheets. Nano Lett 2011;11:5159.
20. Zhang J, Yu X, Han W, Lv B, Li X, Xiao S, et al. Broadband spatial self-phase modulation of black phosphorous. Opt Lett 2016;41:1704.
21. Wu JJ, Tao YR, Wu XC, Chun Y. Nonlinear optical absorption of SnX2 (X = S, Se) semiconductor nanosheets. J Alloys Compd 2017;713:38–45.

22. Huang X, Guan J, Lin Z, Liu B, Xing S, Wang W, et al. Epitaxial growth and band structure of Te film on graphene. Nano Lett 2017;17:4619–23.
23. Guo J, Zhao J, Huang D, Wang Y, Zhang F, Ge Y, et al. Two-dimensional tellurium–polymer membrane for ultrafast photonics. Nanoscale 2019;11:6235–42.
24. Sheik-Bahae M, Said AA, Wei T-H, Hagan DJ, Van Stryland EW. Sensitive measurement of optical nonlinearities using a single beam. IEEE J Quant Electron 1990;26:760.
25. Sheik-bahae M, Said AA, Van Stryland EW. High-sensitivity, single-beam n2 measurements. Opt Lett 1989;14:955.
26. Yang Z, Han L, Qi Y, Ren X, Ud Din SZ, Zhang X, et al. Two-dimensional tellurium saturable absorber for ultrafast solid-state laser. Chin Opt Lett 2021;19:031401.
27. Wang K, Zhang X, Kislyakov IM, Dong N, Zhang S, Wang G, et al. Bacterially synthesized tellurium nanostructures for broadband ultrafast nonlinear optical applications. Nat Commun 2019;10: 3985.
28. Hanlon D, Backes C, Doherty E, Cucinotta CS, Berner NC, Boland C, et al. Liquid exfoliation of solvent-stabilized few-layer black phosphorus for applications beyond electronics. Nat Commun 2015;6:8563.
29. Bao Q, Zhang H, Wang Y, Ni Z, Yan Y, Shen ZX, et al. Atomic-layer graphene as a saturable absorber for ultrafast pulsed lasers. Adv Funct Mater 2019;19:3077–83.
30. Wang G, Wang K, Szydłowska BM, Baker-Murray AA, Wang JJ, Feng Y, et al. Ultrafast nonlinear optical properties of a graphene saturable mirror in the 2 μm wavelength region. Laser Photon Rev 2017;11:1700166.
31. Martinez A, Sun Z. Nanotube and graphene saturable absorbers for fibre lasers. Nat Photonics 2013;7:842–5.
32. Chen Y, Jiang G, Chen S, Guo Z, Yu X, Zhao C, et al. Mechanically exfoliated black phosphorus as a new saturable absorber for both Q-switching and mode-locking laser operation. Opt Express 2015; 23:12823–33.
33. Liao K-S, Wang J, Dias S, James D, Alley NJ, Baesman SM, et al. Strong nonlinear photonic responses from microbiologically synthesized tellurium nanocomposites. Chem Phys Lett 2010; 484:242–6.
34. Qi X, Ma B, Fei X, Liu D-W, Zhai X-P, Li X-Y, et al. Unveiling the dimension-dependence of femtosecond nonlinear optical properties of tellurium nanostructures. Nanoscale Horiz 2021;6: 918–27.

Sara Ali A Aldawood, Aparna Das and Bimal Krishna Banik*

11 Tellurium-induced cyclization of olefinic compounds

Abstract: In this article, we discuss about the importance of Tellurium (Te) in organic synthesis. Tellurium-induced cyclization of alkenyl compounds, as well as alkynyl compounds, are considered for the study. The developments in this area are incorporated in great detail. The mechanism of the reactions does not follow any straightforward process. This study opens up the possibility of stereocontrolled synthesis of complex natural products.

Keywords: alkenyl compounds; alkynyl compounds; cyclization; olefinic compounds; tellurium.

11.1 Introduction

Tellurium (Te) is a metalloid and is a low bandgap semiconductor. Te is useful for various applications. For example, it is used for photovoltaic and thermoelectric applications. Te is used as a coloring agent in glass and ceramics and also in copying machines. Te is also utilized as a vulcanizing agent to make durable products in chemical industry. Te is also used in the semiconductor industry such as in integrated circuits, laser diodes, and sensors, in medical instrumentation, and automobile industry.

Tellurium compounds are also used in organic synthesis. For instance, it uses for carbohydrate synthesis. In organic synthesis, in addition to utilize as reagents for oxidation, tellurium compounds are also described as versatile electrophiles practicable in diverse organic transformations. Tellurium-induced cyclization of alkenyl compounds, as well as alkynyl compounds, are considered for the study. In the following sections, some remarkable and recent, breakthrough in this field is highlighted.

---

*Corresponding author: Bimal Krishna Banik, Department of Mathematics and Natural Sciences, College of Sciences and Human Studies, Prince Mohammad Bin Fahd University, Al Khobar 31952, Kingdom of Saudi Arabia, E-mail: bbanik@pmu.edu.sa
Sara Ali A Aldawood and Aparna Das, Department of Mathematics and Natural Sciences, College of Sciences and Human Studies, Prince Mohammad Bin Fahd University, Al Khobar 31952, Kingdom of Saudi Arabia. https://orcid.org/0000-0002-2502-9446 (A. Das)

As per De Gruyter's policy this article has previously been published in the journal Physical Sciences Reviews. Please cite as: S. Ali A Aldawood, A. Das and B. K. Banik "Tellurium-induced cyclization of olefinic compounds" *Physical Sciences Reviews* [Online] 2022. DOI: 10.1515/psr-2021-0119 | https://doi.org/10.1515/9783110735840-011

11.2 Tellurium-induced cyclization of alkenyl compounds

Developing new methods for tetrahydrofuran derivatives remains an important research area as these compounds are advantageous intermediates in the production of various natural products [1]. One of these examples is the tetrahydrofuran moiety in which an exocyclic double bond is present. This has been used in numerous analyses to assist in synthesizing nonactic acid [2–5].

Jackson et al. showed that the β-dicarbonyl compounds that are substituted with alkenyl can be cyclized with electrophiles of selenium [6]. This study showed the possibility that based on the reaction conditions the product of a C-cyclization (thermodynamic control) or that of an O-cyclization (kinetic control) can be obtained. Later on, other researchers [7–9] covered a comparable iodo-O-cyclization of a series of β-keto esters and ketones substituted with alkenyl. In all of these compounds, the alkenyl chain was at the α-position. The correspondent cyclization of β-enamino esters and ketones promoted by iodine resulted in worthful precursors synthesis for pyrrole nucleous, was reported by Ferraz et al. [10]. Later the same research group examined the behavior of a group of α- and γ-alkenyl-β-keto esters for aryltellurium trichloride [11].

The usage of iodine and aryltellurium tichloride as reagents for cyclization toward β-keto esters substituted with alkenyl were discussed in great detail [11]. The occurrence of this reaction happens through the dicarbonyl compounds that have enolic form. The corresponding five-membered cyclic ethers were obtained in good yields. The results are summarized in Figure 11.1.

Aryltellurium trihalides were employed as electrophilic reagents for the production of organotellurium compounds with greater bioactivity [12–19]. Several studies reported the interactions of aryltellurium trihalides with carbonyl compounds, γ,δ-alkenyl alcohols, acids, and several derivatives of these compounds. For example, the interaction γ,δ-unsaturated alcohols, and ethers **10** with p-alkoxy(alkyl)-phenyltellurium trichlorides **9** in CHCl$_3$ produced derivatives of tetrahydrofuran **11** with an exocyclic aryltelluriium fragment (Figure 11.2). The product yield was very low when the p-methoxyphenyltellurium tribromide was used in the production of tellurium-functionalized furan compounds [20].

Cyclization of aryltellurium trichlorides with sterically hindered alkenols resulted in 2,3,5-substituted tetrahydrofurans and the yield of the product was excellent [21]. The bicyclic furans formation and the furan ring annulation were the consequence of the reaction of γ,δ-alkenols having double carbon-carbon bonds with aryltellurium trihalides [20]. The interaction of o-allylphenols or 2-allylcyclohexanol **13** with p-methoxyphenyltellurium trichloride (**12**) in CHCl$_3$ produced the benzotetrahydrofurans **14** in high yields (Figure 11.3).

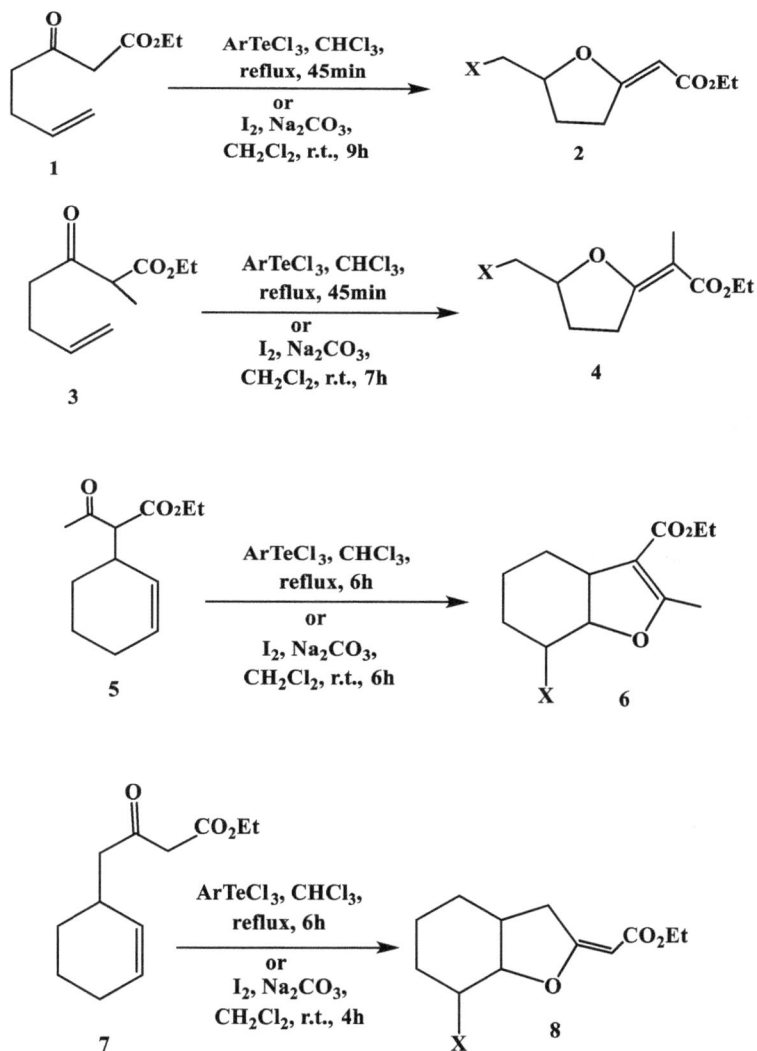

Figure 11.1: Iodine and tellurium-promoted cyclofunctionalization of β-keto esters substituted with alkenyl.

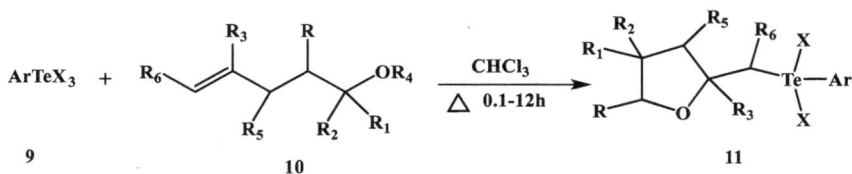

Figure 11.2: Synthesis of tetrahydrofuran derivatives.

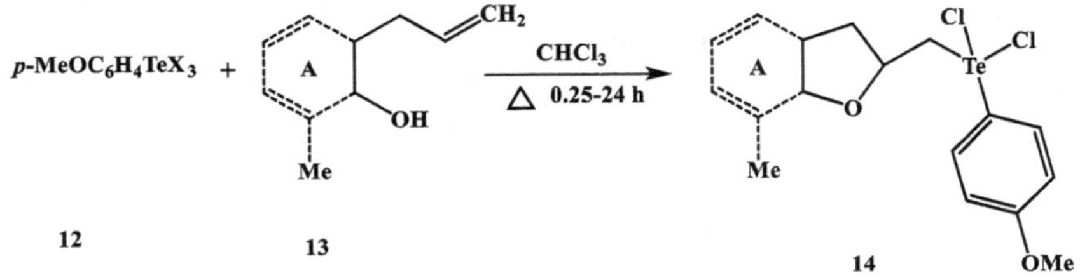

p-MeOC$_6$H$_4$TeX$_3$ + **12** **13** → **14**

X=Cl, Br; A=Cy, Ph, MeC$_6$H$_4$

Figure 11.3: Synthesis of benzotetrahydrofurans.

The γ-lactones substituted with aryldichlorotelluromethyl **16** in high yields was produced when a reaction of γ,δ-unsaturated carboxylic acids **15** with trichloride **12** was performed, as shown in Figure 11.4 [22].

The cyclization of diketone that has enol form and the furan **18** formation was observed as a result of the interaction of trichloride **12** with 2-allyldimedone **17**. Under the same conditions, the α-alkenyl derivatives of acetoacetic ester or acetylacetone **19** were transformed into the representing 2,3,5-substituted derivatives of furan **20** (Figure 11.5) [23].

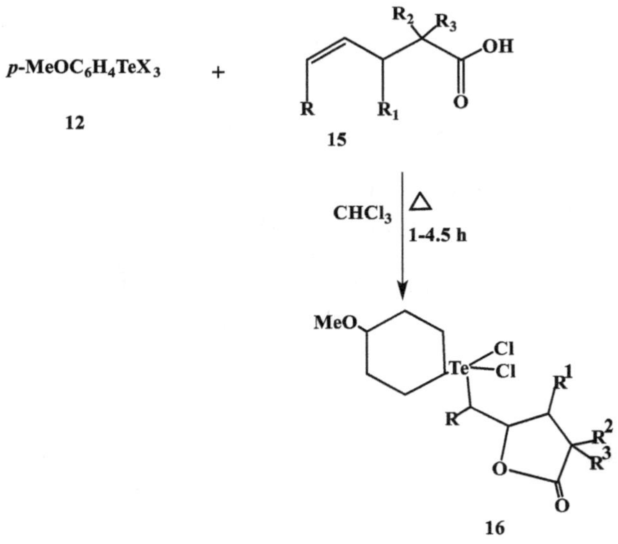

p-MeOC$_6$H$_4$TeX$_3$ +

12

15

CHCl$_3$ $\triangle$ 1–4.5 h

16

R = R = H; R + R$_1$ = (CH$_2$)$_{2\text{-}3}$; R$_2$ = R$_3$ = H, Ph, Me

Figure 11.4: Synthesis of aryldichlorotelluromethyl-substituted γ-lactones.

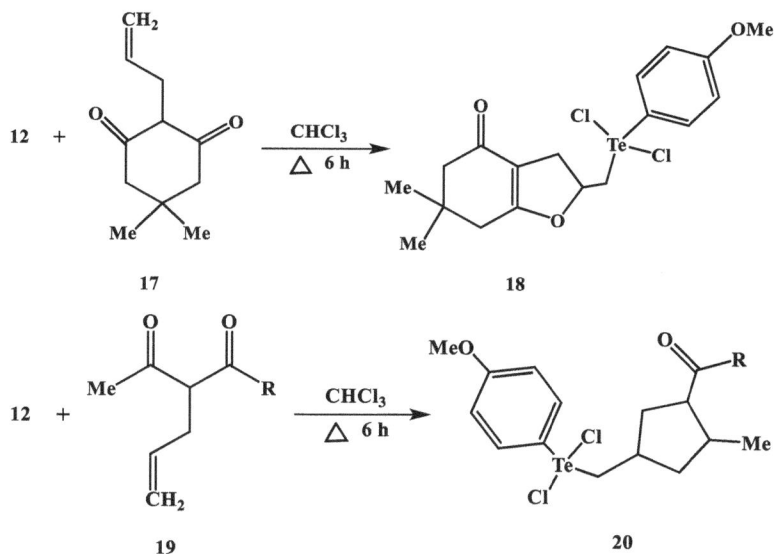

R=Me, OEt

Figure 11.5: Synthesis of 2,3,5-substituted derivatives of furan.

A cyclohexene and ethyl acetoacetate (compound **21**) on appropriate reaction produced condensed furan **22**. The product was with an aryltellurium part [11, 24]. Maintaining the same conditions, the interaction of p-methoxyphenyltellurium trichloride (**12**) with α,β-Diallyl-β-oxoesters **23** in CHCl$_3$ formed dihydrofurans **24** (Figure 11.6). Under the conditions, the γ-alkenyl-substituted β-oxoesters **25** were transformed into tetrahydrofurans **26**. An exocyclic double bond was present in the product (Figure 11.6) [23–25].

Using 5-hexenol (**27**) in the reaction combined with p-methoxyphenyltellurium trichloride (**12**) yielded 2-aryltelluromethyl-substituted pyran **28** with a high yield. 2-Butenyldimedone (**29**) on reacting with trichloride **12** in CHCl$_3$ yielded derivative of pyran **30** in high yield [20, 23]. The schematics of the synthesis are shown in Figure 11.7.

The production of the derivatives of pyran **32** (yield: 84%) and **34** (yield: 65%) was the result of the cyclization of α- and γ-butenyl derivatives **31** and **33,** when these were heated in CHCl$_3$ (Figure 11.8) [24].

Invariantly, aryltellurium trihalides reacting with alkenyloxy compounds gave a cation (telluronium), and consequently, an intramolecular nucleophilic action of hydroxyl oxygen was followed (Figure 11.9).

The oxygen atom reinstated as an additional nucleophilic core with a sulfur or nitrogen atom allowed the formation of sulfur and nitrogen-containing heterocycles. Therefore, N-alkenylthioureas **36** reacting with p-alkoxyphenyltellurium trichlorides

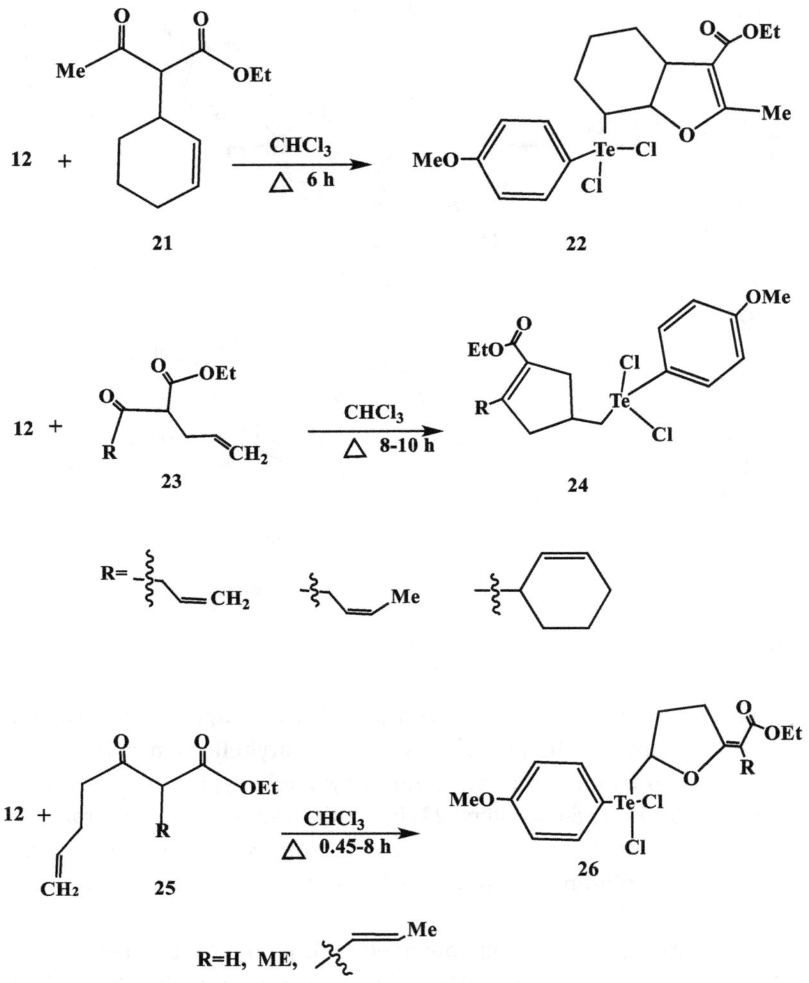

Figure 11.6: Synthesis of 2,5-substituted tetrahydrofurans.

35 made it possible for the formation of thiazolinium chlorides **37**. In the product, an exocyclic tetracoordinated Te was present (Figure 11.10) [26].

The cyclization reaction of S- or N-alkenyl derivatives of thiouracil **40, 38** with p-methoxyphenyltellurium trichloride (**12**) in glacial AcOH produced isomeric thiazolopyrimidines **41, 39**. The yields of products were in range 60–65% (Figure 11.11). The cyclization reaction produced telluro-functionalized selenazolopyrimidinone **41** through introducing 2-(allylselenyl)- 6-methylpyrimidin-4(3H)-one. Te-promoted cyclization reaction of S(Se)-alkenyl derivatives of thio(seleno)-uracil produced thiazolo(selenazolo)-pyrimidinone complex with p-methoxyphenyltellurium trichloride

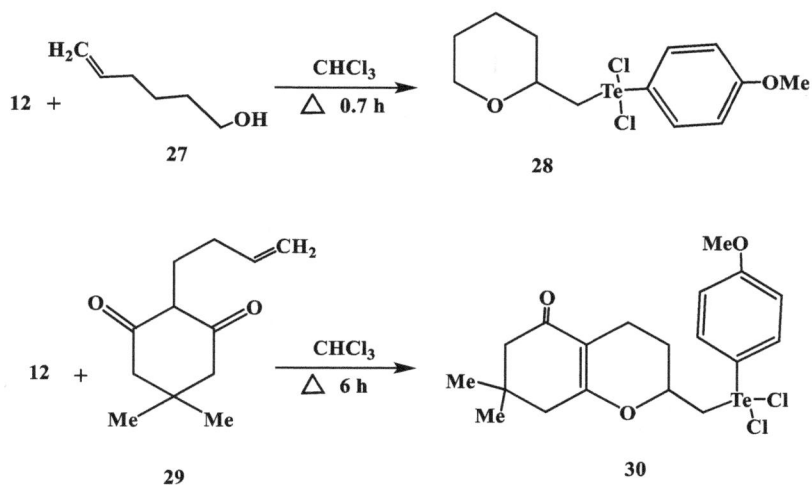

Figure 11.7: Synthesis of pyran derivatives.

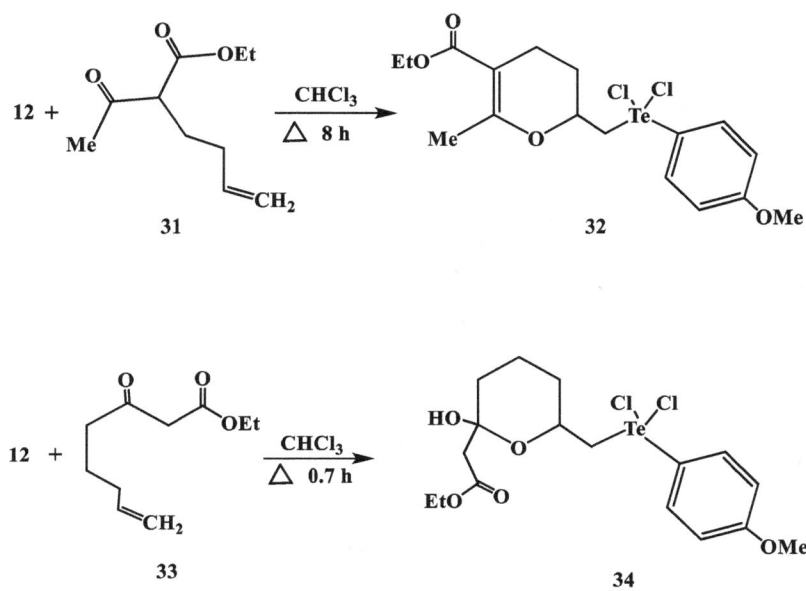

Figure 11.8: Synthesis of pyran derivatives.

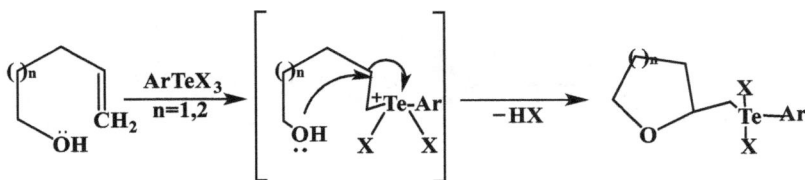

Figure 11.9: Mechanism of the interaction of alkenyloxy compounds with aryltellurium trihalides.

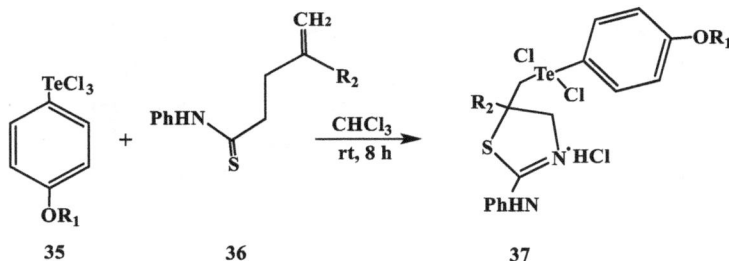

R_1=Me, Et

Figure 11.10: Synthesis of thiazole derivatives.

(12). NMR spectroscopy and mass spectrometry results were supportive to this observation [27].

N- or S-alkenyl derivatives of condensed pyrimidines **42, 44** were cyclized to polycyclic thiazolopyrimidines **43, 45** as a result of interaction with p-alkoxyphenyltellurium trihalides **9 a–d** under similar conditions (Figure 11.12).

The discovery of tellurocyclofunctionalization happened at the same time as the selenocyclofunctionalization [28, 29]. The selenocyclofunctionalization became a crucial synthetic tool [30]. But the tellurium counterpart was neglected.

The information regarding the understanding of the effect of the olefin structure and the Te electrophile for the tellurolactonization of γ,δ,-unsaturated carboxylic acids expanding to benzyl esters of the compound was reported [31]. The diastereomeric lactones mixtures were the results of the reaction, the stereoselectivity of the reaction was also discussed.

The reaction of carboxylic acid **46** with aryltellurium trichloride led to the tellurolactone **47** and the yield of the product was good. At the same time, the analogous benzyl ester failed to produce lactones **47**. The carbon-carbon double bond adducts were produced (**48**) and the compound was having aryltellurium trichlorides (Figure 11.13). The outcome was explained in terms of carboxylic oxygen's lesser nucleophilicity. Perhaps the initial stage consisted of π-complex formation between the electrophile and the double bond. The chloride ion created from the aryltellurium trichloride resulted in the adduct, with the lack of an efficient internal nucleophile.

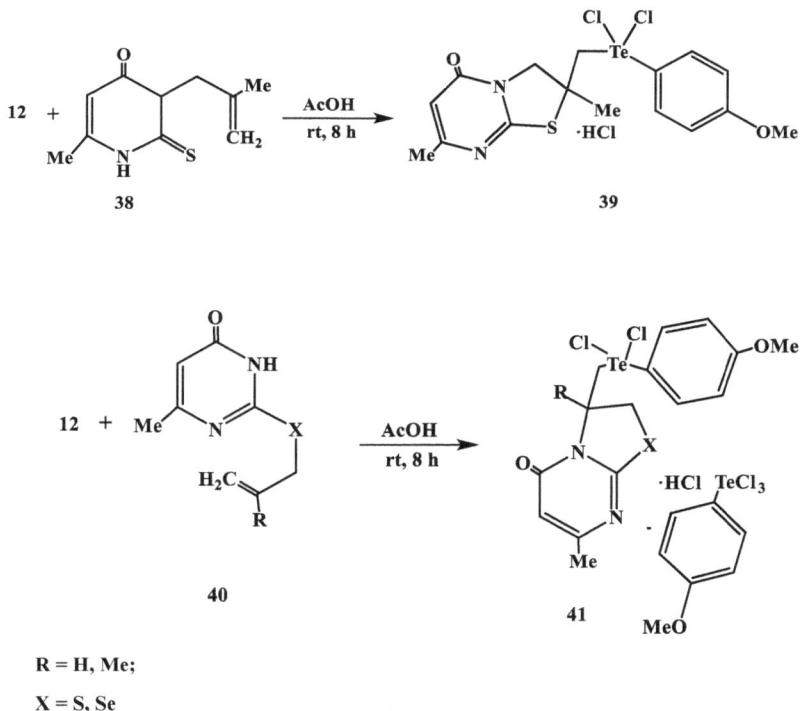

Figure 11.11: Thiazolopyrimidines synthesis.

As expected, the lactone **50** together with the carbon-carbon double bond adduct having hydrochloric acid (**51**) were the result of the reaction of **CH₃OPhTeCl₃** and **PhOPhTeCl₃** with acid **49** (Figure 11.14). During the lactone generation, the hydrochloric acid was released. An attack of the carbon-carbon double bond resulted in a tertiary carbenium ion. The carbenium ion interacts with the chloride ion and produces the compound **51**.

The tellurolactones **50** in good yields were achieved when **CH₃OPhTeCl₃** and **PhOPhTeCl₃** reacted with the benzyl ester since the formation of benzyl chloride was present as a by-product instead of hydrogen chloride. The interaction of aryltelluriumtrichlorides with γ,δ,-unsaturated carboxylic acids and benzylesters are summarized in Table 11.1.

It had been noted that PhTeOSO₂C₆H₄NO₂-4 (**54**) can be achieved *in situ* from 4-nitrobenzenesulfonyl peroxide and Ph₂Te₂. This was also capable of being an electrophilic reagent for the cyclization of unsaturated alcohols, and also for lactonization of unsaturated carboxylic acids (Figure 11.15) [32].

Besides the application of both the aryltellurinic anhydrides and the mixed anhydrides as oxidation reagents in the synthesis of organic compounds, it was employed

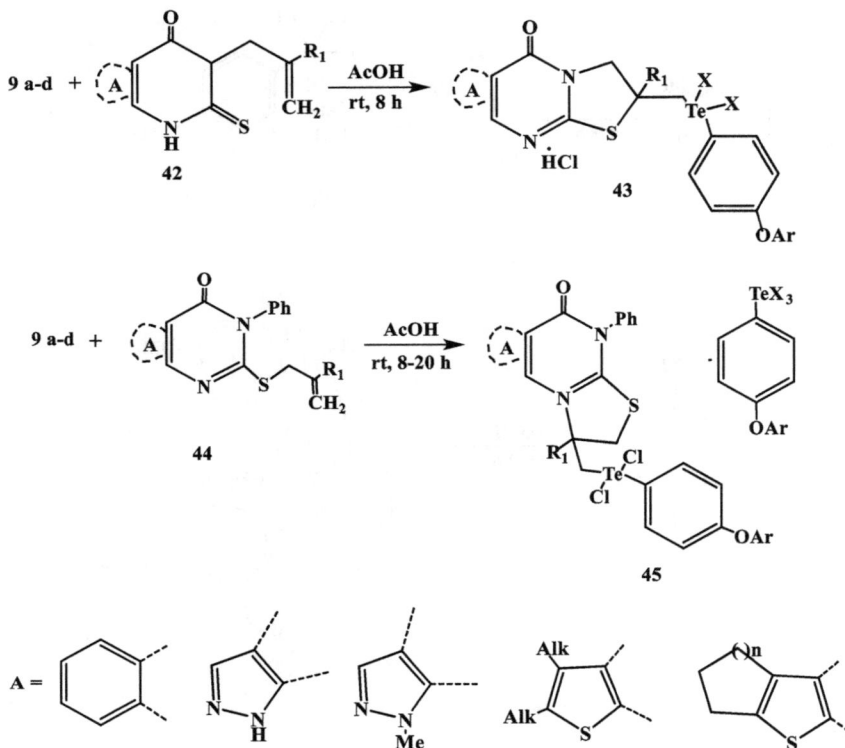

a: Ar = p-MeOC$_6$H$_4$, X= Cl; **b:** Ar = p-MeOC$_6$H$_4$, X = Br

c: Ar = p-EtOC$_6$H$_4$, X = Cl; **d:** Ar = p-EtOC$_6$H$_4$, X = Br

R$_1$=H, Me

Figure 11.12: Synthesis of thiazolopyrimidines.

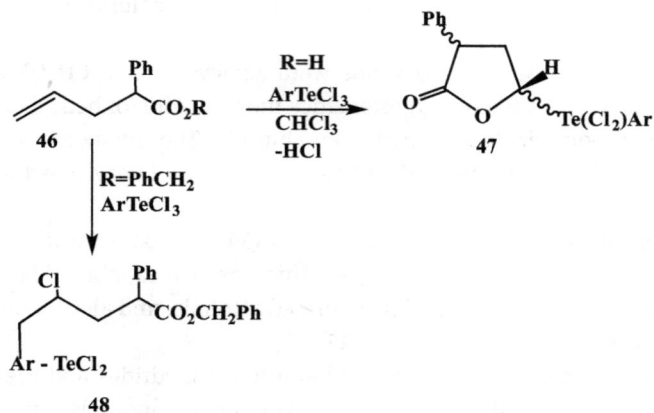

Ar = CH$_3$OPh, PhOPh

Figure 11.13: Synthesis of tellurolactone.

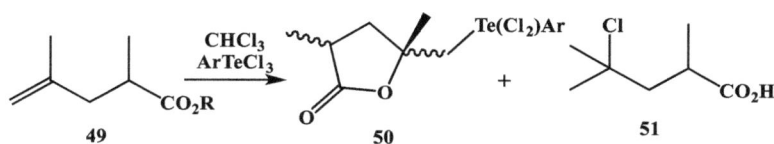

Ar = CH$_3$OPh, PhOPh

Figure 11.14: Synthesis of tellurolactone.

as versatile electrophiles in several organic transformations. For example, hydrox-yalkenes cyclofunctionalization and alkenes acetoxytellurinylation were per-formed [33, 34]. In order to make PhCH$_2$CH(OAc)CH$_2$Te(OAc)$_2$ (C$_6$H$_4$OMe-4) (**58**), the combined anhydride 4-MeOC$_6$H$_4$Te(O)OAc as an intermediate, the reaction of [4-MeOC$_6$H$_4$Te(O)]$_2$O together with PhCH$_2$CH=CH$_2$ the allylbenzene must be held in acetic acid which is at boiling temperature, instead of *vic*-diacetate. As shown in Figure 11.16, when utilizing organic substrates with a suitable nucleophilic group in an appropriate location of the olefinic molecule like in hydroxyalkenes, [ArTe(O)]$_2$O [Ar = Ph, 2-naphthyl, 4-MeOC$_6$H$_4$] it made cyclofunctionalization to 5-, 6-, and 7-membered cyclic ethers [33]. The of hydroxyalkenes cyclofunctionalization by dealing with the combined anhydride PhTe(O)OAc was quicker at room temperature in CHCl$_3$ and BF$_3$·OEt$_2$ was used as Lewis acid catalyst. The generation of 5-membered ring was the preferred option over the 3- or 4-membered rings and the addition interactions were extremely regio- and *trans*-stereoselective [34].

Alkenic carbamates cyclofunctionalization, aminotellurinylation of alkenes, and oxazolidinones synthesis from alkenes were also possible using both the aryltellurinic anhydrides and the mixed anhydrides [35, 36]. Interaction with mixed anhydrides PhTe(O)X [X= OTf, O(O)CCF$_3$, OAc], in CHCl$_3$ or CH$_2$Cl$_2$, a range of olefins went through regio- and stereoselective aminotellurinylation, with excessive amounts of ethyl carbamate, H$_2$NCO$_2$Et, as a nucleophile being present. After reduction with hydrazine hydrate (Figure 11.17a), the tellurinylated products of the reaction were secluded as the equivalent phenyltellurenylated derivatives. The aminotellurinylation was further continued to olefinic carbamates cyclofunctionalization and the nitrogen heterocycles were formed (Figure 11.17b) [36]. Furthermore, while the aminotellurinylation reaction of olefins was executed at an elevated temperature, the 2-oxazolidinone derivatives were achieved right away in a prominent yield. Thus, the reaction provided a basic one-pot synthesis process for compounds of this category from cycloalkenes or alkenes directly (Figure 11.17c) [35, 36].

2-oxazolines synthesis from alkenes and oxazoles synthesis from alkynes and amidotellurinylation of alkenes and alkynes were also possible using both the aryl-tellurinic anhydrides and the mixed anhydrides [37–40]. The mixed anhydride PhTe(O)O(O)CCF$_3$ at room temperature in the presence of BF$_3$·OEt$_2$, produced the alkenes

Table 11.1: Interaction of aryltelluriumtrichlorides with benzylesters and γ,δ-unsaturated carboxylic acids.

Substrates	Electrophiles	Products
	$p-CH_3OPhTeCl_3$	*p*-CH₃OPhTeCl₂
	$p-PhOphTeCl_3$	*p*-PhOPhTeCl₂
	$p-CH_3OphTeCl_3$	*p*-CH₃OPhTeCl₂
	$p-PhOphTeCl_3$	*p*-PhOPhTeCl₂
	$p-CH_3OphTeCl_3$	*p*-CH₃OPhTeCl₂

Table 11.1: (continued)

Substrates	Electrophiles	Products
Ph ... CO₂H	p – PhOphTeCl$_3$	p-**PhOPhTeCl$_2$**
... CO₂H	p – CH₃OphTeCl$_3$	p-**CH$_3$OPhTeCl$_2$**

$$Ph_2Te_2 \ + \ (4\text{-}O_2NC_6H_4SO_2)_2O_2$$

52   **53**

$$PhTeOSO_2\text{-}C_6NC_6H_4NO_2\text{-}4$$

54

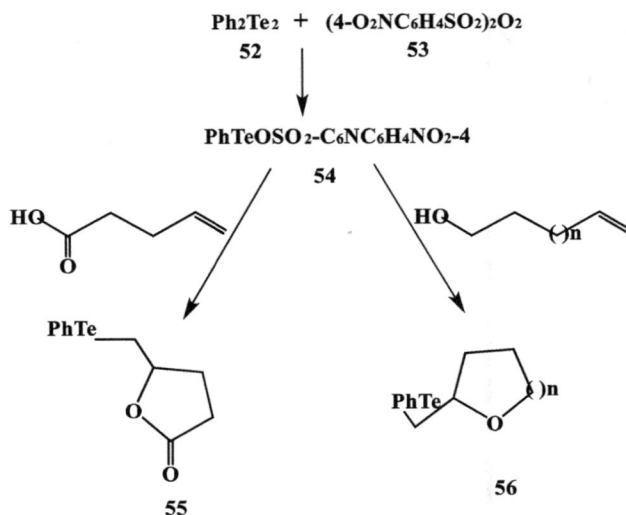

Figure 11.15: Cyclization of unsaturated carboxylic alcohols or acids using $PhTeOSO_2C_6H_4NO_2\text{-}4$ (**54**).

amidotellurinylation in MeCN by functioning as a nucleophile and a solvent as shown in Figure 11.18(a). As nucleophiles in such reactions were the propionitrile and benzonitrile to result in amidotellurinylation products that could be transformed at elevated temperatures into 2-oxazolines following an *in situ* intramolecular substitutions with excellent yields. The reactions were extremely stereo- and regio-selective reactions [37, 38]. Similarly, in MeCN when Brønsted acid (for example, H_2SO_4) was present PhTe(O) OTf goes through (E)-stereoselective amidotellurinylation of alkynes. In the case of terminal alkynes being utilized, the ensuing added products (**71**) were isomerized thermally to the equivalent [(Z)-β-acetamidovinyl] phenyltellurium (IV) oxides (**72**) and was separated as the equivalent tellurides. Comparably, the spontaneous

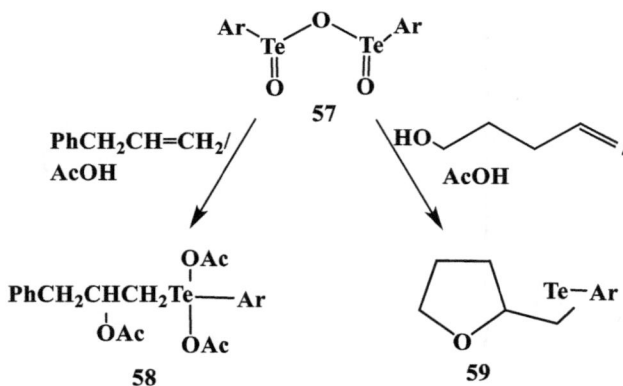

Figure 11.16: Cyclofunctionalization of hydroxyalkenes and acetoxytellurinylation of alkenes and employing aryltellurinic anhydrides, $[ArTe(O)]_2O$ (**57**).

(a)

(b)

(c)

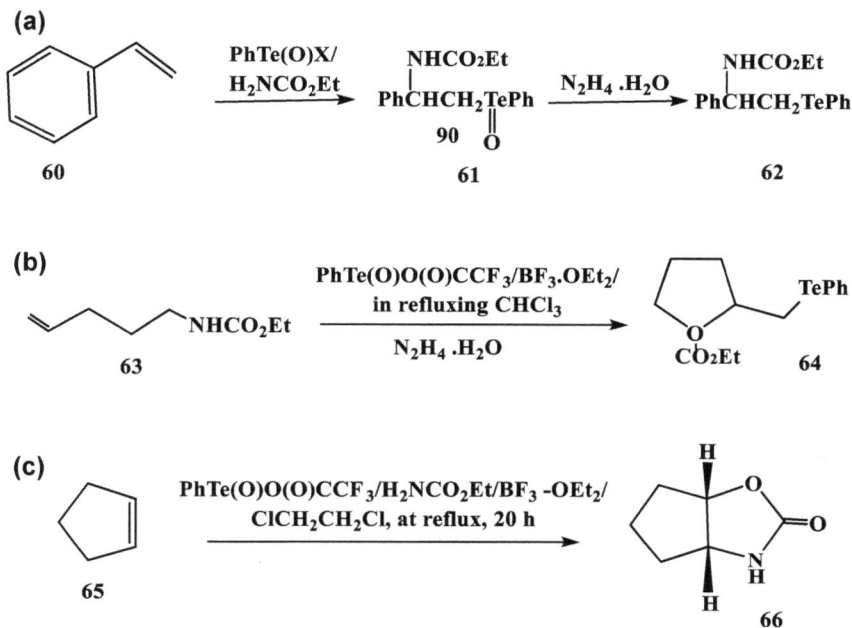

Figure 11.17: (a) Alkenes aminotellurinylation, (b) alkenic carbamates cyclofunctionalization and (c) conversion of alkenes to 2-oxazolidinones.

intramolecular cyclization at a greater temperature partially turned the internal alkynes' products into oxazoles (Figure 11.18b) [39, 40].

During the selective epoxidation of olefins with hydrogen peroxide the catalytic action of a divinylbenzene-styrene copolymer-supported tellurinic acid (**76**) was reported (Figure 11.19) [41]. When preparing a tellurium containing catalyst, its activity potently relies on the level of cross-linking in the divinylbenzene-styrene copolymer employed for its production. Higher conversions were observed with higher cross-linked polystyrene. Withholding of the *cis-trans* geometry in the epoxide product, the oxidation rate depended on olefin structure (for example, the rate was accelerated by increased alkyl substitution) with an oxidation process that is stereospecific. It was worth mentioning that the polymeric structure of the tellurium containing catalyst **76** is obligatory. Also, no epoxidation was observed with free tellurinic acids, for example with 4-MeOC$_6$H$_4$Te(O)OH and C$_6$H$_5$CH$_2$CH$_2$Te(O)OH [41].

It was confirmed that the active catalysts are the corresponding aryltellurinic acids, ArTe(O)OH, developed *in situ* by Ar$_2$Te$_2$ reacting with H$_2$O$_2$ (Figure 11.20) [42]. The best catalyst was discovered to be the unsubstituted phenyltellurinic acid, PhTe(O)OH. For example, the incorporation of Ph$_2$Te$_2$ in 0.20 mol% comparative to substrate boosted a 240-fold enhance in the oxidation rate of NaBr with H$_2$O$_2$, as measured by the bromination of 4-pentenoic acid. And using it provided bromolactonization of a group of alkenoic acids and monobromination of a range of activated aromatic substrates. The rate of reactions was minimized as a result of both electron-withdrawing and electron-donating group on the catalyst's phenyl ring. The stereoelectronic effects

(a)

(b)

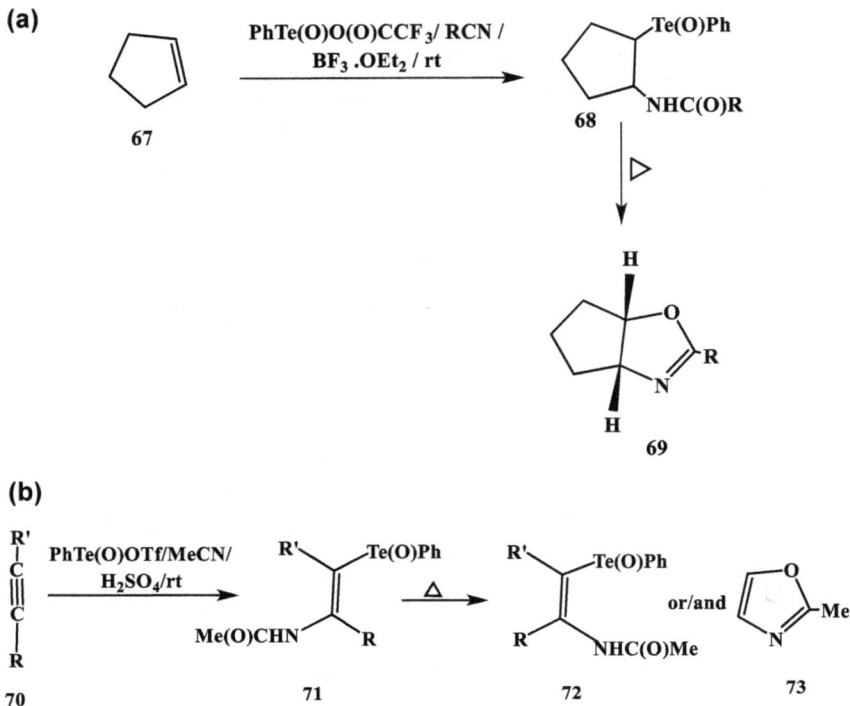

Figure 11.18: (a) Conversion of alkenes to 2-oxazolines and (b) conversion of alkynes to oxazole.

played a role in influencing the bromination rate by regulating (i) the oxidation rate of Ar_2Te_2 to the tellurinic acid, $ArTe(O)OH$, (ii) the reaction rate of NaBr and H_2O_2 with the tellurinic acid, and (iii) the oxidation rate of the conversion of tellurinic acid to the telluronic acid, $ArTe(O)_2$ (OH) (**80**), they were unable to activate the peroxide, so ever since, the latter is working as a catalytic termination stage. Moreover, the balancing act contained relative conversion ratios of the pertellurinic acid, $ArTe(O)OOH$ (**79**) to the

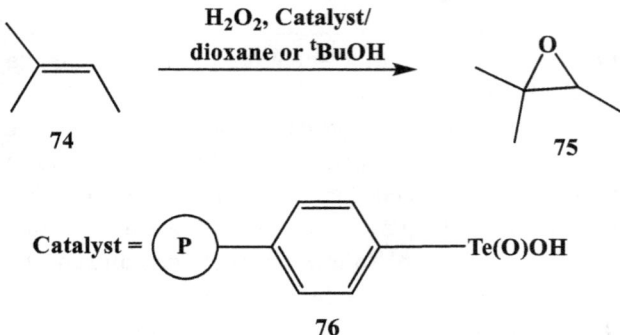

Figure 11.19: Selective epoxidation of olefins with H_2O_2, the copolymer-supported tellurinic acid (**76**) was used as the catalyst.

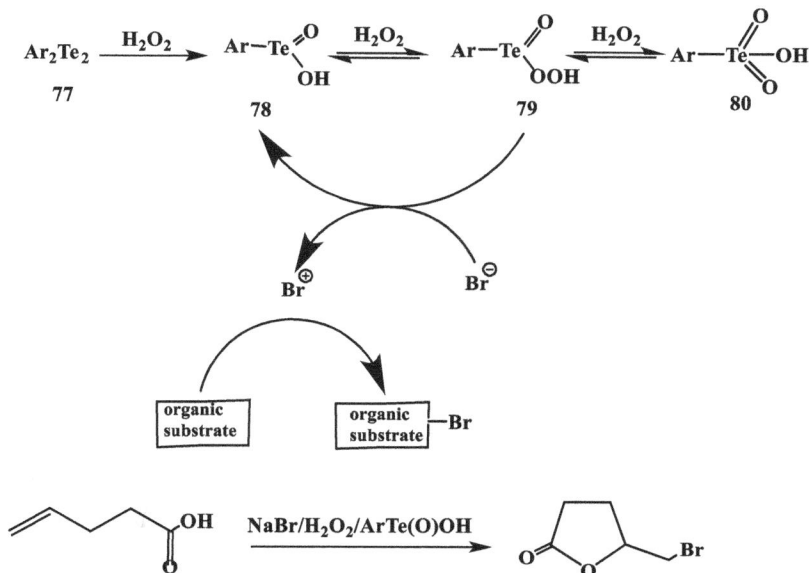

Figure 11.20: Proposed mechanism for the diaryldITelluride/aryltellurinic acid-catalyzed oxidation of halides salts with H_2O_2.

telluronic acid **80** or for the reaction of the bromide with pertellurinic acid **79** (Figure 11.20) [42].

The oxidation of diaryltellurides to either Ar_2TeO or Ar_2TeO/Ar_2TeO_2 mixtures was possible to get under photosensitized conditions with singlet oxygen, utilizing rose bengal, hematoporphyrin, or tetraphenylporphyrin as photosensitizer [43–47]. Noticeably, the generation of diaryltellurones through a nucleophilic pertelluroxide intermediate was observed, $[Ar_2Te^+OO^-]$ [47]. Using various diaryltellurides, Ar_2Te as catalysts, the thiols aerobic oxidation was successfully realized under photosensitized conditions and disulfides were formed [46]. The tellurides with bulkier aryl unit defected in the reaction rate of oxidation and, moreover, it triggered the establishment of certain unidentified by-products. Under identical conditions, employing $(4\text{-MeOC}_6H_4)_2Te$ and rose Bengal or TPP as sensitizers, oxidation of aliphatic dithiols and the formation of cyclic disulfides was attained in reasonable yields (Figure 11.21).

It is a well-known method to make carbo- and heterocyclic ring systems using electrophilic cyclization [48–51]. Studies were reported regarding the application of 1,3-dicarbonyl compounds as furan ring precursors [8, 10, 52–54]. Stefani et al.

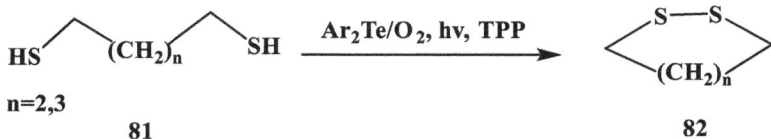

Figure 11.21: Oxidation of aliphatic dithiols and the formation of cyclic disulfides.

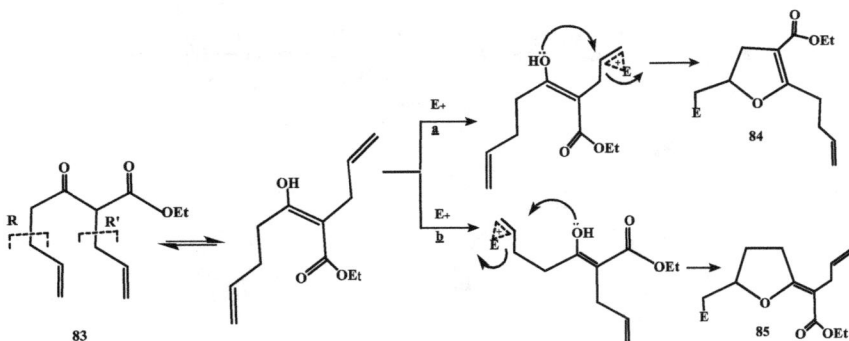

Figure 11.22: Synthesis of telluro 2,3,5-trisubstituted dihydrofurans.

disclosed the ring closure of α,γ-diallyl-β-ketoesters employing Te and Se reagents as electrophiles in keeping with a long-term contribution to the cyclofunctionalization reaction [25]. The interaction of p-methoxyphenyl tellurium trichloride and phenylselenenyl bromide with α,γ-diallyl-β-ketoesters gave seleno- and telurocyclofunctionalized furan derivatives in good yield.

When simply heating the reaction mixture in CHCl$_3$ it was observed that such ring closures, the yield was moderate to good with C$_6$H$_5$SeBr and the yield was high with p-MeOC$_6$H$_4$TeCl$_3$. Dihydrohran derivatives **84** were the preferential products formed through the central allyl group and the enolic OH participation, with disregard to whether or not the ring closure reaction includes a π complex' or a transient "onium" ion (Figure 11.22). The obtained products are shown in Table 11.2.

The addition of ArTeCl$_3$, to alkenes C=C bond bearing a nucleophilic moiety at a appropriate location, is an example of an electrophilic tellurium reagent reaction. This was accompanied by intramolecular entrapping of the telluronium intermediate and this process finally furnished cyclic systems where the tellurium moiety is displayed outer of the freshly formed ring. These interactions were referred to as tellurocyclofunctionalization. It can be either lactonization [22, 29] or etherification [20] (Figure 11.23).

A few other reagents were used for this purpose. PhTeOSO$_2$C$_6$H$_4$NO$_2$-4 [32] for etherifications and lactonizations, and ArTe(O)OAc [36], TeO$_2$/AcOH/LiCl [55] and TeO$_2$/HCl/MeOH [56] for etherifications, were alternative useful reagents for these cyclizations. Further reaction with Bu$_3$SnH was used to accomplish from the cyclic products the reductive elimination of the tellurium moiety [57].

Pd(II) also promoted detellurative carbonylations [58, 59] of vinyl, alkyl, phenyl, and alkynyl tellurides with MeOH/CO/Et$_3$N and produced methyl carboxylates. Only a catalytic amount of Pd(II) was necessary, when carried out by the means of CuCl$_2$ as an oxidant. Substituted butenolides originating from hydroxy vinyl tellurides were also prepared using the same method (Figure 11.24).

Table 11.2: Telluro 2,3,5-trisubstituted dihydrofurans.

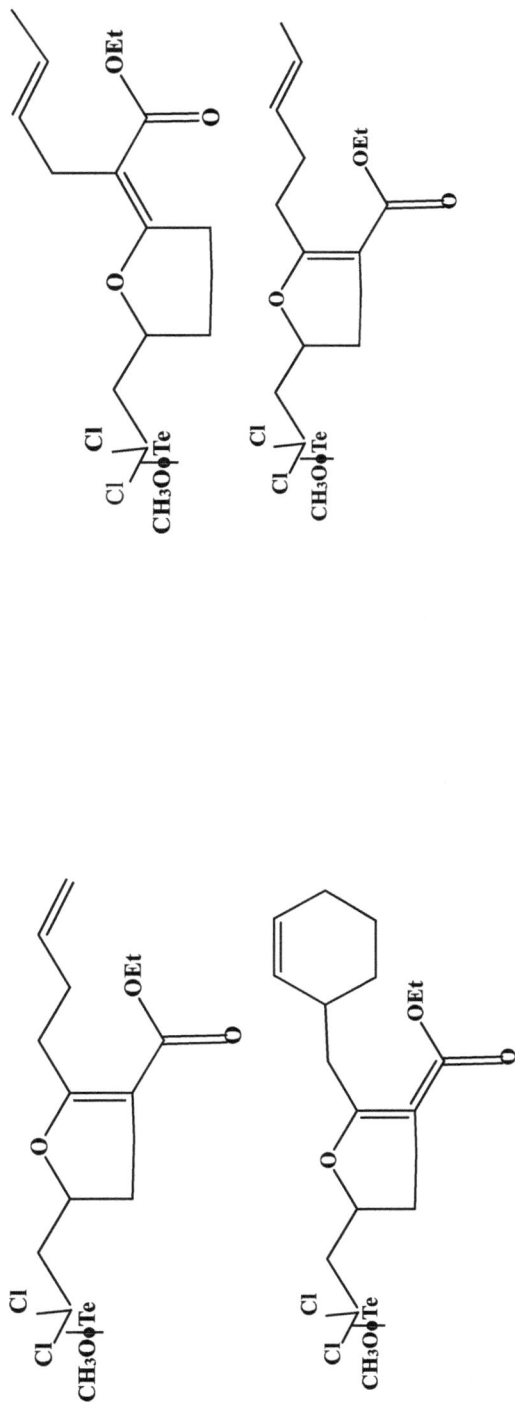

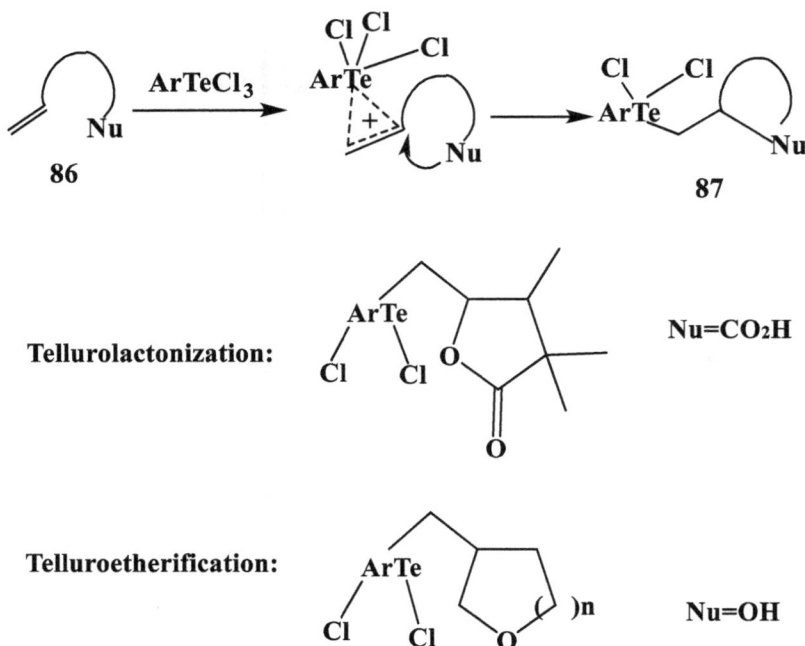

Tellurolactonization:

Nu=CO₂H

Telluroetherification:

Nu=OH

Figure 11.23: Tellurolactonization and telluroetherification.

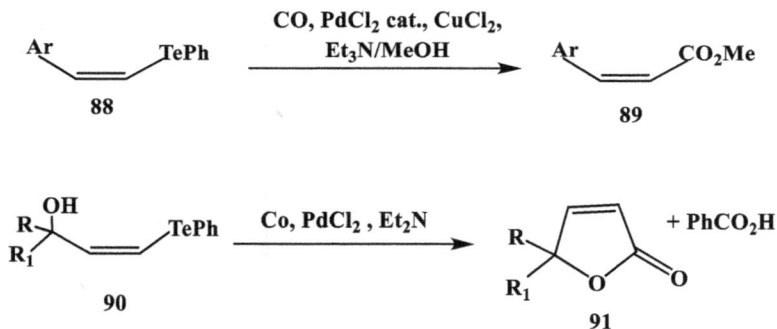

Figure 11.24: Detellurative carbonylation.

The interaction of allylic halides with Li₂Te furnished 1,5-dienes [60] as a result of a coupling process between the appropriate allylic radicals. A radical chain mechanism [61–63] occurred upon irradiation of a mixture comprising an electrophilic olefin, the acetyl derivative of N-hydroxy-2-thiopyridone (**94**) and a alkyl anisyl telluride. This was beneficial in intramolecular cyclizations yielding six-membered rings (Figure 11.25). The similar mix was used in the Te-mediated addition of carbohydrates to olefins.

Telluro methylene and methyl substituted tetrahydrofurans [64] were found from epoxides by the association of radical cyclization and tellurolate-promoted oxirane ring-opening reaction (Figure 11.26).

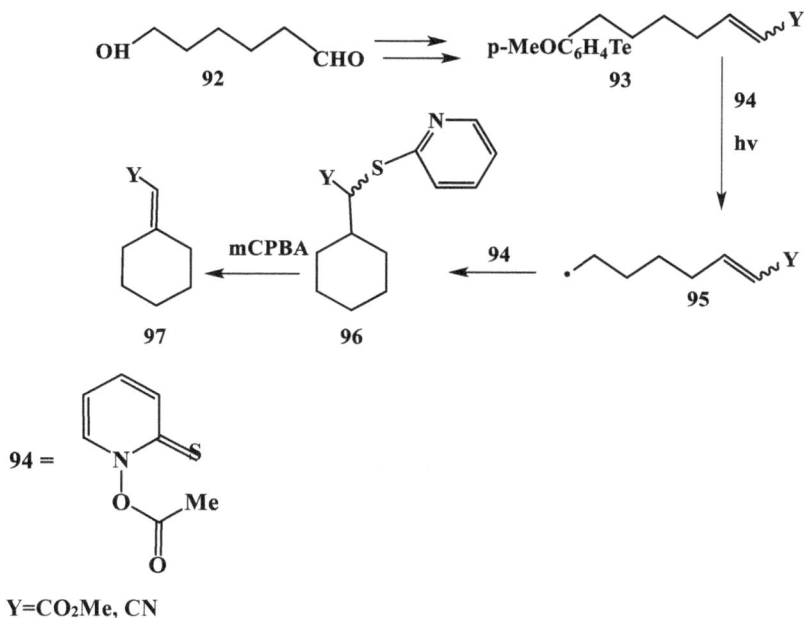

Y=CO₂Me, CN

Figure 11.25: Radical reaction.

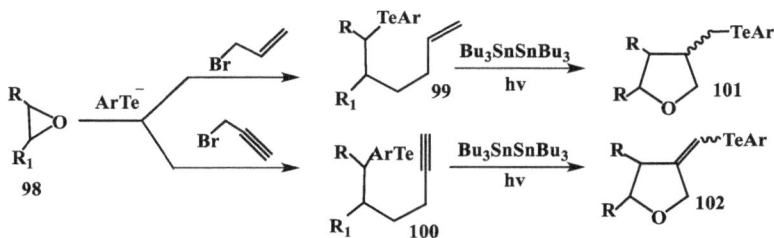

Figure 11.26: Radical cyclizations and tellurolate-promoted oxirane ring-opening.

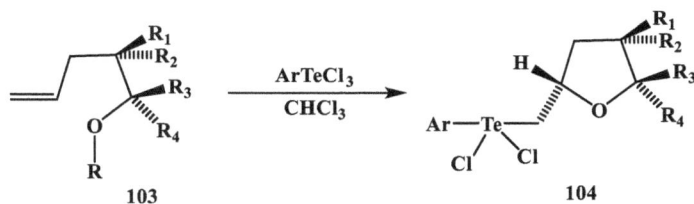

R=H, CH₂C₆H₅; Ar=p-CH₃OC₆H₄, p-C₆H₄OC₆H₄

Figure 11.27: Cyclization of benzylethers and unsaturated alcohols with aryltellurium trichlorides.

Table 11.3: Cyclization of benzylethers and unsaturated alcohols with aryltellurium trichlorides.

Alchohol/Benzy Ether	Tetrahydrofuran

Table 11.3: (continued)

Alchohol/Benzy Ether	Tetrahydrofuran

X= p-CH$_3$OC$_6$H$_4$TeCl$_2$; Y= p-C$_6$H$_4$OC$_6$H$_4$TeCl$_2$.

In the presence of an electrophile, the cyclofunctionalization of double bonds is a very beneficial synthetic methodology. Researchers showed that lithium chloride/ tellurium oxide in acetic acid, aryltellurinic anhydrides, and aryltellurium trihlorides are effective cyclization agents for carbamates, olefinic carboxylic acids, and alcohols [20, 22, 36, 51, 55]. The usage of acidic circumstances to execute the cyclization or the generation of HCl acid during the process was a major setback to these methodologies. The olefinic benzyl ethers cyclization encouraged by aryltellurium trichlorides took place under neutral circumstances in good yield and at reaction times comparable to those found for the corresponding alcohols cyclization was reported (Figure 11.27) [21]. The products are shown in Table 11.3.

The study showed that the time for reaction for certain unsaturated alcohol and the equivalent benzyl ether with p-methoxyphenyltellurium trichloride at 80 °C were nearly the same (Figure 11.27, Table 11.3). The cyclic ether in excellent yield after 5 min was an outcome of the interaction of the p-phenoxyphenyltellurium trichloride with alcohol. The reaction was also performed at 0 °C, but a longer reaction time was necessary. The same product was achieved at room temperature after 30 min of the interaction of the corresponding benzyl ether.

It was reported that diphenylallylacetic acid (**105**) interacts with napthyl tellurenyl iodide, arylselenenyl halides, aryltellurium trichloride or tellurium tetrachloride to produce the corresponding lactones (**106**) (Figure 11.28) [29].

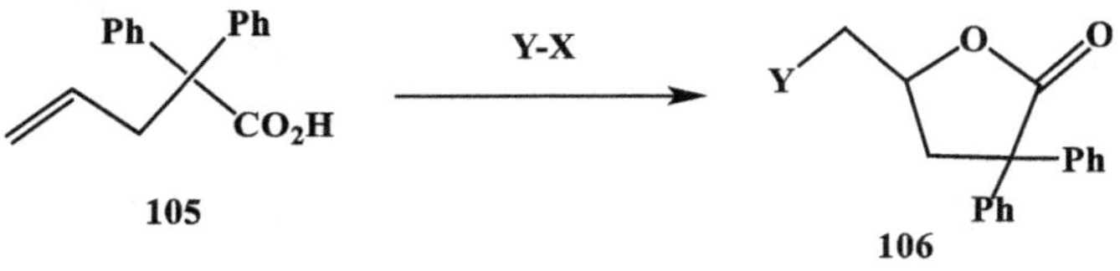

Y = ArSe, ArTeCl₂, Naphtyl-Te, TeCl₃; X= Br, Cl, I, Cl

Figure 11.28: Synthesis of lactones.

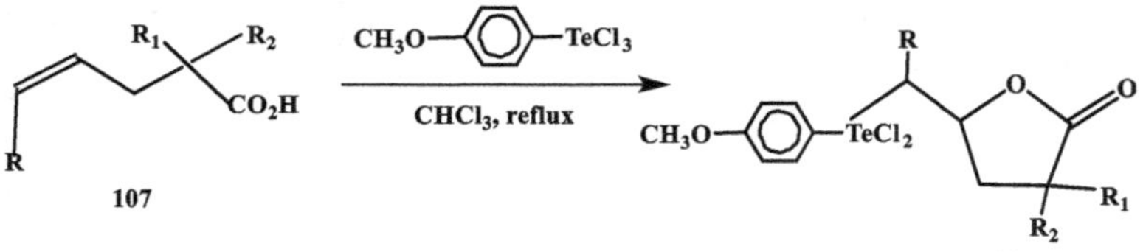

Figure 11.29: Lactonization of γ-δ unsaturated carboxylic acids.

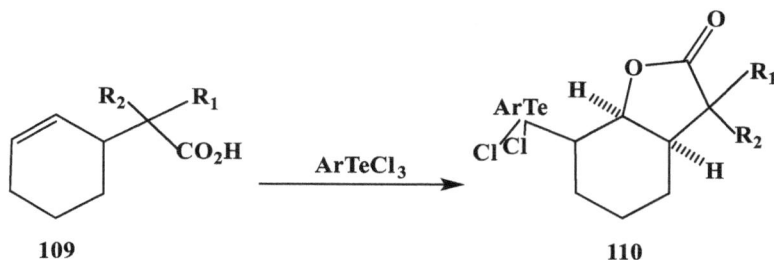

Figure 11.30: Mechanism of lactonization.

The lactonization of γ-δ unsaturated carboxylic acids is shown in Figure 11.29, aryltellurium trichlorides was used for this reaction. The reagents were chosen based on their stability and easiness of formulation when equated to aryltelurenyl halides and Te tetrahalides. The lactonization process proceeded smoothly and in well yields solely by refluxing the unsaturated carboxylic acids (**107**) with an excessive quantity of the Te reagent in chloroform.

Accordant with the mechanisms provided, *cis*-fused lactone was received (Figure 11.30). The lactone ring closure plausibly followed the similar mechanism as the halo and selenolactonization process [29, 65, 66].

An advantageous and extremely effective process for the preparation of β-organotellurobutenolides following the aryltellurenyl halides-induced electrophilic tellurolactonization of α-allenoic acids at modest circumstances was covered [67]. The schematics of the reaction are shown in Figure 11.31. The obtained β-organotellurobutenolides were used as precursors for a variety of butenolide derivatives via a Pd/Cu(I)-catalyzed cross-coupling using terminal alkynes or a substitution process using organocuprate reagent.

Several di- and trisubstituted α-allenoic acids were employed as substrates to provide corresponding β-organotellurobutenolides in better yields within 30 min.

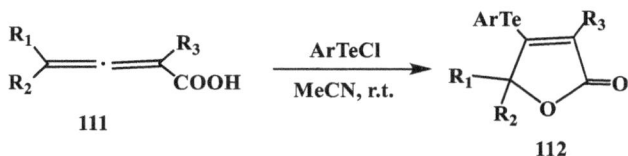

Figure 11.31: Synthesis of substituted β-organotellurobutenolides.

A cyclization product in good yield was obtained by a massive tellurenylating reagent carrying the napthyl group. The allyl group did not react for 2-allyl α-allenoic acid. This indicated that the reactivity of allenyl moiety is better than the individual carbon–carbon double bond. Nonetheless, there was no cyclization product that is spotted with nonsubstituted 2,3-butadienoic acid. A reaction using 2-ally-2,3-butadienoic acid that are monosubstituted as the substrate was somewhat inactive, supplying barely a minimal amount of the desired product.

The tellurolactonization in the case of α-allenoic acids most likely proceeded via the aryltelluro cation's electrophilic attack and this occurs on the central carbon atom of the allenyl moiety. An onium cation **113b** and/or an allyl cation **113a** was formed (Figure 11.32) [68]. The final product was resulted because of the additional cyclization of the intermediate. As the stabilization of the cationic intermediate by substituents was crucial for the formation of **113a** or **113b**, the absence of neighboring group involvement may lead to inactive tellurolactonization of the α-allenoic acids.

The anticipated products **119** and **116** in good yields (Figure 11.33) were produced using vinylic trichlorides **118** and cyclic compounds **115**. These were produced by the tellurocyclization of unsaturated alcohols and the mixing of alkynes to tellurium tetrachloride.

Such lactonization of alkene carboxylic acids [29, 69–71] and unsaturated alcohols's cyclization with electrophiles was reported [20, 72].

The compounding of diphenyl ditelluride/p-nitrobenzenesulfonyl peroxide (NBSP) produced the novel reagent benzenetellurenyl p-nitro benzenesulfonate **120**. It is an exceptional eletrophile for tellurocyclization because as a counter ion it has the feeble nucleophilic nitrobenzenesulfonate [32] (Figure 11.34).

α-Alkenylsubstituted β-dicarbonyl compounds **121**, **123**, and **125** underwent tellurocyclofunctionalization through an exo-mode action of their enolic forms [23] (Figure 11.35).

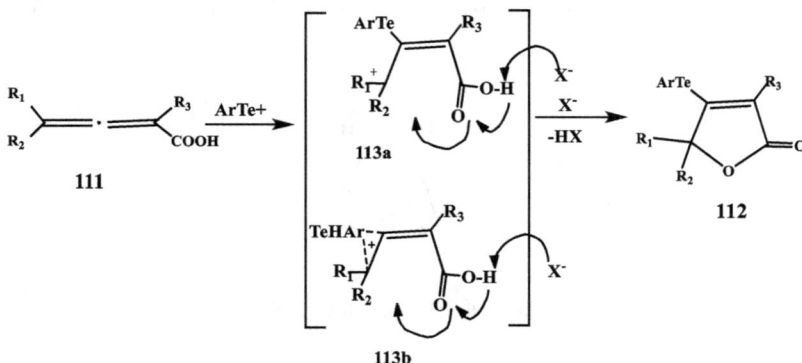

Figure 11.32: Mechanism of tellurolactonization of α-allenoic acids.

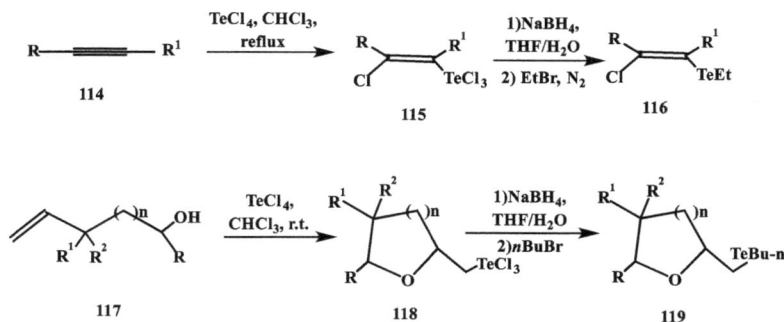

R= Ph; R^1=Ph, Me

n= 1,2; R= H, *i*-prop; R^1=H, Me

Figure 11.33: Synthesis of compounds **116** and **119**.

γ-Alkenyl-substituted β-dicarbonyl **127**, under the similar circumstances, gave 2,5-disubstituted tetrahydrofurans holding exocyclic double bonds (Figure 11.36). After being treated with NaBH$_4$, the products were shrunk to the appropriate size. By the activity with tributyltin hydride, tellurides which were then transformed into Te free methyl derivatives [11].

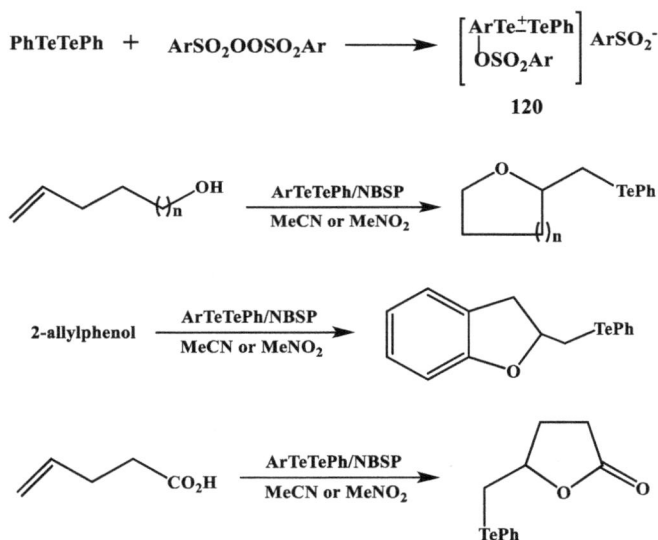

Ar=*p*-NO$_2$C$_6$H$^-$$_4$; n=1, 2

Figure 11.34: Tellurocyclization using NBSP.

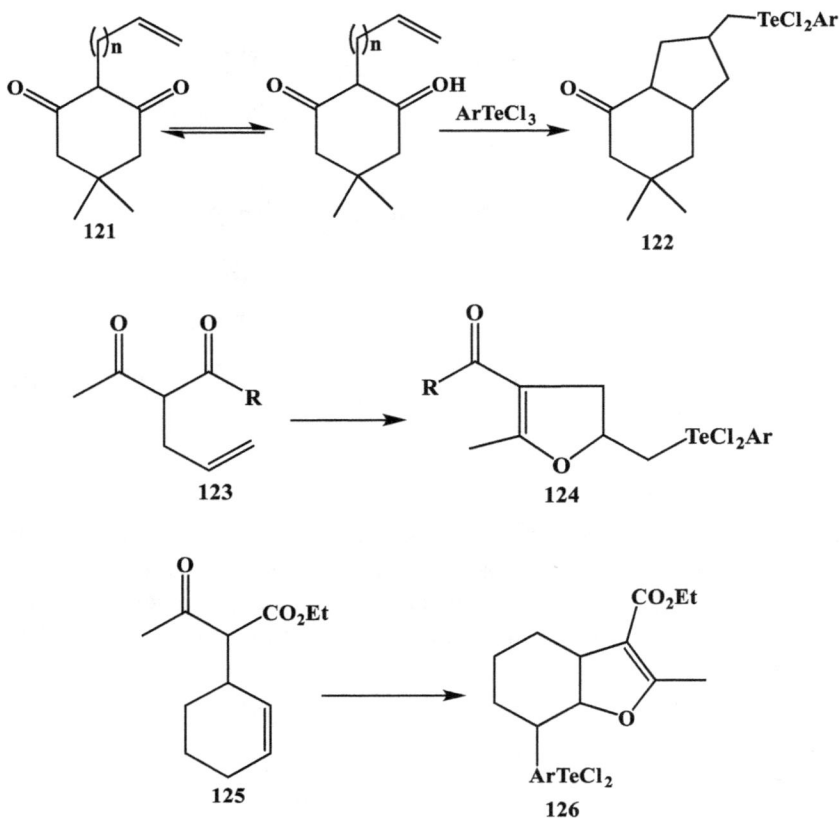

Figure 11.35: Tellurocyclofunctionalization of α-Alkenylsubstituted β-dicarbonyl compounds.

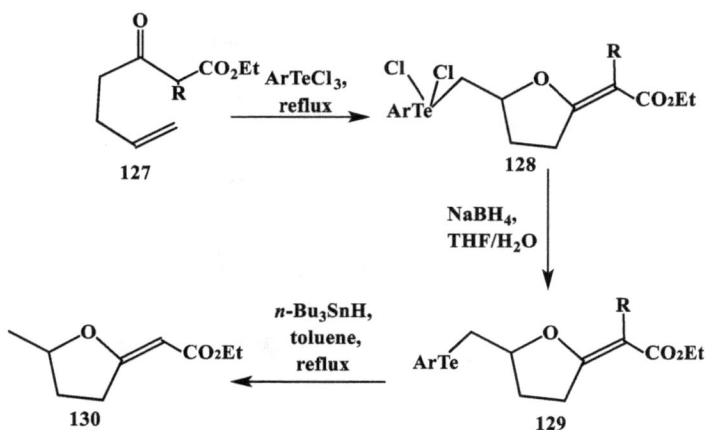

R=H, Me

Figure 11.36: Tellurocyclofunctionalization of γ-Alkenyl-substituted β-dicarbonyl.

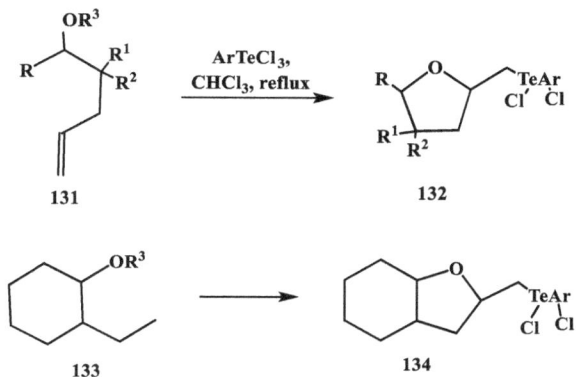

Ar= p-MeOC$_6$H$_4$, p-C$_6$H$_5$OC$_6$H$_4$; R= i-pr; R^1=R^2=Me; R^3 = H, Bz; R= n-Bu; R^1 =H; R^2 = pr; R^3 = H, Bz; R = Ph; R^1 = H; R^2 = Me; R^3 = H, Bz; R = CH$_2$OH. CH$_2$OBz; R^1 =R^2 =H; R^3 =H, Bz; R= H, Bz

Figure 11.37: Ethercyclization with aryltellurium trichlorides.

For ethercyclization with aryltellurium trichlorides, olefinic benzyl ethers **85** and **86** were proven to be appropriate substrates. The reaction times and yields were similar to what has been noticed in the corresponding alcohols's cyclization with low stereoselectivity of the chemical process [21] (Figure 11.37).

The effects of the structure of the substrate in the tellurofunctionalization process of γ,δ-dunsaturated carboxylic acids, and equivalent benzylesters were discussed [31]. Tellurolactone was achieved through the reaction of aryltellurium trichlorides with γ,δ-unsaturated carboxylic acids **135** having monosubstituted carbon–carbon double bond, while the benzyl esters **137** that corresponds provide the extra products of the aryltellurium trichlorides. The 1,1-disubstituted double bond provided a variety of

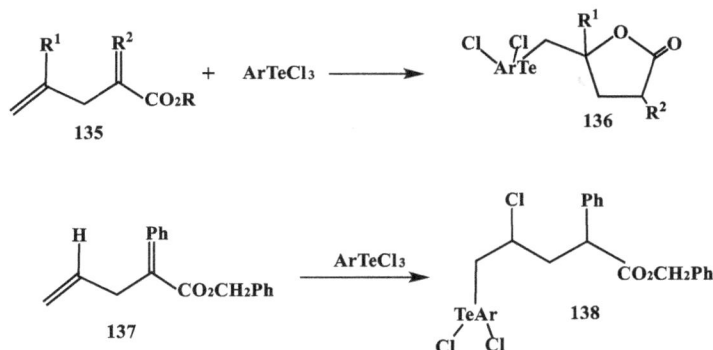

R=H, PhCH$_2$; R$_1$=H, Me; R2=Me, Ph; Ar=p-MeOC$_6$H$_4$, p-C$_6$H$_5$OC$_6$H$_4$

Figure 11.38: Tellurofunctionalization of γ,δ-unsaturated carboxylic acids and equivalent benzylesters.

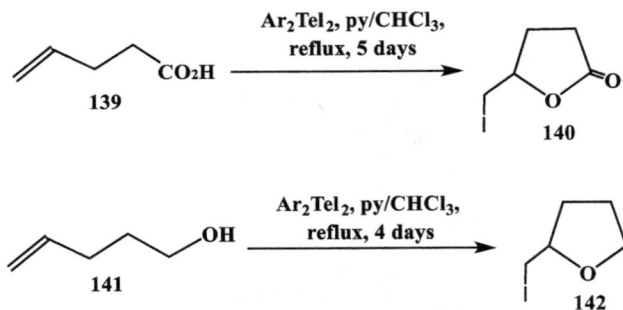

Figure 11.39: Iodocyclization of 4-pentenoic acid and 4-pentenol.

tellurolactones with the HCl adducts to the double bond. At the same time, the equivalent benzyl ester solely provided the tellurolactones (Figure 11.38).

After 5- and 4-days reflux in CHCl₃, in the presence of pyridine, diaryl tellurium diiodides (Ar=p-ClC₆H₄) promoted the iodocyclization of 4-pentenol **141** and 4-pentenoic acid **139** [73] (Figure 11.39).

With 3-butenol and 4-pentenol **145–146**, just bits of bromo tetrahydrofurans were provided and toward 4-pentenoic acid **143**, the diaryl tellurium dibromides were inactive [73] (Figure 11.40).

Under PTC conditions, mixing of o-diethynylbenzene to Na₂Te with hydrazine hydrate produced 3-benzotellurepine **150**. Because the compound is extremely volatile, it was transformed into a more stable dihalide **152** by interaction with Br₂ or SO₂Cl₂. By reduction with Na₂S, the dihalides regenerated the tellurepine [74, 75] (Figure 11.41).

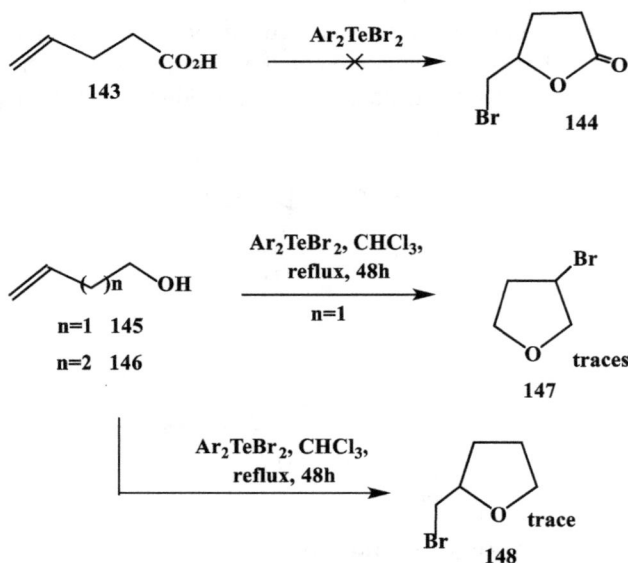

Figure 11.40: Interactions with 4-pentenoic acid, 3-butenol and 4-pentenol.

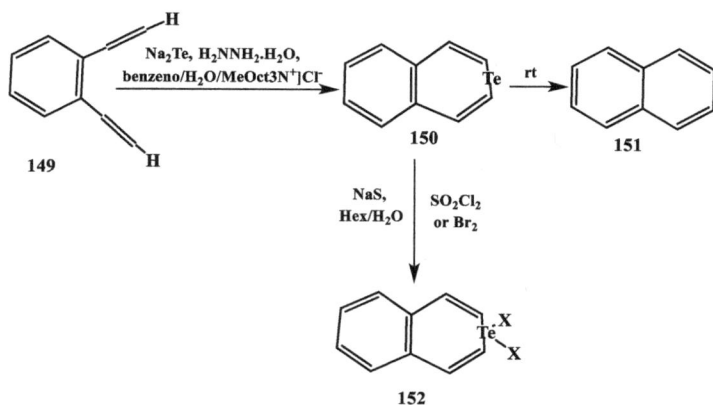

X = Cl, Br

Figure 11.41: Synthesis of tellurepine.

11.3 Tellurium-induced cyclization of alkynyl compounds

To synthesize 1-benzotellurepine **156** and benzotellurochromenes **157** (Figure 11.42), benzo- [b]tellurophene [76] **160** (Figure 11.43) and tellurochromones [77] **162** (Figure 11.44)

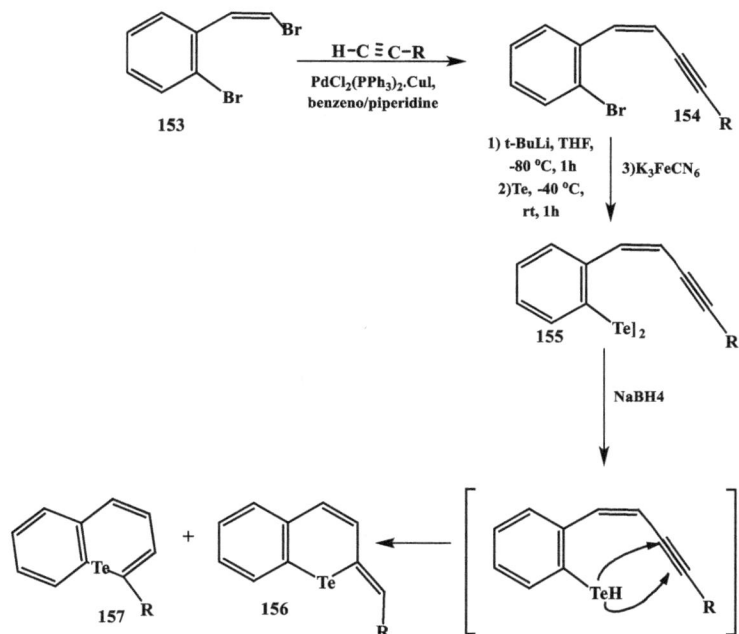

R = Me, n-Bu, t-Bu, n-hex, c-hex, n-oct

Figure 11.42: Synthesize 1-benzotellurepine and benzotellurochromenes.

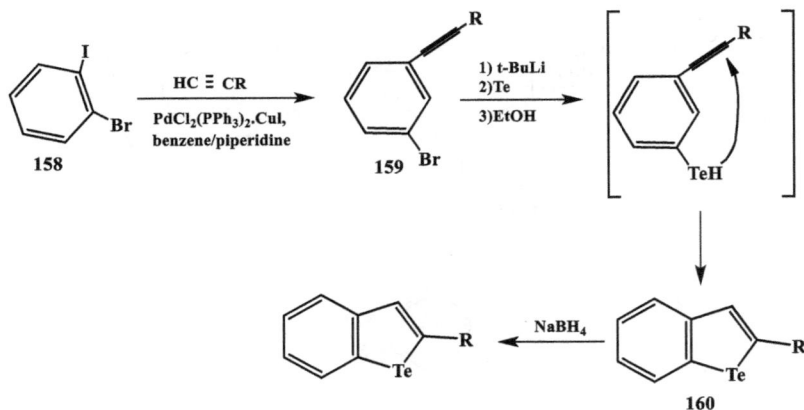

R = Me, *n*-Bu, *t*-Bu, Ph, TMS

Figure 11.43: Synthesis of benzo- [b]tellurophene.

the intramolecular kind of the alkynes's anti-hydrotelluration [74] was used. Coupling reaction of the alkynes with the appropriate *o*-substitutes bromobenzenes **153, 158,** and **161** produced the initial acetylene substrates, the reaction was catalyzed by Pd⁺. The coupling reaction of **161** with trimethylsilylacetylene, which is shown as route **a**, was

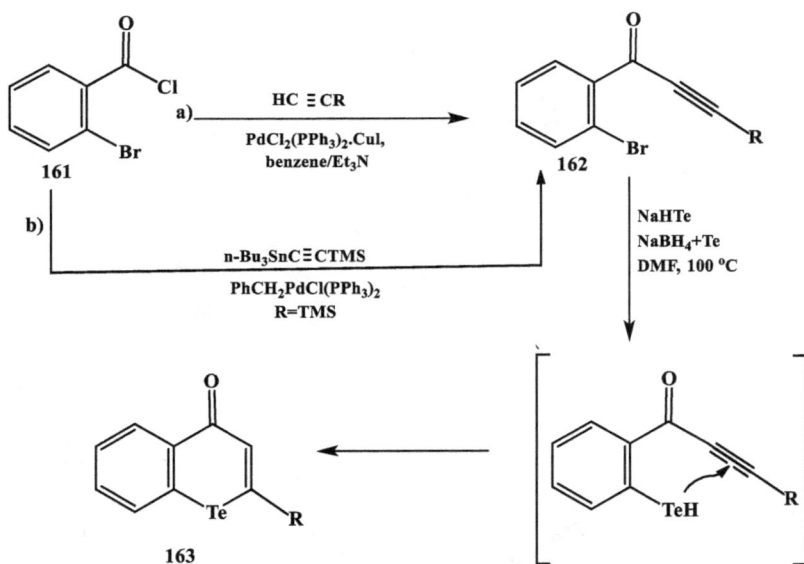

R= Me, n-Bu, t-Bu, n-hex, n-oct, Ph

Figure 11.44: Synthesis of tellurochromones.

discontinued. At the same time, the Stille reaction of **161** with a derivative of tin provided the expected output **162**. The formation of desilylated tellurochromone **163** (RZH) occurred as a result of the reductive elimination of the TMS group.

Figure 11.45 illustrates the synthesis of benzo[c]tellurophenes initiating from *o*-bishalomethyl benzenes [78] **164** [79]. As for R= MeO and H, all transformation efforts to turn diiodide **165** to the tellurophene **167** by straight removal of HI with a base, were unsuccessful. The transition was obtained through the bistrifluoroacetate **166**. The nitro derivative **165** (R=NO$_2$) underwent straightforward dehydroiodination upon re-action with Et$_3$N, as anticipated.

The derivatives **169** and **170** and the nitro-derivative **167** have higher stability than benzotellurophene **167** (R=H) because it can only be stabilized at low temperature when put in benzene solution.

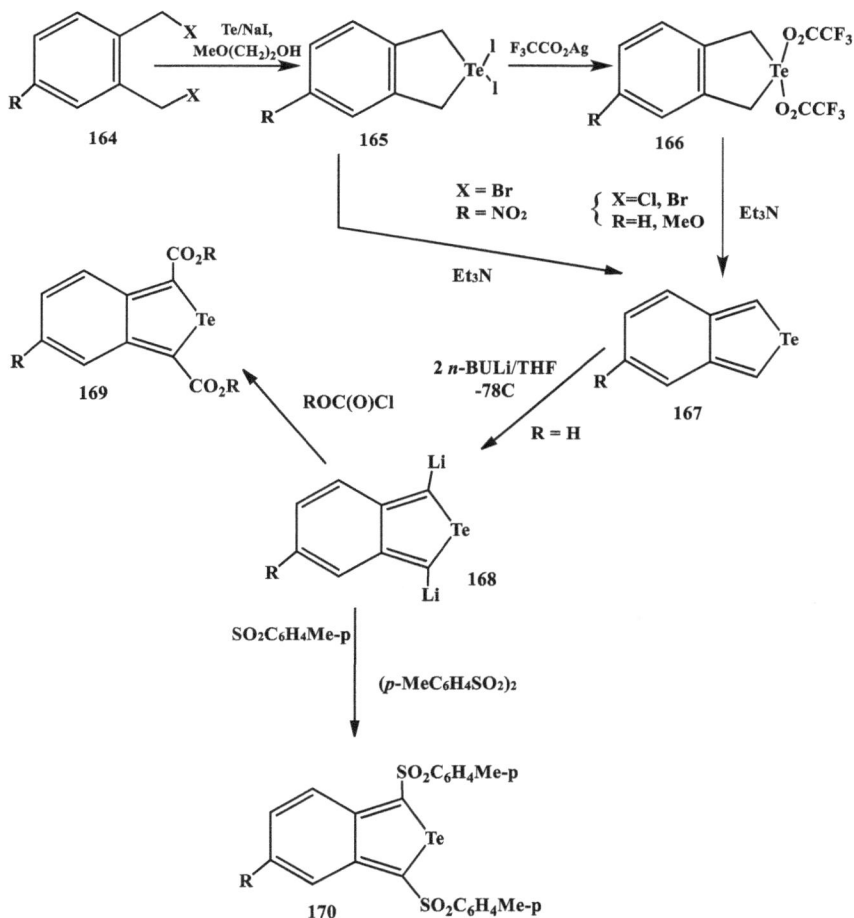

Figure 11.45: Synthesis of benzo[c]tellurophenes.

From the dibromomethyl thiophene derivative **171**, the compound **172** and the compound **173** (the anellated thiophene–tellurophene, the first tellurium containing diheteropentalene) were prepared [80] (Figure 11.46). The interaction of the dimethylacetylene dicarboxylate (DMAD) with dimeric **174** gave the adduct **178**, it was resulted from the summation to the tellurophene part of **173**. The addition of DMAD to the thiophene part gave **177** and was ineffective. The undertook Ac$_2$O-catalyzed Pummerer reaction of **172** provided only the compound **179** and the yield was very low.

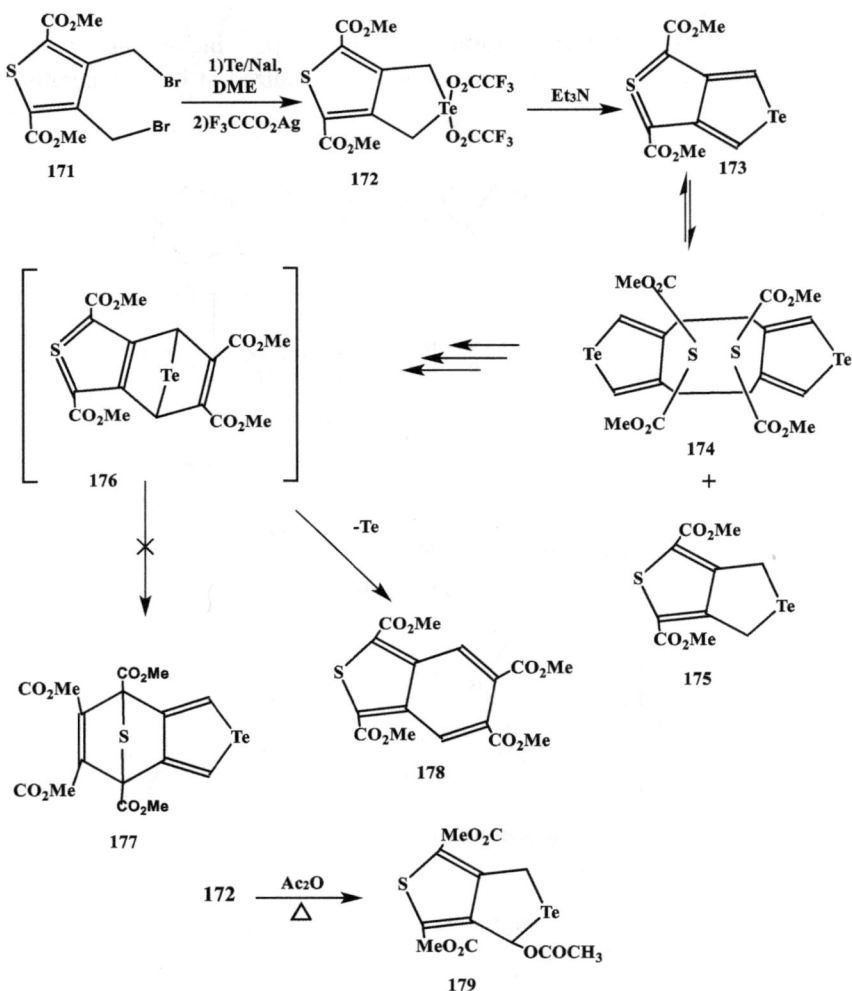

Figure 11.46: Synthesis of compound **179**.

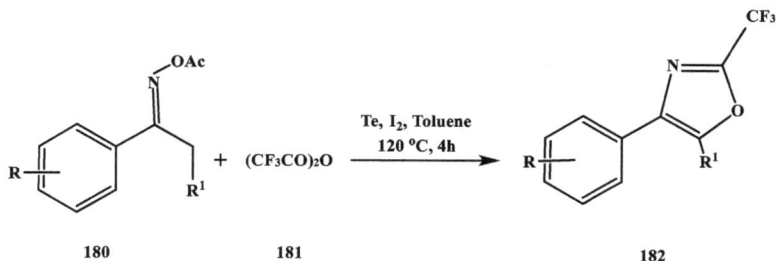

Figure 11.47: Synthesis of 2-(trifluoromethyl) oxazoles.

11.3.1 Other compounds

Te-induced cyclization reaction was also applied for the preparation of other compounds. For instance, through the interaction of trifluoroacetic anhydride with acetophenone oxime acetates, Te-mediated preparation of 2-(trifluoromethyl)- oxazoles was analyzed [81]. This latest tandem cyclization progressed from well to great yields through a SET reduction accompanied by a 5-endo-trig method. Several of the prepared compounds displayed insecticidal and fungicidal activities.

The range and restraints of the oxime acetates's cyclization with trifluoroacetic anhydride were analyzed at optimal reaction conditions (Figure 11.47). The derivatives of acetophenone oxime acetate holding electron-donating substituents such as tert-butyl, methyl, iso-butyl, methylmercapto and methoxy cyclization were achieved with well to outstanding yields. Derivatives of acetophenone oxime acetate holding electron-withdrawing groups such as nitro, ester, fluoro, bromo, and chloro also experienced productive cyclization and produced 2-(trifluoromethyl)oxazoles. The product yield was moderate and no breaking of aryl–halide bond was found.

Figure 11.48 illustrates a reasonable reaction mechanism of the synthesized 2-(trifluoromethyl)oxazoles formation. The electrophilic trifluoroacetylation of acetophenone oxime acetates **180** with **181** gave intermediate **183**, this went through β-H removal to provide trifluoroacetylated enamide derivatives **184**. An alkyl radical intermediate **185** was supplied through a SET reduction of **184** by the reducing Te and accompanying AcO⁻ elimination. It then generated the radical intermediate **187** after isomerization of **185** to an alkoxyl radical **186**, accompanied by a 5-endo cyclization. Lastly, the oxidation of **187** with Te species or I_2 afforded the carbocation intermediate **188**. On deprotonation, the intended compound **182** was formed.

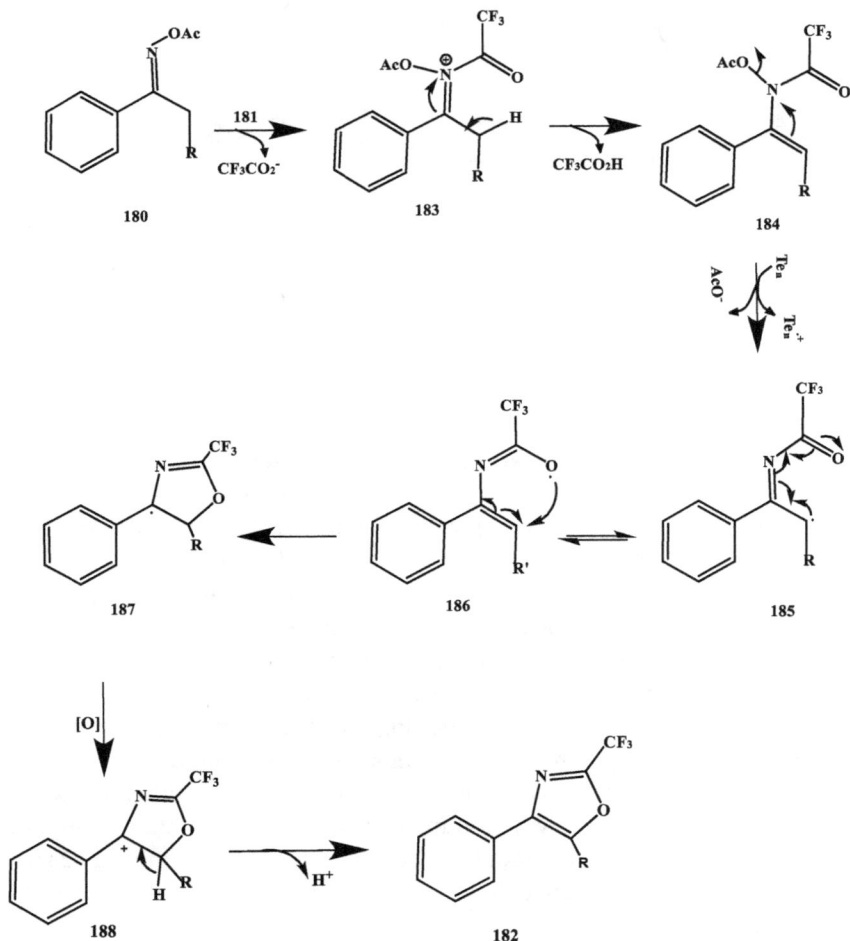

Figure 11.48: Proposed mechanism for the formation of **182**.

11.4 Conclusions

Tellurium is a promising and important functional material due to its unique physical and chemical properties. This article has discussed the importance of Te in organic synthesis. Te-induced cyclization reactions have shown promising results. The unique properties of Te-induced reactions may contribute to the great improvement in this field. The mechanism of many processes as discussed here is complex. Radical as well as ionic mechanisms have been suggested to describe the formation of the product. Clearly, this topic may find application in the stereocontrolled preparation of natural and non-natural products. Studies directed to this area are limited.

Acknowledgements: BKB and AD are thankful to Prince Mohammad Bin Fahd University for support.

References

1. Harmange J-C, Figadère B. Synthetic routes to 2,5-disubstituted tetrahydrofurans. Tetrahedron Asymmetry 1993;4:1711–54.
2. Lee YY, Kim BH. Total synthesis of nonactin. Tetrahedron 1996;52:571–88.
3. Solladie G, Dominguez C. A short asymmetric synthesis of (+)-Nonactic acid and (−)-8-epi-Nonactic acid induced by a chiral sulfoxide group. J Org Chem 1994;59:3898–901.
4. Lygo B. Stereoselective synthesis of (±)-methyl homononactate and (±)-methyl 8-epi-homononactate. Tetrahedron 1988;44:6889–96.
5. Bartlett PA, Meadows JD, Ottow E. Enantiodivergent syntheses of dextro nonactic acid and levo nonactic acid and the total synthesis of nonactin. J Am Chem Soc 1984;106:5304–11.
6. Jackson WP, Ley SV, Morton JA. Cyclisation reactions of alkenyl β-ketoesters involving a novel phenylseleno group migration. Tetrahedron Lett 1981;22:2601–4.
7. Antonioletti R, Bonadies F, Scettri A. A convenient approach to furan derivatives by I2-induced cyclisation of 2-alkenyl substituted 1,3-dicarbonyl compounds. Tetrahedron Lett 1988;29: 4987–90.
8. Antonioletti R, Cecchini C, Ciani B, Magnanti S. Iodoenolcyclization. III. A general approach to tetrasubstituted furans from 2-alkenyl-1,3-dicarbonyl compounds. Tetrahedron Lett 1995;36: 9019–22.
9. Iqbal J, Pandey A. A highly regioselective iodoenoletherification of α-Allyl-β-ketoesters. A convenient route to 2-Alkyl-5-iodomethyl-3-methoxycarbonyl-4, 5-dihydrofurans. Synth Commun 1990;20:665–70.
10. Ferraz HMC, de Oliveira EO, Payret-Arrua ME, Brandt CA. A new and efficient approach to cyclic.beta.-Enamino esters and.beta.-Enamino ketones by iodine-promoted cyclization. J Org Chem 1995;60:7357–9.
11. Ferraz HMC, Sano MK, Scalfo AC. Tellurium and iodine promoted cyclofunctionalization of alkenyl substituted β-keto esters. Synlett 1999;1999:567–8.
12. Silberman A, Kalechman Y, Hirsch S, Erlich Z, Sredni B, Albeck A. The anticancer activity of organotelluranes: potential role in integrin inactivation. Chembiochem 2016;17:918–27.
13. Piovan L, Milani P, Silva MS, Moraes PG, Demasi M, Andrade LH. 20S proteasome as novel biological target for organochalcogenanes. Eur J Med Chem 2014;73:280–5.
14. El Chamy Maluf S, Melo PMS, Varotti FP, Gazarini ML, Cunha RLOR, Carmona AK. Hypervalent organotellurium compounds as inhibitors of P. falciparum calcium-dependent cysteine proteases. Parasitol Int 2016;65:20–2.
15. Cunha RLOR, Gouvêa IE, Feitosa GPV, Alves MFM, Brömme D, Comasseto JV, et al. Irreversible inhibition of human cathepsins B, L, S and K by hypervalent tellurium compounds. Biol Chem 2009;390:1205–12.
16. Cunha RLOR, Urano ME, Chagas JR, Almeida PC, Bincoletto C, Tersariol ILS, et al. Tellurium-based cysteine protease inhibitors: evaluation of novel organotellurium(IV) compounds as inhibitors of human cathepsin B. Bioorg Med Chem Lett 2005;15:755–60.
17. Halpert G, Sredni B. The effect of the novel tellurium compound AS101 on autoimmune diseases. Autoimmun Rev 2014;13:1230–5.

18. Tiekink ERT. Therapeutic potential of selenium and tellurium compounds: opportunities yet unrealised. Dalton Trans 2012;41:6390–5.
19. Cunha RLOR, Gouvea IE, Juliano L. A glimpse on biological activities of tellurium compounds. An Acad Bras Cienc 2009;81:393–407.
20. Comasseto JV, Ferraz HMC, Petragnani N, Brandt CA. Cyclofunctionalization of unsaturated alcohols with aryltellurium trihalides. Tetrahedron Lett 1987;28:5611–4.
21. Comasseto JV, Grazini MVA. Cyclization of olefinic benzyl ethers with aryltellurium trichlorides. Synth Commun 1992;22:949–54.
22. Comasseto JV, Petragnani N. Cyclofunctionalization with aryltellurium trichlorides. Synth Commun 1983;13:889–99.
23. Ferraz HM, Comasseto JV, de Borba EB. Telurociclofuncionalizaqao de compostos 2-alquenil 1,3-dicarbonilicos. Quim Nova 1992;15:298.
24. Ferraz HMC, Sano MK, Nunes MRS, Bianco GG. Synthesis of cyclic enol ethers from alkenyl-β-dicarbonyl compounds. J Org Chem 2002;67:4122–6.
25. Stefani HA, Petragnani N, Brandt CA, Rando DG, Valduga CJ. Seleno and telluro cyclofunctionalization of α,γ-diallyl-β-ketoesters: polysubstituted furan derivatives. Synth Commun 1999;29:3517–31.
26. Kut M, Fizer M, Onysko M, Lendel V. Reactions of N-alkenyl thioureas with p-alkoxyphenyltellurium trichlorides. J Heterocycl Chem 2018;55:2284–90.
27. Kut V, Onysko V, Lendel V, Mykola K. Electrophile cyclization of N(S, Se)-alkenyl derivatives of pyrimidinone with p-metoxyphenyltellurium trichloride. Sci Bull Uzhhorod Univ Chem Ser 2019; 42:63–72.
28. Petragnani N, Comasseto JV.Tellurium reagents in organic synthesis: recent advances. Part 2. Synthesis 1991;1991:897–919.
29. Campos MDM, Petragnani N. Nachbargruppenbeteiligung bei additionsreaktionen, IV. Darstellung von α.α-disubstituierten δ-arylselenenyl-und δ-aryltelluro-γ-valerolactonen. Chem Ber 1960;93: 317–20.
30. Wirth T, Häuptli S, Leuenberger M. Catalytic asymmetric oxyselenenylation–elimination reactions using chiral selenium compounds. Tetrahedron Asymmetry 1998;9:547–50.
31. Moraes DN, Santos RA, Comasseto JV. The influence of the substrate structure in the tellurocyclofunctionalization reaction of g,d-unsaturated carboxylic acids and their corresponding benzyl esters. J Braz Chem Soc 1998;9:397–403.
32. Yoshida M, Suzuki T, Kamigata N. Novel preparation of highly electrophilic species for benzenetellurenylation or benzenesulfenylation by nitrobenzenesulfonyl peroxide in combination with ditelluride or disulfide. Application to intramolecular ring closures. J Org Chem 1992;57: 383–6.
33. Hu NX, Aso Y, Otsubo T, Ogura F. Cyclofunctionalization of hydroxyolefins induced by arenetellurinic anhydride. Tetrahedron Lett 1987;28:1281–4.
34. Hu NX, Aso Y, Otsubo T, Ogura F. Cyclofunctionalization of hydroxy olefins induced by arenetellurinyl acetate. J Org Chem 1989;54:4391–7.
35. Hu NX, Aso Y, Otsubo T, Ogura F. Organotellurium-mediated synthesis of oxazolidin-2-ones from alkenes. J Chem Soc Chem Commun 1987;1447–8. https://doi.org/10.1039/c39870001447.
36. Hu NX, Aso Y, Otsubo T, Ogura F. Organotelluriums. 20. Aminotellurinylation of olefins and its utilization for synthesis of 2-oxazolidinones. J Org Chem 1989;54:4398–404.
37. Hu NX, Aso Y, Otsubo T, Ogura F. Tellurium-based organic synthesis: a novel one-pot formation of 2-oxazolines from alkenes induced by amidotellurinylation. Tetrahedron Lett 1988;29:1049–52.

38. Hu NX, Aso Y, Otsubo T, Ogura F. Organotelluriums. Part 21. Amidotellurinylation of olefins and a novel one-pot synthesis of 4,5-dihydro-oxazoles from olefins. J Chem Soc Perkin 1989;1:1775–80.
39. Fukumoto T, Aso Y, Otsubo T, Ogura F. Stereoselective addition reactions of alkynes with benzenetellurinyl trifluoromethanesulfonate in acetonitrile: organotellurium-mediated one-pot synthesis of oxazoles from internal alkynes. J Chem Soc Chem Commun 1992;1070–2. https://doi.org/10.1039/c39920001070.
40. Fukumoto T, Aso Y, Otsubo T, Ogura F. Syntheses of β-amidovinyltellurides and oxazoles by addition reactions of alkynes with benzenetellurinyl trifluoromethanesulfonate in acetonitrile. Heteroat Chem 1993;4:511–6.
41. Brill WF. A site isolated tellurium oxidation catalyst having no soluble analog. J Org Chem 1986;51: 1149–50.
42. Alberto EE, Muller LM, Detty MR. Rate accelerations of bromination reactions with NaBr and H2O2 via the addition of catalytic quantities of diaryl ditellurides. Organometallics 2014;33:5571–81.
43. Oba M, Endo M, Nishiyama K, Ouchi A, Ando W. Photosensitized oxygenation of diaryl tellurides to telluroxides and their oxidizing properties. Chem Commun 2004;35:1672–3.
44. Oba M, Okada Y, Nishiyama K, Ando W. Aerobic photooxidation of phosphite esters using diorganotelluride catalysts. Org Lett 2009;11:1879–81.
45. Okada Y, Oba M, Arai A, Tanaka K, Nishiyama K, Ando W. Diorganotelluride-catalyzed oxidation of silanes to silanols under atmospheric oxygen. Inorg Chem 2010;49:383–5.
46. Oba M, Tanaka K, Nishiyama K, Ando W. Aerobic oxidation of thiols to disulfides catalyzed by diaryl tellurides under photosensitized conditions. J Org Chem 2011;76:4173–7.
47. Oba M, Okada Y, Endo M, Tanaka K, Nishiyama K, Shimada S, et al. Formation of diaryl telluroxides and tellurones by photosensitized oxygenation of diaryl tellurides. Inorg Chem 2010;49:10680–6.
48. Meinwald J. Modern synthetic reactions. J Chem Educ 1965;42:A910.
49. Bartlett PA, Richardson DP, Myerson J. Electrophilic lactonization as a tool in acyclic stereocontrol. Synthesis of serricornin. Tetrahedron 1984;40:2317–27.
50. Gyu Kim Y, Cha JK. Stereoselective synthesis of hydroxy-substituted tetrahydrofurans. Tetrahedron Lett 1988;29:2011–3.
51. Cardillo G, Orena M. Stereocontrolled cyclofunctionalizations of double bonds through heterocyclic intermediates. Tetrahedron 1990;46:3321–408.
52. Ley SV, Lygo B, Molines H, Morton JA. Synthesis of (cis-6-methyltetrahydropyran-2-yl)acetic acid involving the use of an organoselenium-mediated cyclization reaction. J Chem Soc Chem Commun 1982;1251–2. https://doi.org/10.1039/c39820001251.
53. Minami I, Yuhara M, Watanabe H, Tsuji J. A new furan annelation reaction by the palladium-catalyzed reaction of 2-alkynyl carbonates or 2-(1-alkynyl)oxiranes with β-keto esters. J Organomet Chem 1987;334:225–42.
54. Tiecco M, Testaferri L, Tingoli M, Bartoli D, Balducci R. Ring-closure reactions initiated by the peroxydisulfate ion oxidation of diphenyl diselenide. J Org Chem 1990;55:429–34.
55. Bergman J, Engman L. Oxidative cyclization of some.gamma.- and.delta.-hydroxy olefins induced by tellurium dioxide. J Am Chem Soc 1981;103:5196–200.
56. Engman L. Alkoxytellurination of olefins for the preparation of bis (.beta.-alkoxyalkyl) ditellurides and (.beta.-alkoxyalkyl) tellurium trichlorides. Organometallics 1989;8:1997–2000.
57. Comasseto JV, Ferraz HMC, Brandt CA, Gaeta KK. Reduction of tellurium – carbon bonds of tellurolactones and telluroethers. Tetrahedron Lett 1989;30:1209–12.
58. Ohe K, Takahashi H, Uemura S, Sugita N. Carbonylation of aryl- and vinyl-tellurium compounds with carbon monoxide in the presence of palladium(II) salts. J Organomet Chem 1987;326:35–47.
59. Ohe K, Takahashi H, Uemura S, Sugita N. Palladium(II) chloride catalyzed carbonylation of organic tellurides with carbon monoxide. J Org Chem 1987;52:4859–63.

60. Clive DLJ, Anderson PC, Moss N, Singh A. New method for coupling allylic halides: use of telluride(2-) ion species. J Org Chem 1982;47:1641–7.
61. Barton DH, Ramesh M. Tandem nucleophilic and radical chemistry in the replacement of the hydroxyl group by a carbon-carbon bond. A concise synthesis of showdomycin. J Am Chem Soc 1990;112:891–2.
62. Barton DHR, Ozbalik N, Sarma JC. The role of organic tellurides as accumulators and exchangers of carbon radicals. Tetrahedron Lett 1988;29:6581–4.
63. Barton DHR, Dalko PI, Géro SD. Preparation of six membered carbocycles by aryl-tellurium mediated free-radical cyclisation. Tetrahedron Lett 1991;32:4713–6.
64. Kanda T, Sugino T, Kambe N, Sonoda N. A new preparative method of organoalkali and organoalkaline-earth metals using metal-tellurium exchange reactions. Phosphorus Sulfur Silicon Relat Elem 1992;67:103–6.
65. Clive DLJ, Russell CG, Chittattu G, Singh A. Cyclofunctionalisation of unsaturated acids with benzeneselenenyl chloride: kinetic and thermodynamic aspects of the rules for ring closure. Tetrahedron 1980;36:1399–408.
66. Nicolaou KC. Organoselenium-induced cyclizations in organic synthesis. Tetrahedron 1981;37:4097–109.
67. Xu Q, Huang X, Yuan J. Facile synthesis of β-organotellurobutenolides via electrophilic tellurolactonization of α-allenoic acids. J Org Chem 2005;70:6948–51.
68. Macomber RS, Krudy GA, Seff K, Rendon-Diaz-Miron LE. Sulfur- and selenium-promoted cyclization of allenic phosphonates and phosphinates to substituted 1,2-oxaphosphol-3-enes: stereochemical consequences at phosphorus. The crystal and molecular structure of (Z)-3,5-di-tert-butyl-2-methoxy-4-(phenylseleno)-1,2-oxaphosphol-3-ene 2-oxide. J Org Chem 1983;48:1425–30.
69. Arnold RT, Campos Mde M, Lindsay KL. Participation of a neighboring carboxyl group in addition reactions. I. The mechanism of the reaction of bromine with γ, δ-unsaturated acids and Esters1. J Am Chem Soc 1953;75:1044–7.
70. Rowland RL, Perry WL, Friedman HL. Mercurial diuretics. III. Mercuration of allylacetic acid and related compounds. J Am Chem Soc 1951;73:1040–1.
71. de Moura Campos M, Petragnani N. Organic tellurium compounds– IV: vinylic and ethynylic tellurium derivatives. Tetrahedron 1962;18:527–30.
72. Adams R, Roman FL, Sperry WN. The structure of the compounds produced from olefins and mercury salts: mercurated dihydrobenzofurans. J Am Chem Soc 1922;44:1781–92.
73. Leonard KA, Zhou F, Detty MR. Chalcogen(IV)–Chalcogen(II) redox cycles. 1. Halogenation of organic substrates with dihaloselenium(IV) and -tellurium(IV) derivatives. Dehalogenation of vicinal dibromides with diaryl tellurides. Organometallics 1996;15:4285–92.
74. Sashida H, Ito K, Tsuchiya T. Synthesis of the first examples of 1-benzotellurepines and 1-benzoselenepines. J Chem Soc Chem Commun 1993;1493–4. https://doi.org/10.1039/c39930001493.
75. Sashida H, Ito K, Tsuchiya T. Studies on seven-membered heterocycles. XXXV. Synthesis of the group 16 1-benzoheteroepines involving the first examples of 1-benzotellurepine and 1-benzoselenepine rings. Chem Pharm Bull 1995;43:19–25.
76. Sashida H, Sadamori K, Tsuchiya T. A convenient one-pot preparation of benzo[b]-tellurophenes,-selenophenes, and -thiophenes from o-bromoethynylbenzenes. Synth Commun 1998;28:713–27.
77. Sashida H. An alternative facile preparation of telluro- and selenochromones from o-bromophenyl ethynyl ketones. Synthesis 1998;1998:745–8.
78. Huang Z, Lakshmikantham MV, Lyon M, Cava MP. Synthesis and isolation of some benzo[c] tellurophenes. J Org Chem 2000;65:5413–5.

79. Ziolo RF, Günther WHH. The synthesis and characterization of α- and β-1, 1-diiodo-3,4-benzo-1-telluracyclopentane, $C_8H_8TeI_2$. J Organomet Chem 1978;146:245–51.
80. Rajagopal D, Lakshmikantham MV, Mørkved EH, Cava MP. Generation of the first tellurium-containing diheteropentalene. Org Lett 2002;4:1193–5.
81. Luo B, Weng Z. Elemental tellurium mediated synthesis of 2-(trifluoromethyl)oxazoles using trifluoroacetic anhydride as reagent. Chem Commun 2018;54:10750–3.

Devalina Ray*, Aparna Das, Suman Mazumdar and Bimal K. Banik*

12 Tellurium-induced functional group activation

Abstract: Tellurium-chemistry comprises of vibrant and innovative prospects in major area of research and development. The function of Tellurium in organic synthesis remained underexplored till date. Moreover, the reactivity of Tellurium as Lewis acid or electrophilic reagents to activate functional group conceptually remains as an ever-demanding area to be investigated extensively. In this context, the present compilation portrays a detailed study on the reactivity of organotellurium compounds as catalyst, reagent, and sensors to explore the reactions occurring specifically through functional group activation.

Keywords: catalysis; chalcogen; lewis acid; organic synthesis; Tellurium.

12.1 Introduction

Chalcogen bond induced reactions are the non-covalent interactions between Lewis acidic or electrophilic chalcogen centers with the Lewis basic or nucleophilic atoms of the substrates (Figure 12.1). In general, non-covalent organocatalysis are predominantly overtaken by hydrogen bonding (HB), involving (thio)urea as backbones which has been used as efficient catalysts [1]. The non-covalent organocatalysis is under epitomized till date.

Basically, the elements of group 16 such as sulfur, selenium and tellurium derivatized organic molecules correspond to the electron donating character which prefers to act as Lewis bases in electrophilic reactions through formation of covalent bonds [2]. However, the reverse approach of the chalcogens to act as intermolecular Lewis acid electrophilic center might not be a common or well-known approach and was documented for the first time in 2017 [3].

The anisotropic distribution of electrons in chalcogen atom instigates an electrostatic attraction between chalcogen-based Lewis acids (ChLA) and Lewis bases (LB) which generates a zone with positive electrostatic potential known as s-hole, [4] in the elongation of the R-Ch axis. The electron donation occurs from the LB into the σ^* orbital

*Corresponding authors: Devalina Ray**, Amity Institute of Biotechnology, Amity University, Sector 125, 201313, Noida, India, E-mail: dray@amity.edu; and **Bimal K. Banik,** Department of Mathematics and Natural Sciences, College of Sciences and Human Studies, Prince Mohammad Bin Fahd University, Al Khobar, Kingdom of Saudi Arabia, E-mail: bimalbanik10@gmail.com
Aparna Das, Department of Mathematics and Natural Sciences, College of Sciences and Human Studies, Prince Mohammad Bin Fahd University, Al Khobar, Kingdom of Saudi Arabia
Suman Mazumdar, Department of Scientific and Industrial Research, Ministry of Science & Technology, Government of India, Technology Bhawan, New Mehrauli Road, 110016, New Delhi, Delhi, India

As per De Gruyter's policy this article has previously been published in the journal Physical Sciences Reviews. Please cite as:
D. Ray, A. Das, S. Mazumdar and B. K. Banik "Tellurium-induced functional group activation" *Physical Sciences Reviews*
[Online] 2022. DOI: 10.1515/psr-2021-0221 | https://doi.org/10.1515/9783110735840-012

Figure 12.1: Non-covalent interactions in chalcogen.
(a) Interaction of lewis base with σ* orbital of chalcogen. (b) Telluride as chalcogen bond donor.

of ChLA which is assumed to be crucial for strong ChB complexes (Figure 12.1a) [5]. In this regard, tellurides serves as efficient chalcogen bond donor via non-covalent interactions (Figure 12.1b).

Certain factors attribute to the strength of chalcogen bond namely the Lewis basicity of the atom in substrate and the chalcogen type (Te > Se > S) and the favorable LB-Ch-R bond angle of 180°. Despite of having considerable preferences in terms of a linear interaction angles of 180° [5, 6] and manipulation of binding strength by variation of various parameters, majority of the results from chalcogen binding is limited to their application in solid-state and supramolecular chemistry.

Furthermore, organotellurium compounds have been rarely employed as catalysts for functional group activation through non-covalent bond interaction, although they have been predicted to have stronger Lewis acidity according to chalcogen bonding theory. The present chapter deals with exploration of the role of tellurium compounds primarily in electrophilic reactions along with various other related transformations.

12.1.1 Lewis acidity in Organotellurium

Tellurium is having remarkably varied properties compared to its related group elements, as the heaviest element of group 16. With the increase in atomic weight of chalcogen, the disproportionation of ChX_2 (Ch = Chalcogen and X = Halogen) into Ch_2X_2 and ChX_4 increases. The first crystallography for organotellurium (II) and (IV) complex possessing intramolecular coordination of nitrogen to tellurium was developed by McWhinnie and coworkers [7]. It was interesting to observe that several Te^{2+} cations can coordinate to s-bond donors such as 2,2′-bipyridyl for successful trapping and isolation of intermediates [8]. Highly bulky groups or strong electron donors were utilized for successful isolation of tellurenyl (HTe^+) cations [9] (Figure 12.2).

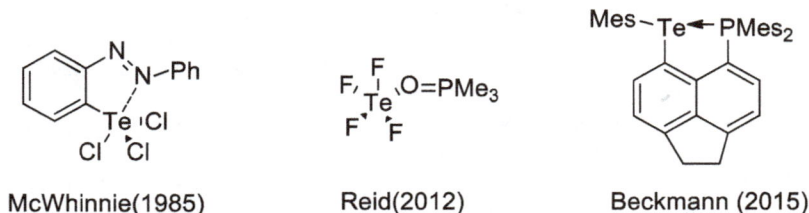

McWhinnie(1985)   Reid(2012)   Beckmann (2015)

Figure 12.2: Lewis acidic Te(II) and Te(IV) compounds.

Stabilized Lewis acid-base adducts of tellurium might be formed with phosphines, amines or oxygen. Apart from having high polarizability, Te possesses the property of three-centre-four-electron bonding. Furthermore, the coordination properties of neutral tellurium are well explored.

Telluranes (IV) (D) cannot act as donor ligands in transition metal complexes whereas tellurides (II) (C) have the tendency to act as L-ligand donors [10]. Thus, this contrasting behaviour suggests that a declination in electron donation capacity occurs for tetravalent tellurium species possessing single lone pair. These electrostatic characteristics further predominate in cationic tellurium which fails to coordinate with metals by donation of lone pair to form the complex E. Thus, the cationic tellurium functions as strong Lewis acid and further coordinates to form hypervalent adducts (F) [11].

However, the advancement towards metallated hypervalent main group compounds such as metalated tellurium complexes still remain underexplored. In this regard, Gabbaï et al introduced a Te–Pt complex (1), where the hypervalent tetra-coordinated Te is covalently bonded to octahedral Pt [12]. Thereafter, the same group explained the successful introduction of tellurium centre in the coordination sphere of Pd(II) complex as s-acceptors (2) (Figure 12.3). Te telluronium ion was assumed to act as Z-ligand s-acceptor accommodating a d-electron pair from the palladium atom, although it is a group 16 element having a lone pair. The formation of a metal-telluronium bond was assessed through the interaction of telluronium ligand with the metal center supported by auxiliary donor groups.

Structural analysis of cationic tellurium complex (2) discloses the involvement of nitrogen atoms in ligation of both quinolinyl motifs to the palladium. The tellurium

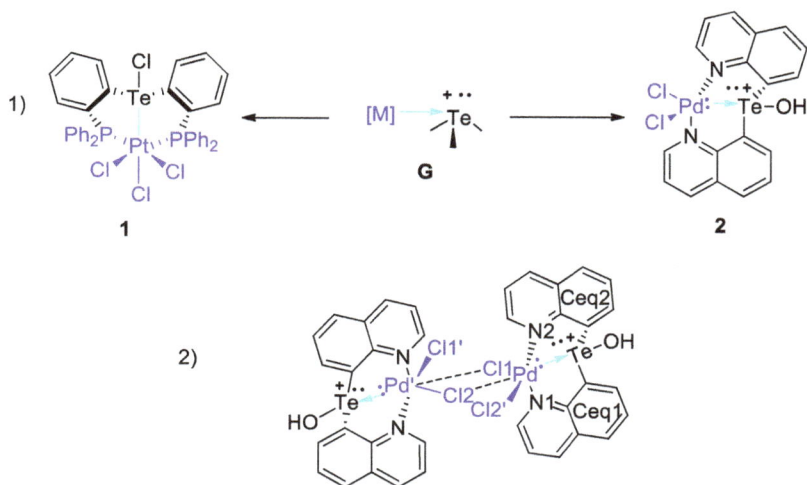

Figure 12.3: Metal coordinated lewis acidic tellurium complex.

atom is situated at 2.7823(8) Å from the palladium center in this arrangement. This Pd–Te distance exceeds the sum of the covalent radii (2.56–2.77 Å) [9] by only 0.4–8.7%. The tellurium atom is also bonded to hydroxide ligand (Te–O 1.941(4) Å), that comes from the basic conditions applied during its synthesis. As indicated by the Pd–Te–O (169.3(2)°) and Ceq1–Te–Ceq2 (91.8(3)°) angles, a seesaw geometry was adopted by tellurium atom in the complex where the orientation of palladium is trans with respect to hydroxide ligand. Additionally, the palladium atom is having a defined coordination sphere which can accomodate dual quinolinyl nitrogen atoms (N1 and N2) along with two chloride ligands (Cl1 and Cl2), which is responsible for the square plane formation while the bridging chloride (Pd–Cl2′ 3.221 Å, Cl2′-PdTe 174.50°) involves another molecule of the complex.

Although the Pd–Cl2′ bond was found to be more than Pd–Cl1 (2.3228(14) Å) and Pd–Cl2 (2.3100(15) Å), the existence of additional ligand clearly provided the clue that the palladium coordination sphere is intermediate between square pyramidal (t = 0.003) and octahedral which indicates a tetravalent state. Overall, it can be concluded that a Pd-Te interaction exists in the complex **2** (Figure 12.3).

F. P. Gabbaı and coworkers reported a bidentate Lewis acidic complex consisting of a boryl and a telluronium moiety that possess a strong fluoride binding affinity as evident from anion complexation studies [13]. The exceptional attraction of bidentate telluronium borane for fluoride may be demonstrated through the generation of chelated complex B–F → Te facilitating a strong lone-pair of F with σ* of Te–C donor-acceptor interaction Scheme 12.1. The heavier chalcogenium centres was analyzed to be better anion-binding sites and provides a new direction towards the synthesis of polydentate Lewis acids with increased anion affinities. The electrostatic interactions resulting k of the polarizability and relative electropositivity of Ch atom is the dictating factor in addition to the ability to donate electron from the filled-orbital of the donor to σ*-orbital of Ch–C bond. Naturally the donor–acceptor interactions incline in the order S < Se < Te leading to the complexes of type F (Figure 12.4). The short anion–cation contacts revealed from crystal structures of the salts of telluronium ions accounts for its Lewis acidity. The synthesis of tellurium complex was carried out in three steps with good yield (Scheme 12.1).

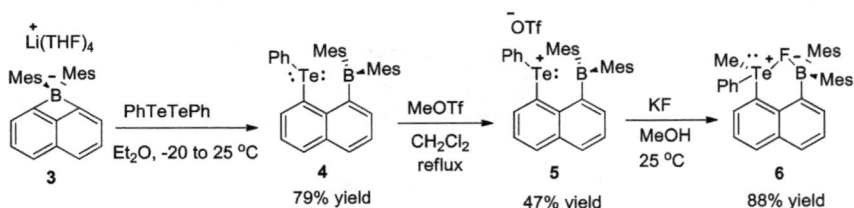

Scheme 12.1: Synthetic approach towards the chalcogenium borane salts.

Figure 12.4: Neutral and cationic Te(II) and Te(IV) species.

The existence of chalcogen bonding in the solid state can be proved by certain evidence whereas there are less reports on the strength of these interactions in the solution phase. In this context, M. S. Taylor and coworkers interpretated the association constants for benzotelluradiazoles with various Lewis bases such as -Cl, -Br, -I, -NO₃ and quinuclidine anions [14]. The existence of chlcogen bond interactions between tellurium species and the Lewis bases was determined by ^{1}H and ^{19}F NMR along with UV–visible and nano-ESI mass spectroscopy.

The free energy changes with variations in solvent, acceptor and donor in chalcogen bonds are reflected from the analytical data. A linear free energy relationship was obtained for the donor ability of chalcogen bond and electrostatic potential at the tellurium center.

The tellurium containing analogue of 1,2,5-chalcogenadiazoles easily proceeds with self-association via N···Te interactions particularly in the solid state (Figure 12.5) [15]. The Lewis basic solvents [16] and anions [17] successfully led to cocrystallization (Figure 12.3). The solution-phase association constants were documented for the Lewis acidic dicyanotelluradiazole 1a with I⁻ (Ka = 6.8 × 105 M⁻¹ in DCM) and PhS⁻ (Ka = 7.4 × 104 M⁻¹ in THF) [18]. The functionalization in fused phenyl ring provides options to manipulate the donor ability of tellurium by varied effects of substituent. The strength of chalcogen bond donor was fine-tuned through structure–activity relationships analyzing the effects of substitution in the phenyl ring.

The Lewis base–benzotelluradiazole complexes formed were further confirmed was by NMR spectroscopy and nanoelectrospray ionization mass spectrometry (nanoESI-MS). The interaction of two was analyzed with quinuclidine in solvent to assess the effect of solvent. The trend in chalcogen bond donor and acceptor capacity were observed through quantum chemical calculations. It indicated single point chalcogen bonding interactions of benzotelluradiazoles that can lead to higher association constants as 105 M⁻¹ and the change in solvent, acceptor, and substitution pattern produces remarkable difference in strength.

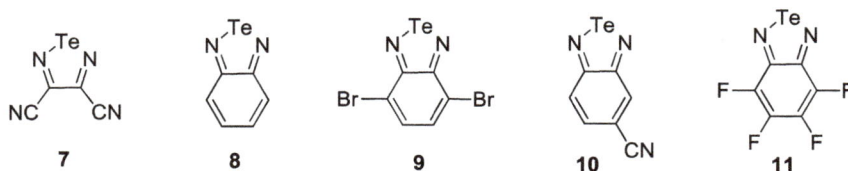

Figure 12.5: Various types of 1, 2, 5-benzotelluradiazoles.

Beckmann and Woollins developed a peri-substituted system involving phosphorus-tellurium interaction through space [19]. The effect of substituents at the phosphorus and tellurium centres along with the metal-coordination (Pt, Au) has been successfully demonstrated using NMR, single-crystal X-ray diffraction, and advanced density functional theory studies. Due to stability issues, phosphorus–tellurium complexes are very limited in number [20].

A variety of phosphorus–tellurium peri-functionalized systems R′Te–Acenap–PR$_2$ (R′ = Ph, p-An, Nap, Mes, Tips; R = ^{i}Pr, Ph) can be readily synthesized from organo-tellurium monohalides by reacting with lithiated R$_2$p–Acenap. The phosphorus–tellurium peri-substituted system exhibits high spin–spin coupling constants through space which can act as the driving factor for three-center four-electron type interactions. Due to the absence of lone pair in the oxidized phosphorus atoms containing sulfur and selenium, strong p–Te coupling was no longer observed. The gold complex however, furnished large p–Te couplings probably due to overlap of a tellurium lone pair orbital with a p–Au bonding orbital (Figure 12.6). It was speculated that there might be a specific role of donor–acceptor interaction in dictating the coupling constants in overlapping lone pair orbitals whereas coupling through bond might have minor contribution.

12.1.1.1 Organotelluride-induced activation

Organochalcogenides were introduced primarily as activators in halide abstraction reactions, where strong Lewis basic anions were taken into account. Furthermore, the coordination of chalcogen bond donor (actually ChLA) with neutral atom is much weaker. In this regard, there are limited reports on the activation of carbonyl compounds [20] and reduction of quinolines [21, 22].

Recently, Stefan M. Huber and coworkers investigated the first chalcogen bond-assisted activation of nitro group in the Michael addition of 5-methoxyindole with trans-β-nitrostyrene (Scheme 12.2) [23]. In this method, the dicationic and bidentate organochalcogenide preferentially selenium and tellurium-based compounds were utilized to obtain the highest Lewis acidity (Scheme 12.3). The better activity of cationic chalcogen bonding catalysts with tellurium based chalcogen bond donors was

Figure 12.6: Donor acceptor orbital overlap in Tellurides.

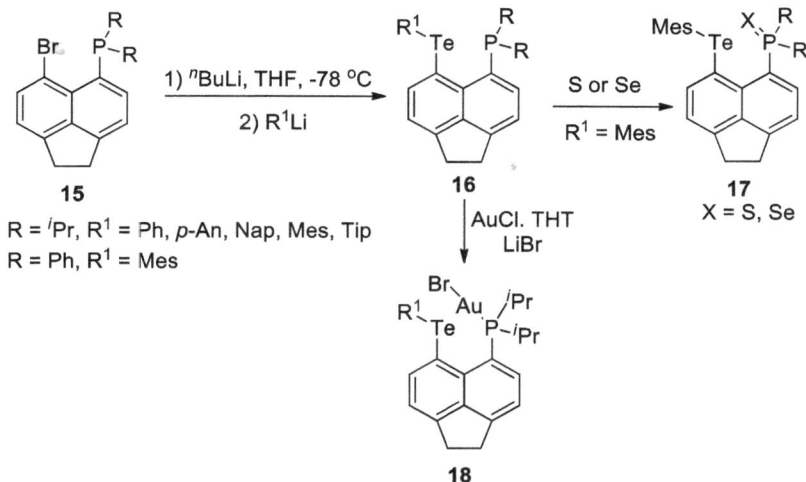

Scheme 12.2: Synthetic route to phosphorus tellurium peri substituted systems.

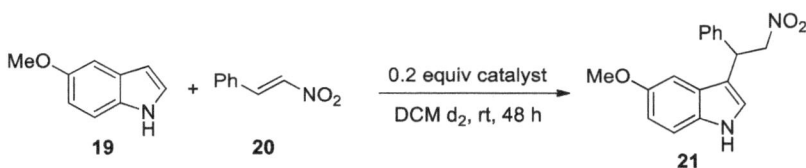

Scheme 12.3: Michael addition of 5-methoxyindole to β-nitrostyrenes.

established over the neutral version. Cationic framework was accomplished with the introduction of triazolium units because the neutral version of these frameworks are stable compounds. Further synthesis of charged species was operationally straight forward and could be attained by easy alkylation. The other substituents of chalcogens were methodically chosen as phenyl to avoid probable dealkylation. The competiting Lewis acidities of chalcogen bond donors (ChLA) were analyzed by titration. Furthermore, the probable mechanism of activation with the bidendate catalyst was verified with DFT calculations.

This report displays the first dicationic tellurium-based chalcogen bond donors for functional group activation of the substrate. The exceptional strength of Lewis acidity for dicationic Te compound coordinated to triflate as well as water was supported by X-ray structure (Figure 12.7). Additionally, the counter anions and the cationic tellurium counterpart commonly possess close contacts.

The same research group demonstrated the first oragonotellurium-induced activation of the carbonyl group for an α, β-unsaturated carbonyl compound in the Michael addition reaction of 1-methylindole (**1**) with trans-Crotonophenone (Table 12.1) [24].

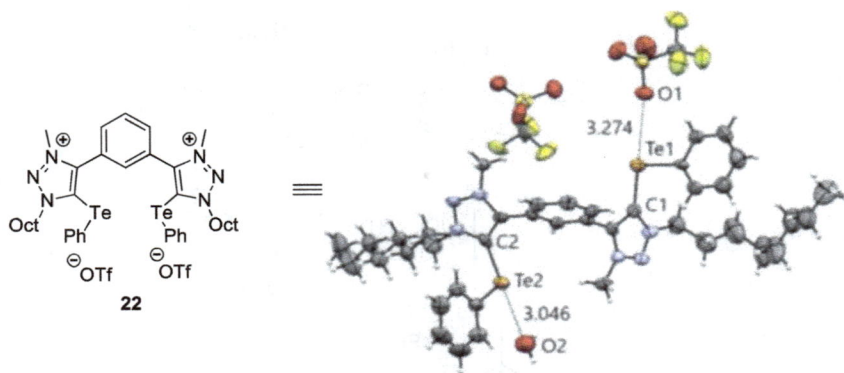

Figure 12.7: Crystal structure of organotelluride.

Table 12.1: Organochalcogenide-catalyzed Michael addition.

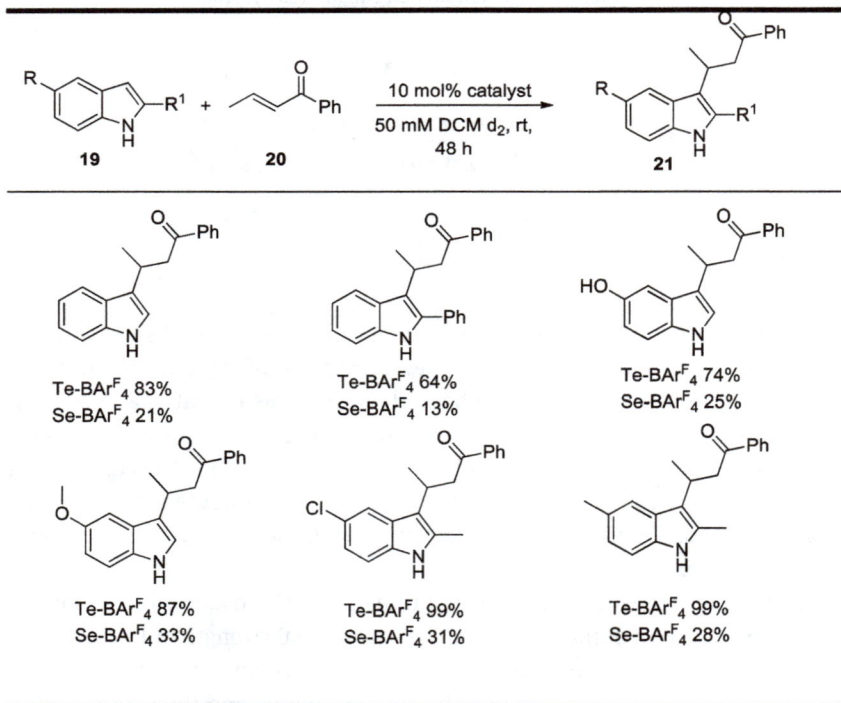

1-methylindole was employed to prevent unnecessary complications because of the involvement of the acidic *N*-proton in reaction.

The initial experiments were done with bis(triazolium)benzene derivatives as the structural motif as the catalysts which have proved to be active in their nitro-Michael reaction previously (Figure 12.8). Additionally, the change in chalcogens as well as the influence of the counter anion for the compounds of tellurium and selenium was extremely important as these factors have been rarely addressed till date. To rule out any other activation such as halogen bond activations, reference compounds were tested for Lewis acidity. Among the dicationic and neutral analogues of bis(triazolium) benzene derivatives with central chalcogen atoms, the charged version was found to have prominent activity. Additionally, the tellurium analogues have enhanced activity as compared to the selenium catalysts as was evident from yield of the products. The experimental outcomes thus supported the previous reports with the increasing order of chalcogen bond donor activity as S < Se < Te. The activity of counter anion for the dicationic chalcogen bond donors was also analyzed in the carbonyl activation. The coordinating power of the anions was supposed to decrease following the trend $NTf_2^- > OTf^- \approx BF_4^- > BAr_F^-$ so that the Lewis acidic character of chalcogens get a better exposure to substrates. However, the OTf or the BF_4 derivatives of selenium failed to show significant activity. The substrate scope was assessed with both the Te and Se analogues to compare the activity of both the catalysts in each case (Table 12.1). The dicationic organotellurium was proved to be the most efficient catalyst for all the substrates. The products were obtained in 64–99% yield with Te catalyst while 13–33% yield was obtained with Se analogue.

Tellurium tetrachloride was introduced as an efficient Lewis acid catalyst by H. Tani and coworkers for dithioacetalization under mild condition [25]. A series of aldehydes and aliphatic ketones underwent protection in presence of lesser amount of tellurium tetrachloride as catalyst into the corresponding dithioacetals at room temperature in good yields.

The synthetic approach for the formation of dithioacetals involves condensation of carbonyl compounds with thiols in presence of acid-catalysts. Several acid catalysts have been documented for this purpose including aluminum chloride, [26] boron trifluoride etherate, [27] tungstophosphoric acid, [28] iodine, [29] bronsted acidic ionic liquid, [30] ferric chloride, [31] p-dodecylbenzenesulfonic acid, [32] yttrium triflate, [33] hafnium trifluoromethanesulfonate, [34] and vanadyl triflate [35]. Mixed reagent

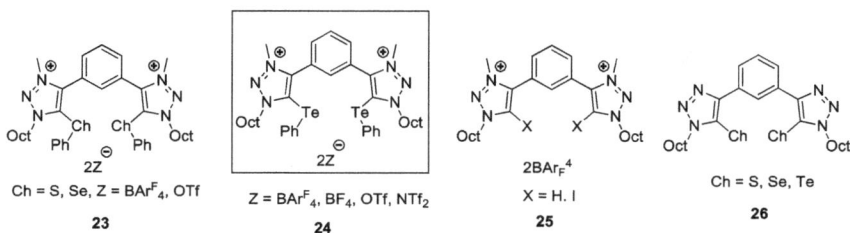

Figure 12.8: Dicationic and neutral organochalcogenides and organohalides.

systems such as silica gel-supported perchloric acid, [36] *p*-toluenesulfonic acid on silica gel, [37] silica gel supported sulfamic acid, [38] immobilized scandium(III) triflate in ionic liquids, [39] and anhydrous cobalt(II) bromide dispersed on silica gel [40] have also been explored. Some of these reported procedures suffer limitations related to low chemoselectivity or yield, vinyl sulfide formation from carbonyl compound, requirement of expensive reagent in stoichiometric amounts, harsh reaction conditions, and tedious work-up.

The aldehydes were reacted with either alkanethiols or alkanedithiols in the presence of tellurium tetrachloride in 1,2-dichloroethane (DCE) as solvent for 2–3 h at room temperature to afford the corresponding dithioacetals in good yields through thioacetalization of carbonyl compound (Scheme 12.4). Aliphatic ketones were smoothly reacted to produce corresponding dithioacetals, however, aromatic ketones failed to react and was found intact in the reaction mixture. The generalization of the reaction was done with diversified substrates that afforded products in good yields. The selectivity in functional group protection was demonstrated through various examples.

The ketoaldehyde (**30**) was protected with 1,3-propanedithiol (**31**) under optimized conditions where the aldehyde group was selectively thioacetalized in presence of the keto group to afford the product (**32**) in 76% yield. Furthermore, when 9-methyl-A5(te)-octalin-1,6-dione (Wieland-Miescher ketone) **33** underwent reaction with 1,3-propanediitiol, the α, β-unsaturated carbonyl group was selectively dithioacetalized over the saturated carbonyl to give the product (**34**) in 68% yield (Scheme 12.5).

The vinyl sulfides formed through the α-proton were not obtained in this protocol. The non-existence of vinyl sulfide can be logically explained through the reduced ability of tellurium tetrachloride toward effective complexation with the *in situ*

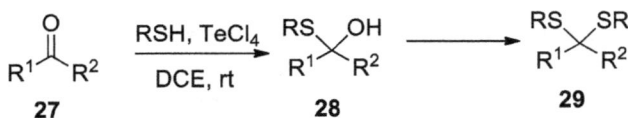

Scheme 12.4: Tellurium-catalyzed dithioacetalization.

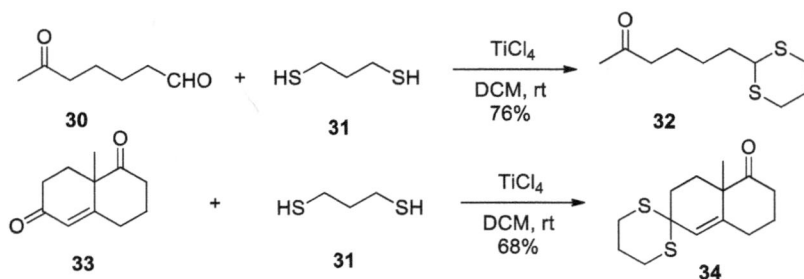

Scheme 12.5: Selective dithioacetalization of ketoaldehyde and diketone.

generated hemithioacetal intermediates (**28**) which further facilitates the attack of residual thiol instead of carbocation formation by cleavage of thiol functional group. Tellurium tetrachloride efficiently differentiates aldehyde from keto carbonyl group due to the steric difference as well as lower amount of tellurium catalyst in the reaction medium (Scheme 12.4).

S. M. Huber and coworkers introduced chalcogen bond donors containing bidentate cationic tellurium as catalysts for removal of chloride in 1-chloroisochroman (Scheme 12.6). Tellurium-based catalysts displayed much improved activity over selenium or sulfur analogues which showed lesser activity in the standard reaction conditions. Tellurium-variants rate accelerated the rate of the reaction by [40] as compared to non-chalogenated reference compounds.

The weak counter cation raised the activity of the catalysts. However, tetrafluoroborate had reduced activity as it got involved in the side reaction namely fluoride transfer. Stability of the catalyst was confirmed through a fluoro-tagged variant. The synthesis of catalysts was carried out following reported procedure where the starting material 1,3-diethynyl-5-fluorobenzene (**39**) was synthesized using Sonogashira coupling of 1,3-dibromo-3-fluorobenene (**38**) followed by deprotection of trimethylsilyl group in acetylene (Scheme 12.7). Thereafter, the alkyne-azide click reaction afforded the substituted triazole compounds (**40**) which further react with diaryl chalcogenide to form neutral chalcogenated product (**41**). Subsequent methylation and anion exchange led to the desired dicationic chalcogenides (**43**).

N. Petragnani and coworkers revealed the solvent-mediated variation in product formation for the telluroxide-catalyzed oxidation of thioamides [41]. It was observed that the telluroxides supported on polymer can undergo reaction with thioamides in acidic solvent at room temperature to afford 1,2,4-thiodiazoles. It also yielded 4.5-hydrothiazole when thiourea is used as substrates (Scheme 12.8). However, the application of polar non-acidic solvents such as dichloromethane, chloroform and methanol under similar reaction condition led to dehydrosulfurization followed by nitrile formation.

The two different mechanistic pathways have been proposed for the nitriles and thiadiazole which basically involves similar adduct. It was proposed that in non-acidic

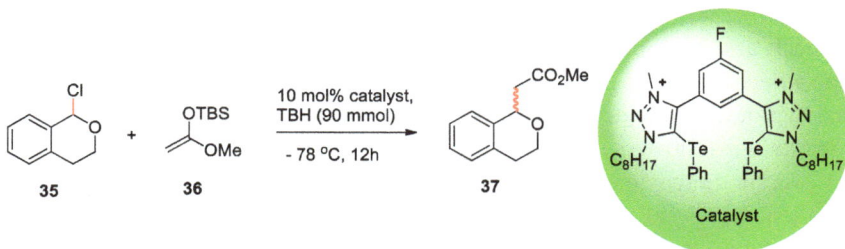

Scheme 12.6: Organotelluride-catalyzed substitution reaction in 1-chloroisochroman.

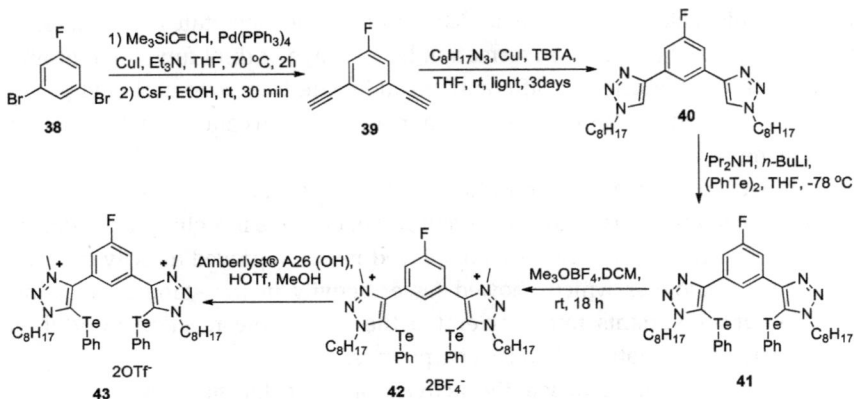

Scheme 12.7: Synthesis of phenyl linked dicationic tellurides.

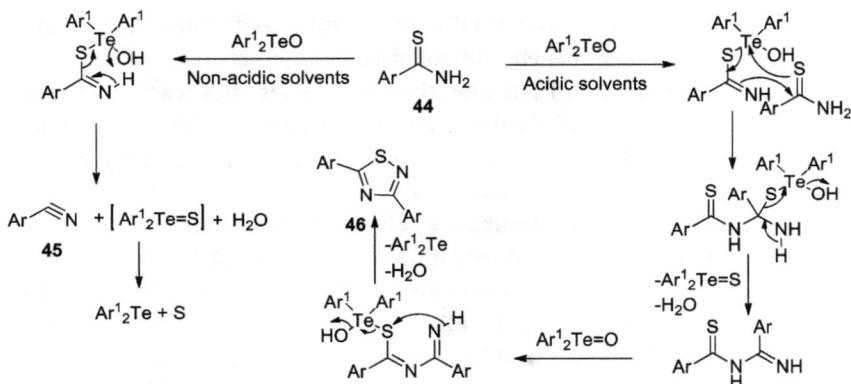

Scheme 12.8: Polymeric supported telluroxide-catalyzed oxidation of thioamide.

reaction medium, the intermediate adduct undergoes elimination to form nitrile whereas in acidic medium a second molecule of thiamide reacts with the intermediate leading to the generation of highly electrophilic iminium adduct with further formation of thiadiazoles.

M. Albeck and coworkers explored $TeCl_4$-catalyzed oligomerization and polymerization where the lewis acidic tellurium facilitated the reaction of phenylethylenes and benzyl chloride analogues [42]. The termination step in the polymerization of vinylic derivatives followed Friedel craft reaction. Being a source of $TeCl_3^+$, $TeCl_4$ promotes cationic polymerization. Trans-stilbene has been marked as inert toward polymerization, however, $TeCl_4$ proved to have exceptional efficiency in this regard over all other catalysts. The chain termination in this case proceeds through internal Friedel craft reaction providing 5-membered fused ring system (**50**, Scheme 12.9a). The polymerization of substituted benzyl chloride (**51**) proceeded with Friedel Crafts reaction through the TeCl$_4$ catalyzed functional group activation to form the anionic TeCl$_5$ and benzyl cation (**52**, Scheme 12.9b). The progression occurs with the polymerization of the benzylic cation followed by termination of polymeric cation with chloride ion to form the polymer **56** (Scheme 12.9b). TeCl$_4$, being a catalyst, gets regenerated in the reaction medium by combination of H$^+$ and TeCl$_5^-$.

The same group established a facile conversion of a variety of cycloheptatrienes (**57**) to benzylic halides (**58**) through TeCl$_4$-catalyzed rearrangement (Scheme 12.10) [43]. Two different mechanistic pathways have been proposed for the reaction where two different carbenium intermediate formed by activation of double bond and electrophilic attack of TeCl$_4$ to cycloheptatriene.

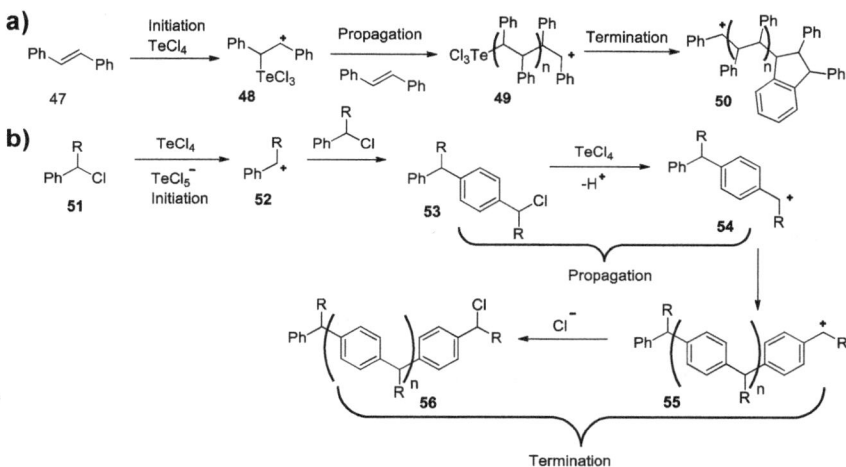

Scheme 12.9: TeCl$_4$ catalyzed oligo- and polymerization.

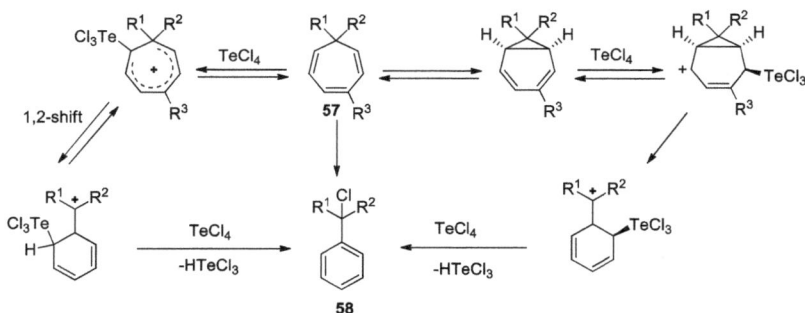

Scheme 12.10: TeCl$_4$-catalyzed rearrangement of cycloheptatriene.

12.2 Conclusions

Organotellurides have emerged as highly efficient compounds in various areas of organic synthesis, organocatalysis, and anion recognition through non-covalent chalcogen bonding interaction. These interactions take the Lewis acidic centre of chalcogens into account for activation of specific atom or functional groups. Being a heaviest and non-radioactive chalcogen, tellurium was found to be most efficient catalyst than others for certain organic transformations due to its ability to act as stronger Lewis acid. The advancement in the field of organotellurides through non-covalent chalcogen bond interaction has led to the significant expansion of its utility in various research areas.

Acknowledgments: DR and BKB are grateful to Amity University, Noida, UP, India and Prince Mohammad Bin Fahd University, Saudi Arabia for support.

References

1. a) Taylor MS, Jacobsen EN. Asymmetric catalysis by chiral hydrogen-bond donors. Angew Chem Int Ed 2006;45:1520–43. Angew. Chem. 2006;118:1550–73.
    b) Doyle AG, Jacobsen EN. Small-molecule H-bond donors in asymmetric catalysis. Chem Rev 2007; 107:5713–43.
    c) Connon SJ. Organocatalysis mediated by (Thio)urea derivatives. Chem Eur J 2006;12:5418–27.
    d) Alemn J, Parra A, Jiang H, Jørgensen KA. Squaramides: bridging from molecular recognition to bifunctional organocatalysis. Chem Eur J 2011;17:6890–9.
2. a) Vedejs E, Denmark SE. Lewis base catalysis in organic synthesis. KGaA: Verlag GmbH & Co; 2016.
    b) Lenardão EJ, Santi C, Sancineto L. New frontiers in organoselenium compounds. Cham: Springer International Publishing; 2018.
3. a) Wonner P, Vogel L, Düser M, Gomes L, Kniep F, Mallick B, et al. Carbon–halogen bond activation by selenium-based chalcogen bonding. Angew Chem Int Ed 2017;56:12009–12. Angew Chem 2017; 129:12172–6.
    b) Wonner P, Vogel L, Kniep F, Huber SM. Catalytic carbon–chlorine bond activation by selenium-based chalcogen bond donors. Chem Eur J 2017;23:16972–5.
    c) Benz S, López-Andarias J, Mareda J, Sakai N, Matile S. Catalysis with chalcogen bonds. Angew Chem Int Ed 2017;56:812–5.
4. a) Murray JS, Lane P, Clark T, Politzer P. Sigma-hole bonding: molecules containing group VI atoms. J Mol Model 2007;13:1033–8.
    b) Angyan JG, Poirier RA, Kucsman A, Csizmadia IG. Bonding between nonbonded sulfur and oxygen atoms in selected organic molecules (a quantum chemical study). J Am Chem Soc 1987;109: 2237–45.
    c) Burgi HB, Dunitz JD. Fractional bonds: relations among their lengths, strengths, and stretching force constants. J Am Chem Soc 1987;109:2924–6.
    d) Murray JS, Lane P, Politzer P. Expansion of the σ-hole concept. J Mol Model 2009;15:723–9.
5. a) Weiss R, Schlierf C, Schloter K. Toward metallocyclopropenium ions: redox cleavage of diorganyldichalcogenides by trichlorocyclopropenium salts. J Am Chem Soc 1976;98:4668.

b) Rosenfield RE, Parthasarathy R, Dunitz JD. Directional preferences of nonbonded atomic contacts with divalent sulfur. 1. Electrophiles and nucleophiles. J Am Chem Soc 1977;99:4860–2.

6. a) Row TNG, Parthasarathy R. Directional preferences of nonbonded atomic contacts with divalent sulfur in terms of its orbital orientations. 2. Sulfur···sulfur interactions and nonspherical shape of sulfur in crystals. J Am Chem Soc 1981;103:477–9.

   b) Politzer P, Murray JS, Clark T, Resnati G. The σ-hole revisited. Phys Chem Chem Phys 2017;19: 32166–78.

7. a) McWhinnie WR. Intra-molecular co-ordination - a route to novel organo tellurium compounds. Phosphorus Sulfur Relat Elem 1992;67:107.

   b) Ahmed MAK, McWhinnie WR, Hamor TA. Tellurated azobenzenes: the crystal and molecular structure of (2-phenylazophenyl-C,7N')tellurium(IV) trichloride. J Organomet Chem 1985;281:205.

   c) Granger P, Chapelle S, McWhinnie WR, Al-Rubaie A. A 125Te NMR study of the exchange reaction between diarylditellurides. J Organomet Chem 1981;220:149.

8. a) Reeske G, Cowley AH. Direct reactions of tellurium tetrahalides with chelating nitrogenligands. Trapping of TeI2 by a 1,2-bis(arylimino)acenaphthene (aryl-BIAN) ligand and C–H activation of an α,α′-diiminopyridine (DIMPY) ligand. Chem Commun 2006:4856.

   b) Kozma A, Petuskova J, Lehmann CW, Alcarazo M. Synthesis, structure and reactivity of cyclopropenyl-1-ylidene stabilized S(ii), Se(ii) and Te(ii) mono- and dications. Chem Commun 2013; 49:4145.

   c) Dube JW, Hanninen MM, Dutton JL, Tuononen HM, Ragogna PJ. Homoleptic pnictogen-chalcogen coordination complexes. Inorg Chem 2012;51:8897.

   d) Magdzinski E, Gobbo P, Martin CD, Workentin MS, Ragogna PJ. The syntheses and electrochemical studies of a ferrocene substituted diiminopyridine ligand and its P, S, Se, and Te complexes. Inorg Chem 2012;51:8425.

   e) Martin CD, Ragogna PJ. Reactions of diiminopyridine ligands with chalcogen halides. Inorg Chem 2012;51:2947.

   f) Dutton JL, Ragogna PJ. Dicationic tellurium analogues of the classic N-heterocyclic carbene. Chem Eur J 2010;16:12454.

   g) Dutton JL, Farrar GJ, Sgro MJ, Battista TL, Ragogna PJ. Lewis base sequestered chalcogen dihalides: synthetic sources of ChX2 (Ch=Se, Te; X=Cl, Br). Chem Eur J 2009;15:10263.

   h) Dutton JL, Tuononen HM, Ragogna PJ. Tellurium(II)-centered dications from the pseudohalide "Te(OTf)2". Angew Chem Int Ed 2009;48:4409.

   i) Dutton JL, Martin CD, Sgro MJ, Jones ND, Ragogna PJ. Synthesis of N,C bound sulfur, selenium, and tellurium heterocycles via the reaction of chalcogen halides with –CH3 substituted diazabutadiene ligands. Inorg Chem 2009;48:3239.

9. a) Sugamata K, Sasamori T, Tokitoh N. Generation of an organotellurium(II) cation. Eur J Inorg Chem 2012;2012:775.

   b) Beckmann J, Finke P, Heitz S, Hesse M. Aryltellurenyl cation [RTe(CR'2)]+ stabilized by an N-heterocyclic carbene. Eur J Inorg Chem 2008;51:1921.

   c) Boyle PD, Cross WI, Godfrey SM, McAuliffe CA, Pritchard RG, Sarwar S, et al. Synthesis and characterization of Ph4Te4I4, containing a Te4 square, and Ph3PTe(Ph)I. Angew Chem Int Ed 2000; 39:1796.

   d) Liu L, Zhu D, Cao LL, Stephan DW. N-heterocyclic carbene stabilized parent sulfenyl, selenenyl, and tellurenyl cations (XH+, X = S, Se, Te). Dalton Trans 2017;46:3095.

   e) Beckmann J, Bolsinger J, Duthie A, Finke P, Lork E, Luedtke C, et al. Mesityltellurenyl cations stabilized by triphenylpnictogens [MesTe(EPh3)]+ (E = P, As, Sb). Inorg Chem 2012;51:12395.

10. a) Hope EG, Levason W. Recent developments in the coordination chemistry of selenoether and telluroether ligands. Coord Chem Rev 1993;122:109–70.

    b) Gysling HJ. The ligand chemistry of tellurium. Coord Chem Rev 1982;42:133–244.

    c) Singh AK. Synthesis and characterization of Ph4Te4I4, containing a Te4 square, and Ph3PTe(Ph)I. Focus Organomet Chem Res 2005;79:109.

d) Singh AK, Sharma S. Recent developments in the ligand chemistry of tellurium. Coord Chem Rev 2000;209:49–98.

e) Murray SG, Hartley FR. Coordination chemistry of thioethers, selenoethers, and telluroethers in transition-metal complexes. Chem Rev 1981;81:365–414.

11. a) Lin T-P, Nelson RC, Wu T, Miller JT, Gabba FP. Lewis acid enhancement by juxtaposition with an onium ion: the case of a mercury stibonium complex. Chem Sci 2012;3:1128–36.

b) Lin T-P, Wade CR, Prez LM, Gabba FP. A mercury → antimony interaction. Angew Chem 2010;122: 6501–4; Angew Chem Int Ed 2010;49:6357–60.

c) Wade CR, Gabba FP. Two-electron redox chemistry and reversible umpolung of a gold–antimony bond. Angew Chem 2011;123:7507–10; Angew Chem Int Ed 2011;50:7369–72.

d) Wade CR, Lin T-P, Nelson RC, Mader EA, Miller JT, Gabba FP. Synthesis, structure, and properties of a T-shaped 14-electron stiboranyl-gold complex. J Am Chem Soc 2011;133:8948–55.

e) Wade CR, Ke I-S, Gabba FP. Sensing of aqueous fluoride anions by cationic stibine–palladium complexes. Angew Chem 2012;124:493–6; Angew Chem Int Ed 2012;51:478–81.

12. Lin T-P, Gabba FP. Two-electron redox chemistry at the dinuclear core of a TePt platform: chlorine photoreductive elimination and isolation of a TeVPtI complex. J Am Chem Soc 2012;134:12230–8.

13. Lin T-P, Gabbai FÅP. Telluronium ions as s-acceptor ligands. Angew Chem Int Ed 2013;52:3864–8.

14. Garrett GE, Gibson GL, Straus RN, Seferos DS, Taylor MS. Chalcogen bonding in solution: interactions of benzotelluradiazoles with anionic and uncharged Lewis bases. J Am Chem Soc 2015;137:4126–33.

15. a) Cozzolino AF, Vargas-Baca I. The supramolecular chemistry of 1,2,5-chalcogenadiazoles. J Organomet Chem 2007;692:2654–7.

b) Berionni G, Pégot B, Marrot J, Goumont R. Supramolecular association of 1,2,5-chalcogenadiazoles: an unexpected self-assembled dissymetric [Se···N]2 four-membered ring. Cryst Eng Commun 2009;11:986–8.

c) Chivers T, Gao X, Parvez M. Preparation, crystal structures, and isomerization of the tellurium diimide dimers RNTe(μ-NR')2TeNR (R = R' = tBu; R = PPh2NSiMe3, R' = tBu, tOct): X-ray structure of the telluradiazole dimer [tBu2C6H2N2Te]2. Inorg Chem 1996;35:9–15.

d) Alcock NW. Bonding and structure: structural principles in inorganic and organic chemistry. Harlow, UK: Ellis Horwood, Ltd.; 1990.

16. a) Cozzolino AF, Britten JF, Vargas-Baca I. The effect of steric hindrance on the association of telluradiazoles through Te–N secondary bonding interactions. Cryst Growth Des 2005;6:181–6.

b) Cozzolino AF, Whitfield PS, VargasBaca I. Supramolecular chromotropism of the crystalline phases of 4,5,6,7-tetrafluorobenzo-2,1,3-telluradiazole. J Am Chem Soc 2010;132:17265–70.

17. Semenov NA, Pushkarevsky NA, Beckmann J, Finke P, Lork E, Mews R, et al. Tellurium–nitrogen π-heterocyclic chemistry – synthesis, structure, and reactivity toward halides and pyridine of 3,4-dicyano-1,2,5-telluradiazole. Eur J Inorg Chem 2012;2012:3693–703.

18. Semenov NA, Lonchakov AV, Kushkarevsky NA, Suturina EA, Korolev VV, Lork E, et al. Coordination of halide and chalcogenolate anions to heavier 1,2,5-chalcogenadiazoles: experiment and theory. Organometallics 2014;33:4302–14.

19. Nordheider A, Hupf E, Chalmers BA, Knight FR, Buhl M, Mebs S, et al. Peri-substituted phosphorus–tellurium systems – an experimental and theoretical investigation of the P···Te through-space interaction. Inorg Chem 2015;54:2435.

20. a) Davis R, Patel L. Chapter 5: chalcogen–phosphorus (and heavier congener) compounds. In: Devillanova FA, du Mont W-W, editors. Handbook of chalcogen chemistry: new perspectives in sulfur, selenium and tellurium, 2nd ed. Cambridge, UK: RSC; 2013, vol 1.

b) Corbridge DEC. Phosphorus world-chemistry, biochemistry and technology. Boca Raton, FL: CRC Press; 2005:747 p.

c) Allen FH. The Cambridge Structural Database: a quarter of a million crystal structures and rising. Acta Crystallogr 2002;B58:380–8.

21. a) Benz S, López-Andarias J, Mareda J, Sakai N, Matile S. Catalysis with chalcogen bonds. Angew Chem Int Ed 2017;56:812–5; Angew Chem 2017;129:830–3.

    b) Benz S, Mareda J, Besnard C, Sakai N, Matile S. Catalysis with chalcogen bonds: neutral benzodiselenazole scaffolds with high-precision selenium donors of variable strength. Chem Sci 2017;8:8164–9.

22. a) Wonner P, Steinke T, Huber SM. Activation of quinolines by cationic chalcogen bond donors. Synlett 2019;30:1673–8.

    b) Wang W, Zhu H, Liu S, Zhao Z, Zhang L, Hao J, et al. Chalcogen–chalcogen bonding catalysis enables assembly of discrete molecules. J Am Chem Soc 2019;141:9175–9.

23. Wonner P, Dreger A, Vogel L, Engelage E, Huber SM. Chalcogen bonding catalysis in a nitro–Michael reaction. Angew Chem Int Ed 2019;58:16923.

24. Wonner P, Steinke T, Vogel L, Stefan M, Huber. Carbonyl activation by selenium- and tellurium-based chalcogen bonding in a Michael addition reaction. Chem Eur J 2020;26:1258–62.

25. Tani H, Masumoto K, Inamasu T. Tellurium tetrachloride as a mild and efficient catalyst for dithioacetalization. Tetrahedron Lett 1991;32:2039–42.

26. Ong BS. Mild and efficient chemoselective protection of aldehydes as dithioacetals employing – bromosuccinimide. Tetrahedron Lett 1980;21:4225.

27. Fieser LF. Preparation of ethylenethioketals. J Am Chem Soc 1954;76:1945.

28. Firouzabadi H, Iranpoor N, Amani K. Heteropoly acids as heterogeneous catalysts for thioacetalization and transthioacetalization reactions. Synthesis 2002:59–60. https://doi.org/10.1055/s-2002-19300.

29. Firouzabadi H, Iranpoor N, Hazarkhani H. Iodine catalyzes efficient and chemoselective thioacetalization of carbonyl functions, transthioacetalization of O,O- and S,O-acetals and acylals. J Org Chem 2001;66:7527–9.

30. Hajipour AR, Azizi G, Ruoho AE. An efficient method for chemoselective thioacetalization of aldehydes in the presence of a catalytic amount of acidic ionic liquid under solvent-free conditions. Synlett 2009:1974–8.

31. Lai J, Du W, Tian L, Zhou C, She X, Tang S. Fe-catalyzed direct dithioacetalization of aldehydes with 2-chloro-1,3-dithiane. Org Lett 2014;16:4396–9.

32. Dong D, Ouyang Y, Yu H, Liu Q, Liu J, Wang M, et al. Chemoselective thioacetalization in water: 3-(1,3-dithian-2-ylidene) pentane-2,4-dione as an odorless, efficient, and practical thioacetalization reagent. J Org Chem 2005;70:4535–7.

33. De SK. Yttrium triflate as an efficient and useful catalyst for chemoselective protection of carbonyl compounds. Tetrahedron Lett 2004;45:2339–3241.

34. Wu Y-C, Zhu J. Hafnium trifluoromethanesulfonate (hafnium triflate) as a highly efficient catalyst for chemoselective thioacetalization and transthioacetalization of carbonyl compounds. J Org Chem 2008;73:9522–4.

35. De SK. Vanadyl triflate as an efficient and recyclable catalyst for chemoselective thioacetalization of aldehydes. J Mol Catal Chem 2005;226:77–9.

36. Rudrawar S, Besra RC, Chakraborti AK. Perchloric acid adsorbed on silica gel (HClO4-SiO2) as an extremely efficient and reusable catalyst for 1,3-dithiolane/dithiane formation. Synthesis 2006:2767–71.

37. Ali MH, Gomes MG. A simple and efficient heterogeneous procedure for thioacetalization of aldehydes and ketones. Synthesis 2005:1326–32.

38. Aoyama T, Suzuki T, Nagaoka T, Takido T, Kodomari M. Silica-gel supported sulfamic acid (SA/SiO$_2$) as an efficient and reusable catalyst for conversion of ketones into oxathioacetals and dithioacetals. Synth Commun 2013;43:553–66.

39. Kamal A, Chouhan G. Chemoselective thioacetalization and transthioacetalization of carbonyl compounds catalyzed by immobilized scandium(III) triflate in ionic liquids. Tetrahedron Lett 2003; 44:3337–40.
40. Patney HK. Anhydrous cobalt(II) bromide dispersed on silica gel: a mild and efficient reagent for thioacetalisation of carbonyl compounds. Tetrahedron Lett 1994;35:5717–8.
41. a) Petragnani N, Comasseto JV. Tellurium reagents in organic synthesis; recent advances. Part 2. Synthesis 1991;11:897–919.
b) Hu NX, Aso Y, Otsubo T, Ogura F. Polymer-supported diaryl selenoxide and telluroxide as mild and selective oxidizing agents. Bull Chem Soc Jpn 1986;59:879–84.
42. Albeck M, Tamari T. TeCl₄ as a catalyst in cationic oligomerisations and polymerisations. J Organomet Chem 1982;238:357–62.
43. Albeck M, Tamari T, Sprecher T. Formation of benzylic chlorides by rearrangement of cycloheptatrienes with tellurium tetrachloride. J Org Chem 1983;48:2276–8.

Index

https://doi.org/10.1515/9783110735840-013

www.ingramcontent.com/pod-product-compliance
Lightning Source LLC
Chambersburg PA
CBHW080924220326
41598CB00034B/5675